The Ethics of Information

The Ethics of Information

Luciano Floridi

OXFORD
UNIVERSITY PRESS

OXFORD
UNIVERSITY PRESS

Great Clarendon Street, Oxford, OX2 6DP,
United Kingdom

Oxford University Press is a department of the University of Oxford.
It furthers the University's objective of excellence in research, scholarship,
and education by publishing worldwide. Oxford is a registered trade mark of
Oxford University Press in the UK and in certain other countries

© Luciano Floridi 2013

The moral rights of the author have been asserted

First published 2013
First published in paperback 2015

All rights reserved. No part of this publication may be reproduced, stored in
a retrieval system, or transmitted, in any form or by any means, without the
prior permission in writing of Oxford University Press, or as expressly permitted
by law, by licence or under terms agreed with the appropriate reprographics
rights organization. Enquiries concerning reproduction outside the scope of the
above should be sent to the Rights Department, Oxford University Press, at the
address above

You must not circulate this work in any other form
and you must impose this same condition on any acquirer

Published in the United States of America by Oxford University Press
198 Madison Avenue, New York, NY 10016, United States of America

British Library Cataloguing in Publication Data

Data available

Library of Congress Cataloging in Publication Data

Data available

ISBN 978-0-19-964132-1 (Hbk.)
ISBN 978-0-19-874805-2 (Pbk.)

Links to third party websites are provided by Oxford in good faith and
for information only. Oxford disclaims any responsibility for the materials
contained in any third party website referenced in this work.

Contents

Preface xii
Acknowledgements xvii
List of most common acronyms xx
List of figures xxi
List of tables xxii

Chapter 1—Ethics after the information revolution 1

    Summary 1
1.1 Introduction: the hyperhistorical predicament 3
1.2 The zettabyte era 5
1.3 ICTs as re-ontologizing technologies 6
1.4 The global infosphere, or how information is becoming our ecosystem 8
1.5 The metaphysics of the infosphere 10
1.6 The information turn as the fourth revolution 13
1.7 The evolution of inforgs 14
    Conclusion 17

Chapter 2—What is information ethics? 19

    Summary 19
2.1 Introduction: a unified model of information ethics 20
2.2 First stage: IE as an ethics of informational resources 21
2.3 Second stage: IE as an ethics of informational products 23
2.4 Third stage: IE as an ethics of the informational environment 24
2.5 Fourth stage: IE as a macroethics 25
    Conclusion 28

Chapter 3—The method of abstraction 29

    Summary 29
3.1 Introduction: on the very idea of levels of abstraction 30
3.2 The definition of a level of abstraction 31
3.3 Abstraction and ontological commitment 34
3.4 An application of the method of abstraction: telepresence 36
    3.4.1 Presence as epistemic failure 37
    3.4.2 Presence as successful observation 41
    3.4.3 Virtual pornography 44
    3.4.4 An objection against presence as successful observation 46

3.4.5 Informational privacy: from intrusion to abduction	49
3.4.6 The method of abstraction and telepresence	51
Conclusion	51

Chapter 4—Information ethics as e-nvironmental ethics — 53

Summary	53
4.1 Introduction: the foundationalist problem	54
4.2 Classic macroethics and ICTs ethical problems	58
4.3 An informational model of macroethics	61
4.4 From computer ethics to information ethics	63
4.5 Information ethics as a patient-oriented and ontocentric theory	65
4.5.1 Uniformity of Being	65
4.5.2 Uniformity of nothingness, non-Being, or metaphysical entropy	65
4.5.3 Uniformity of change	67
4.5.4 The ontological equality principle	68
4.6 The normative aspect of information ethics: four ethical principles	70
4.6.1 The non-monotonic nature of goodness and its resilience	72
4.7 Information ethics as a macroethics	74
4.7.1 IE is a complete macroethics	74
4.7.2 IE and other non-standard ethics	76
4.7.3 IE and virtue ethics	77
4.7.4 IE and deontologism	77
4.7.5 IE and consequentialism	77
4.8 Case analysis: three examples	79
4.8.1 Vandalism	80
4.8.2 Bioengineering	82
4.8.3 Death	82
Conclusion	84

Chapter 5—Information ethics and the foundationalist debate — 86

Summary	86
5.1 Introduction: looking for the foundations of computer ethics	87
5.2 The 'no resolution approach': CE as not a real discipline	88
5.3 The professional approach: CE as a pedagogical methodology	90
5.4 Theoretical CE and the uniqueness debate	92
5.5 The radical approach: CE as a unique discipline	93
5.6 The conservative approach: CE as applied ethics	94
5.7 The innovative approach: information ethics as the foundation of CE	97
Conclusion	99

Chapter 6—The intrinsic value of the infosphere 102
    Summary 102
    6.1 Introduction: an object-oriented model of moral action 103
    6.2 The role of information in ethics 109
    6.3 An axiological analysis of information 112
        6.3.1 A critique of Kantian axiology 114
        6.3.2 A patient-oriented approach to axiology 115
        6.3.3 IE's axiological ecumenism 122
    6.4 Five objections 124
        6.4.1 The need for an ontology 124
        6.4.2 How can an informational entity have 'a good of its own'? 125
        6.4.3 What happened to Evil? 128
        6.4.4 Is there a communication problem? 130
        6.4.5 Right but irrelevant? 132
    Conclusion 133

Chapter 7—The morality of artificial agents 134
    Summary 134
    7.1 Introduction: standard vs. non-standard theories of agents and patients 135
    7.2 What is an agent? 138
        7.2.1 An effective characterization of agents 140
        7.2.2 Examples 141
    7.3 What is a moral agent? 146
    7.4 Mindless morality 148
        7.4.1 The teleological objection 148
        7.4.2 The intentional objection 149
        7.4.3 The freedom objection 149
        7.4.4 The responsibility objection 150
    7.5 The morality threshold 152
    7.6 A concrete application 153
        7.6.1 Codes of ethics 155
        7.6.2 Censorship 156
    7.7 The advantage of extending the class of moral agents 157
    Conclusion 159

Chapter 8—The constructionist values of *homo poieticus* 161
    Summary 161
    8.1 Introduction: reactive and proactive macroethics 162
    8.2 The scope and limits of virtue ethics as a constructionist ethics 164
    8.3 Why information ethics cannot be based on virtue ethics 166
    8.4 Ecopoiesis 168

8.5 Poiesis in the infosphere	169
8.5.1 Interfaces	169
8.5.2 Open source	170
8.5.3 Digital arts	172
8.5.4 The construction of the self	173
8.5.5 Virtual communities	174
8.5.6 Constructionism on the web	175
8.6 *Homo poieticus*	175
Conclusion	177
Chapter 9—Artificial evil	**180**
Summary	180
9.1 Introduction: the nature of evil	180
9.2 Nonsubstantialism: a deflatory interpretation of the existence of evil	183
9.3 The evolution of evil and the theodicean problem	185
9.4 Artificial evil	187
Conclusion	191
Chapter 10—The tragedy of the Good Will	**194**
Summary	194
10.1 Introduction: modelling a Good Will	194
10.2 The tragic and the scandalous	197
10.3 The IT-heodicean problem	200
10.4 Cassandra's predicament	201
10.5 Escaping the tragic condition	204
10.5.1 The Copenhagen Consensus: using information to cope with information	206
Conclusion	208
Chapter 11—The informational nature of selves	**210**
Summary	210
11.1 Introduction: Plato and the problem of the chariot	211
11.2 Egology and its two branches	213
11.3 Egology as synchronic individualization	215
11.4 A reconciling hypothesis: the three membranes model	217
11.4.1 Phase one: the corporeal membrane and the organism	219
11.4.2 Phase two: the cognitive membrane and the intelligent animal	220
11.4.3 Phase three: the consciousness membrane and the self-conscious mind	220

11.5	ICTs as technologies of the self	221
	11.5.1 Embodiment: from dualism to polarism	221
	11.5.2 Space: the detachment between location and presence	222
	11.5.3 Time: the detachment between outdating and ageing	222
	11.5.4 Memories and interactions: fixing the self	223
	11.5.5 Perception: the digital gaze	223
11.6	The logic of realization	225
11.7	From the egology to the ecology of the self	226
	Conclusion	227

Chapter 12—The ontological interpretation of informational privacy — 228

	Summary	228
12.1	Introduction: the dearest of our possessions	229
12.2	Informational privacy and computer ethics	229
12.3	Informational privacy as a function of ontological friction	231
12.4	Ontological friction and the difference between old and new ICTs	232
12.5	Informational privacy in the re-ontologized infosphere	235
	12.5.1 Empowering the information agent	236
	12.5.2 The return of the (digital) community	236
	12.5.3 Assessing theories of privacy	240
	12.5.4 The ontological interpretation of informational privacy and its value	242
12.6	Informational privacy, personal identity, and biometrics	246
12.7	Four challenges for a theory of informational privacy	249
	12.7.1 Non-Western approaches to informational privacy	249
	12.7.2 Individualism and the anthropology of informational privacy	251
	12.7.3 The scope and limits of informational privacy	253
	12.7.4 Public, passive, and active informational privacy	256
12.8	Non-informational privacies	257
	Conclusion	258

Chapter 13—Distributed morality — 261

	Summary	261
13.1	Introduction: the basic idea of distributed morality	262
13.2	The old ethical scenario without distributed morality	263
13.3	The new ethical scenario with distributed morality	265
13.4	Some examples of distributed morality	267
	13.4.1 The shopping Samaritan: (RED)	268
	13.4.2 Plastic fidelity: the Co-operative Bank	268
	13.4.3 The power of giving: JustGiving	268
	13.4.4 Socially oriented capitalism: peer-to-peer lending	269

13.5 The big challenge: harnessing the power of DM	269
13.6 Distributed morality and the enabling infraethics	272
Conclusion	274
Chapter 14—Information business ethics	**277**
Summary	277
14.1 Introduction: from information ethics to business ethics	279
14.2 The informational analysis of business	280
14.3 The WHI ethical questions: what, how, and impact	284
14.4 Normative pressure points	285
14.5 The ethical business	287
Conclusion	290
Chapter 15—Global information ethics	**292**
Summary	292
15.1 Introduction: from globalization to information ethics	293
15.1.1 Contraction	293
15.1.2 Expansion	294
15.1.3 Porosity	294
15.1.4 Telepresence	295
15.1.5 Synchronization	295
15.1.6 Correlation	295
15.2 Globalizing ethics	296
15.3 Global-communication ethics vs. global-information ethics	296
15.4 Global-information ethics and the problem of the lion	297
15.5 Global information-ethics and its advantages	299
15.6 The cost of a global-information ethics: postulating the ontic trust	300
Conclusion	303
Chapter 16—In defence of information ethics	**306**
Summary	306
16.1 Introduction: addressing the sceptic	306
16.2 IE is an ethics of news	307
16.3 IE is too reductivist	308
16.4 IE fails to indicate what information constitutes an individual	309
16.5 IE's de-anthropocentrization of the ethical discourse is mistaken	312
16.6 IE is inapplicable	313
16.7 IE is supererogatory	314
16.8 IE is hypermoralistic	316
16.9 IE's measure of intrinsic moral value is insufficiently clear and specific	317

16.10	IE's inference from moral value to moral respect is incorrect	317
16.11	IE's negative argument for the intrinsic moral goodness of Being is incorrect	319
16.12	IE's claim to be universal is unclear and possibly contradictory	320
16.13	IE's egalitarianism is untenable	322
16.14	IE commits the naturalistic fallacy	323
16.15	IE's account of intrinsic value is incorrect	324
16.16	IE is counterintuitive	325
16.17	IE's adoption of LoA is mistaken	326
16.18	IE's interpretation of artificial agents as autonomous moral agents is mistaken	327
16.19	IE is too conservationist	327
16.20	IE is pantheistic or panpsychistic	328
	Conclusion	329
Epilogue		331
References		334
Index		353

Preface

What I am concerned with is knowledge only—that we should think correctly and so far arrive at some truth, however unimportant: I do not say that such knowledge will make us more useful members of society. If anyone does not care for knowledge for its own sake, then I have nothing to say to him: only it should not be thought that a lack of interest in what I have to say is any ground for holding it untrue.

G. E. Moore, *Principia Ethica*, Moore (1993), p. 115.

The information revolution has been changing the world profoundly and irreversibly for some time now, at a breath-taking pace, and with an unprecedented scope. It has made the creation, processing, management, and utilization of information vital issues, and brought enormous benefits as well as opportunities. However, it has also greatly outpaced our understanding of its nature, implications, and consequences, and raised conceptual issues that are rapidly expanding and evolving.[1] They are also becoming increasingly serious. Today, philosophy faces the challenge of providing a foundational treatment of the phenomena and the ideas underlying the information revolution, in order to foster our understanding and guide both the responsible construction of our society and the sustainable management of our natural and synthetic environments. In short, we need to develop a *philosophy of information*.

The philosophy of information investigates the conceptual nature and basic principles of information, including its ethical consequences (Floridi, 2011a). It is a thriving new area of research that intersects with, and complements, other classic areas of philosophical investigation, especially epistemology, metaphysics, logic, philosophy of science, philosophy of language and mind, and ethics. It is based on two simple ideas. Information is something as fundamental and significant as knowledge, being, validity, truth, meaning, mind, or good and evil, and so equally worthy of autonomous, philosophical investigation. But it is also a more impoverished concept, in terms of which the others can be expressed, interrelated, and investigated philosophically.

As the reader will see in Chapter 2, I interpret *information ethics* (IE) as the branch of the philosophy of information that investigates, in a broad sense, the ethical impact of Information and Communication Technologies (ICTs) on human life and society. ICTs have profoundly affected many aspects of the human condition, including the nature of communication, education, work, entertainment, industrial production and business, health care, social relations, and armed conflicts. They have had a radical and

---

[1] See The European Group on Ethics in Science and New Technologies, *Opinion no. 26: Ethics of Information and Communication Technologies* (2012) and the UNESCO Observatory on the Information Society, Information for All Programme (IFAP) Information Society Observatory <http://ifap-is-observatory.ittk.hu/taxonomy/term/165>.

widespread influence on our moral lives and on contemporary ethical debates. Examples come readily to mind, from privacy and freedom of expression to Wikileaks, from the digital divide to a dystopian 'surveillance society', from artificial companions to drones and cyberwar. Indeed, the ethical problems raised by ICTs are ubiquitous in our society and in contemporary culture, and often lie behind debates in medical ethics, environmental ethics, neuroethics, and bioethics. As I shall argue in this book, they actually invite us to reconsider some fundamental tenets in our moral theories.

The ethical issues brought about by ICTs constitute a complicated and potentially confusing scenario, not least because it is in constant and rapid evolution. A simple analogy may help to make sense of the current situation. Our technological tree has been growing its far-reaching branches much more widely, rapidly, and chaotically than its conceptual, ethical, and cultural roots. The lack of balance is obvious and a matter of daily experience in the life of millions of people.[2] The risk is that, like a tree with weak roots, further and healthier growth at the top will be impaired by a fragile foundation at the bottom. As a consequence, today, any information society faces the pressing task of equipping itself with a shareable and sustainable information ethics. It is not by chance that IFAP (UNESCO intergovernmental council for the Information For All Programme) is drafting the UNESCO 'Code of Ethics for the Information Society', which is expected to lead to the UNESCO 'Declaration on Infoethics in Cyberspace'. Applying the previous analogy, while technology keeps growing bottom-up, it is high time we start digging deeper, top-down, in order to expand and reinforce our conceptual understanding of the foundations of our information ethics and of the moral implications and impact of ICTs. A better philosophical grasp will be essential, if we wish to have a better chance of anticipating difficulties, identifying opportunities, and resolving moral conflicts and dilemmas (I shall return to this analogy in the conclusion of Chapter 16).

The task of this book is to contribute to such conceptual foundations of IE as a new area of philosophical research. It does so systematically (conceptual architecture is pursued as a valuable feature of philosophical thinking) rather than exhaustively, by pursuing three goals.

The first goal is meta-theoretical. The book describes the information revolution, the role and nature of information ethics after it, its method of levels of abstraction, its problems, and the foundational debate about computer and information ethics. These are the topics of the first part, which comprises Chapters 1 to 5.

The second goal is introductory. In Chapters 6 to 11, the book explores the nature of informational entities and the infosphere as patients (i.e. receivers) of moral actions, the nature of moral agents (i.e. senders), our constructionist values as human agents responsible for the well-being of our environments and their inhabitants, good and evil

---

[2] See e.g. the i2010—Annual Information Society Reports, <http://eur-lex.europa.eu/LexUriServ/LexUriServ.do?uri=CELEX:52007DC0146:EN:NOT>.

in the infosphere, the difficulties encountered by good moral agents, and the informational interpretation of our selves.

The third goal is constructive. In Chapters 12 to 15, the book answers questions about privacy, morality in distributed systems, the relation between information and business ethics, and the global nature of an information ethics for all. The final chapter provides a defence of IE from some recurrent criticisms. The sixteen chapters are strictly related, so I have added internal references whenever it might be useful.

This book is the second volume of a tetralogy, entitled *Principia Philosophiae Informationis*. Although entirely independent of volume one, entitled *The Philosophy of Information* (Floridi, 2011a), the two are complementary. The essential message from volume one is quite straightforward. Semantic information is well-formed, meaningful, and truthful data; knowledge is relevant semantic information properly accounted for; humans are the only known semantic engines and conscious informational organisms who can design and understand semantic artefacts and thus develop a growing knowledge of reality; and reality is the totality of information (notice the crucial absence of 'semantic'). Against this background, this second volume investigates the foundations of the ethics of informational organisms (inforgs) like us, which flourish in informational environments (infospheres) and are responsible for their construction and well-being. In short, it is about the ethics of inforgs in the infosphere. Thus, in a classic Kantian move, we are shifting from theoretical to pragmatic philosophy.

Like volume one, this too is a German book, written from a post-analytic–continental divide perspective. Unlike volume one, however, it is much less neo-Kantian than I expected it to be. Instead, the careful reader will easily place this work in the tradition linking Platonism—from Plato and Plotinus to Augustine and G. E. Moore, the latter being the philosopher who has most influenced my ethical thinking—and Spinozism.[3] Apparently, there are also some spiritual overtones and connections to Confucianism, Buddhism, Taoism, and Shintoism.[4] They were unplanned and they are not based on any intended study of the corresponding sources. I was made aware of such connections by other philosophers, while working on the articles that led to this book. Once we grasp them, ideas have their own way of leading us by the hand to unknown places we might not have meant to visit. Some books write their authors.

Regarding the style and structure of this book, as I wrote in the preface of volume one, I am painfully aware that this second volume too is not a page-turner, to put it mildly, despite my attempts to make it as interesting and reader-friendly as possible. It will require not only patience and time but also an open mind, three scarce resources. For in a decade or so of debates, I have been made fully aware that some of the ideas I defend in the following pages are controversial. Ethics is often considered a strictly human business, in which deeply seated intuitions are treated as the ultimate criteria to assess the value of moral ideas. Ethics is obsessed with religious beliefs, psychologistic

---

[3] I owe this insight to Hongladarom (2008).
[4] See e.g. Ess (2008) and Herold (2005).

introspection, and a reliance on often ungrounded, idiosyncratic intuitions or very foggy ideas (intentionality being one of them). It is still largely centred on a stand-alone, Cartesian-like, ratiocinating, human individual—a vision which is in turn based on what might be called, borrowing a technical expression from mathematics, *degenerate* epistemology (Floridi, 2012b)—when the world has in fact moved towards hybrid, distributed, and multi-agent systems (there is probably more 'moral agency' occurring at the level of governments, non-governmental organizations, parties, groups, companies, and so forth, than in the life of millions of individuals). A pinch of serious computer science and *rigorous* philosophy can provide a great counterbalance. After all, artificial agents tell us as much about ourselves as about our artefacts. In almost any other intellectual investigation, we are ready to follow our reasoning even when it clashes with our intuitions. *Fiat veritas, pereat intuitio*, to paraphrase the equivalent idea in ethics that there should be justice, even if the world were to perish because of it. We accept that the earth is spinning around its axis at the speed of 1040 miles (1670 km) per hour; that it orbits around the sun at the speed of about 66000 miles (107000 km) per hour; that we are the product of natural evolution; that space is not Euclidean; that only about 4 per cent of the Universe is made of normal matter (the stuff of stars, planets and people); that our bodies are mostly water; that bananas are not fruit; that whales are not fish; and so forth. In ethics, however, where there is so much disagreement, so much more uncertainty, and so fewer universal standards shared unanimously and uncontroversially, we seem to believe that the subjectively counterintuitive is simply equivalent to the objectively silly, the immoral, or a combination of both. This book is not meant to irritate sensitivities about what may count as a moral patient worthy of some respect, what may count as a moral agent accountable for its actions, and what may count as the ethical environment within which morally loaded interactions take place, or what it means to lead an ethical (not just moral, see Section 16.8) life. But it does run against some dogmas, in the Greek sense of the word, as established beliefs or theory, the sort of dogmas against which Sextus Empiricus wrote. Our ethical intuitions are whence we inevitably depart, but they are not where we should necessarily arrive. To those who may find the conclusions reached in this book preposterous, I can only ask them to be tolerant. I never meant to step on their toes.

As in the previous volume, two features that I thought might help the reader to access the contents of this book more easily are the summaries and conclusions at the beginning and the end of each chapter, and some redundancy. Regarding the first feature, I know it is slightly unorthodox, but the solution, already adopted in volume one, of starting each chapter with a 'Previously in Chapter x...', should enable the reader to browse the text, or fast-forward entire sections of it, without losing the essential plot. Science-fiction fans, who recognize the reference to *Battlestar Galactica*, may consider this second volume as the equivalent of season two.

Regarding the second feature, while editing the final version of the book, I decided to retain in the chapters some repetitions and some rephrasing of recurrent themes, whenever I thought that the place where the original content had been introduced was too distant, either in terms of pages or in terms of theoretical context. If every now and then the reader experiences some *déjà vu* it is not a bug, it is a feature, and I hope it will be to the advantage of clarity.

A final word now on what the reader will not find in the following pages. This is not an introduction to computer or information ethics for the general reader, and it does not seek to provide an exhaustive investigation of all the moral issues debated under such labels. It is not a textbook on professional ethics, on computer science and its ethical implications, or on the sort of topics discussed in undergraduate courses entitled 'computers in society' either. The reader interested in such topics might wish to look at *The Cambridge Handbook of Information and Computer Ethics* (Floridi, 2010d) and at *Information—A Very Short Introduction* (Floridi, 2010c). This book also avoids any discussion, insofar as this is possible given its topic, of political issues. One of the OUP anonymous reviewers rightly spotted that 'liberalism [is] at the heart of [Floridi's] "distributed constructionism"', but it would have been too messy and too long to include here a full elaboration of such constructive liberalism. The interested reader may wish to know that this is volume three of the tetralogy, entitled *The Policies of Information*.

In short, this book is about the roots of some of the ethical problems of our time, not their leaves. It is about the digital seeds we are sowing and what we shall reap, as Shakespeare suggests at the beginning of Chapter 1.

Acknowledgements

I could not have worked on such a long-term project without dividing it into some feasible and much smaller tasks. I must also confess that I was surprised by the fact that they still fitted together in the end. I hope this is not the symptom of a stubbornly closed mind. All the chapters were planned as conference papers or (sometimes inclusive or) journal articles. The bibliographic details are in the list of references below. This way of working was laborious, but it also seemed inevitable, given the innovative nature of the field. It did require a perseverance and commitment that I hope were not ill-exercised. I wished to test the ideas presented in this volume as thoroughly as possible, and publishing the articles gave me the opportunity and privilege to enjoy a vast amount of feedback, from a very large number of colleagues and anonymous referees. If I do not thank all of them here, this is not for lack of manners or mere reason of space, but because the appropriate acknowledgements can be found in the corresponding, published articles.

There are, however, some people who played a significant role throughout the project (including volume one) and during the revisions of the final text. Kia first of all. She and I have been married for as long as I have been working on this book. Without her, I would never have had the confidence to undertake such a task or the spiritual energy to complete it. She has made our life blissful, and I am very grateful to her for the endless hours we spent talking about the topics of this book, for all her sharp suggestions, and for her lovely patience with an utterly obsessed husband. The last paragraph in Section 12.6 is dedicated to her. I owe to a former colleague in Oxford, Jeff Sanders, a much better understanding of the more technical aspects of the method of abstraction in computer science. Some of the ideas presented in Chapter 3 were developed and formulated in close collaboration with him, and he should really be considered a co-author of it (see References below), although not someone co-responsible for any potential shortcomings. Jeff and I also co-authored several articles of which Chapters 4, 6, and 10 are revised versions. I learnt much from our collaboration, and I am very grateful to him for his permission to reproduce some of our work here. Mariarosaria Taddeo, Matteo Turilli, and other members of the IEG, the interdepartmental research group on the philosophy of information at Oxford, were very generous with their time and provided numerous opportunities for further reflection on virtually any topic discussed in this book. Massimo Durante helped me in understanding several implications of the ideas developed in this book, when we were working on the Italian translations of a collection of articles on information ethics which formed a sort of prequel to this volume, Floridi (2009b). Anthony Beavers and Marty J. Wolf carefully read and commented on a final draft of this book and helped me to improve it enormously. Many conversations and debates with Terry Bynum, Charles Ess, Jim Moor, Ugo Pagallo, Julian Savulescu, and Jeroen Van den Hoven have influenced my work, perhaps not deeply enough, according to some critics. The CEPE (Computer Ethics Philosophical Enquiries) meetings, organized by the International Society for Ethics and Information Technology, and the CAP (Computing and Philosophy) meetings, organized by the International Association for Computing and Philosophy, provided stimulating and fruitful venues to test some of the ideas presented in this and in the previous volume. I have learnt a lot from the colleagues I met there, though I wish I had been able to learn more. Peter Momtchiloff was pivotal for the realization of both volumes—both because of his timely invitation to publish them with OUP, and because of his support and firm patience, when it seemed that I was never going to complete them. The anonymous reviewers appointed by OUP were incredibly helpful with their comments and suggestions. I was impressed especially by one of them for her or his careful reading and very insightful recommendations. I wish I could thank her or him in person.

Penny Driscoll, my personal assistant, very kindly and most skilfully proofread the final version of the manuscript, making it much more readable. She also provided some very helpful philosophical feedback on the final version of the book. Without her exceptional support and impeccable managerial skills I could not have completed this project.

Finally, I would like to thank the University of Bari, Hertfordshire and the University of Oxford for having provided me with the time to pursue my research at different stages during the past fourteen years. The final writing effort was made possible thanks to two Arts and Humanities Research Council grants during the academic years 2010–11 and 2011–12.

References

The 16 chapters constituting the book are based on the following publications:

Chapter 1: Floridi (2005c, 2007b); Floridi (2007a); Floridi (2008c, 2008a, 2008b); Floridi (2010c)
Chapter 2: Floridi (2005c, 2007b, 2008a, 2008b)
Chapter 3: Floridi and Sanders (2004b); Floridi (2005b, 2008d, 2010a); Floridi (2011a); Floridi (2012f, 2012g, forthcoming-b)
Chapter 4: Floridi (1999a)
Chapter 5: Floridi and Sanders (2002, 2004a)
Chapter 6: Floridi (2003)
Chapter 7: Floridi and Sanders (2004c)
Chapter 8: Floridi and Sanders (2005)
Chapter 9: Floridi and Sanders (1999, 2001)
Chapter 10: Floridi (2006b)

Chapter 11: Floridi (2011c)
Chapter 12: Floridi (2005a, 2006a)
Chapter 13: Floridi (forthcoming-a)
Chapter 14: Floridi (2010b)
Chapter 15: Floridi (2002); Taddeo and Floridi (2007)
Chapter 16: Floridi (2008f, 2008e); Floridi (2012f, 2012e)

I am grateful to the following publishers and organizations for permission to reproduce extracts from my publications listed above: the American Philosophical Association (APA), the Association for Computing Machinery (ACM), Cambridge University Press, Elsevier, Etica & Politica, the European Centre for International Political Economy (ECIPE), Jones and Bartlett, Peter Lang, MIT Press, Palgrave MacMillan, The Royal Society, Springer, SUNY Press, Taylor & Francis, John Wiley & Sons, Wiley-Blackwell.

List of most common acronyms

A	agent, Alice
B	agent, Bob
AA	Artificial Agent
AI	Artificial Intelligence
CE	Computer Ethics
ICTs	Information and Communication Technologies
IE	Information Ethics
LoA	Level of Abstraction
P	patient, Peter
PI	Philosophy of Information

The use of Alice as synonymous with Agent is not merely an Oxford-related reference. The reader acquainted with the literature on quantum information will recognize in Alice, Bob, Carol, and Dave the more memorable and concrete placeholder names for otherwise abstract agents, see <http://en.wikipedia.org/wiki/Alice_and_Bob>. Whenever I need an abstract patient, I shall refer to him as P for Peter.

List of figures

1	A typical information life-cycle	4
2	The 'external' R(esource) P(roduct) T(arget) model	20
3	The 'internal' R(esource) P(roduct) T(arget) model	27
4	The scheme of a theory	35
5	The SLMS scheme with ontological commitment	35
6	An example of levels of abstraction	41
7	A model of telepresence	43
8	A model of telepistemics as backward presence	48
9	The model of a moral action as a binary relation	106
10	The informational model of a moral action	108
11	The logical relations between classes of moral agents and patients	136
12	MENACE (Matchbox Educable Noughts and Crosses Engine)	143
13	Example: Songs of the Korean War	224
14	The logic of realization	226
15	CCTV image of the four London terrorists as they set out from Luton	237
16	The old ethical scenario without distributed morality	263
17	The old ethical scenario with moral thresholds and morally negligible actions	265
18	The new ethical scenario of distributed morality	266
19	EU population accessing the Internet, away from home or work	271
20	The relational analysis of business	282
21	The concentric ring model of business	282
22	The whirlpool model of business	283
23	The WHI ethical questions: what, how, and impact	284

List of tables

1	The four ethical principles of IE	71
2	The definition of an agent at LoA_1	140
3	Examples of agents	142
4	The principles guiding ethical behaviour in the ACM code of ethics	155
5	Interfaces as theatre	170
6	The definition of evil action	183
7	A taxonomy of agents	187

1
Ethics after the information revolution

> If you can look into the seeds of time,
> And say which grain will grow and which will not,
> Speak then to me, who neither beg nor fear
> Your favours nor your hate.
>
> Shakespeare, *Macbeth*, Act I, Scene III, 59–62.

SUMMARY

In this introductory chapter, I shall discuss some conceptual undercurrents, which seem to be flowing beneath the surface of what we call the information revolution. I shall introduce a few key concepts and themes that will recur in different places throughout this book. Yet no technological determinism is endorsed—a mythical *bête noire* long exorcised and dead, even assuming it ever was alive. Rather, the idea is to look at the informational seeds that might grow into realities. As any gardener knows, Paul of Tarsus was a bit too radical when he stated that 'a man reaps what he sows' (*Galatians* 6:7). If that were true, my garden would not be so disappointing. However, there is no denying that logic plays a big role in our actions, for the latter imply consequences the likelihood of which we must consider. Going back to the initial analogy and the conceptual undercurrents, I do not intend to suggest that we are passive passengers on the history train which we cannot control. We are more like sailors (Plato's Greek word for this is *kybernetes*, steersman or governor, with cybernetics meaning 'the study of self-governance' in *The Laws*; Wiener knew his classics) on a boat that needs to be steered in the right direction, negotiating the aforementioned currents, while navigating according to plans.

In the following pages, I shall focus, more generally, on the potential impact of information and communication technologies (ICTs) on our lives. But since there would be no merit in explaining the obvious, I shall avoid calling attention to a large variety of issues, such as privacy and ownership, spamming, viruses, the importance of semantic tagging, online shopping and virtual communities, or the fairly new phenomena of information warfare and massive open online course (MOOC), which I presume to be well known to the reader. Nor will I try to steal ideas from those who know better than I do the future development of the actual technologies.[1] I will, instead, stick to what philosophers do better, *conceptual design*, and seek to synthesize the silent *Weltanschauung* that might be dawning on us. Before doing so, let me add a final clarification.

By 'conceptual design' I mean to refer to a *constructionist* (not a *constructivist*) philosophy that can explain (better: account for) our semantic artefacts and design or re-purpose those needed by our new infosphere. In the past, I have spoken of such a philosophical task in terms of *conceptual engineering* (Floridi, 2011b). However, I recently came to realize that the very word 'engineering' may generate confusion. True, both Descartes and Wittgenstein were engineers. Yet Carnap (who seems to have been the first to coin the expression (Carus, 2007)) and his followers attached to it a linguistic value (conceptual engineering as language engineering) that I do not share. Moreover, the expression has mechanical and deterministic overtones that I would be reluctant to endorse but that are difficult to shake off. I now find it rather clunky, and hence I much prefer speaking of *conceptual design*, especially in view of the fact that design is neither discovery nor invention, nor a mere matter of tinkering, fixing, or improving, but indeed the art of implementing requirements and exploiting constraining affordances intelligently and teleologically, in order to build artefacts in view of a specific goal. Philosophy as conceptual design[2] is therefore a realistic philosophy, which treats semantic artefacts as mind- *and* reality-co-dependent, in the same way as a house is not a representation but the outcome of a specific architectural design both constrained and afforded by the building materials.

I take it to be a foundationalist enterprise (something not very fashionable these days). Again, relying on the previous analogy and the conceptual undercurrents, I do not take philosophy to be in the business of repairing but rather in that of building the raft while swimming, to paraphrase Neurath. The emphasis is on the radical and difficult nature of the philosophical task ahead of us, not on any anti-foundationalist suggestion. Understanding philosophy as conceptual design means giving up not on its foundationalist vocation, but rather on the possibility of outsourcing its task to any combination of logico-mathematical and empirical approaches.[3] At the same time, understanding philosophy as conceptual design enables one to avoid epistemic relativism at the expense of representationalism. For the equations in front of us are rather

---

[1] See e.g. Microsoft-Research (2005); *Nature* (2006); O'Reilly (2005).

[2] To the reader who may wish to build a network of references and 'embed' the constructivism advocated in my philosophy of information into a more contextualized set of influences and perspectives, I would highly recommend reading Deleuze and Guattari (1994).

[3] I know this was not Neurath's intention when he first introduced the metaphor of the raft in the 1930s. As he later wrote:

> There is no way of taking conclusively established pure protocol sentences as the starting point of the sciences. No *tabula rasa* exists. We are like sailors who must rebuild their ship on the open sea, never able to dismantle it in dry-dock and to reconstruct it there out of the best materials. Only the metaphysical elements can be allowed to vanish without trace. Vague linguist conglomerations always remain in one way or another as components of the ship (Neurath 1959, p. 201).

simple: we can either embrace a representationalist epistemology, which can avoid relativism by dropping the constructionist stance; or we can accept the fact that we are in charge of our conceptual constructions, some of which are very ill-conceived (astrology, homeopathy, Othello's understanding of Desdemona's behaviour, etc.) while others are increasingly successful in making sense of the world (astrophysics, medicine, the perfect understanding between Romeo and Juliet, etc.), but then constructionism without relativism becomes possible only by unveiling representationalist epistemology as another ill-conceived artefact. I shall return to this topic in Chapter 3.

1.1 Introduction: the hyperhistorical predicament

History has many metrics. Some are natural and circular, relying on recurring seasons and planetary motions. Some are social or political and linear, being determined, for example, by the succession of Olympic Games, or the number of years since the founding of the city of Rome (*ab urbe condita*), or the ascension of a king. Still others are religious and have a V-shape, counting years before and after a particular event, such as the birth of Christ. There are larger periods that encompass smaller ones, named after influential styles (Baroque), people (the Victorian era), particular circumstances (the Cold War), or some new technology (the Nuclear age). What all these and many other metrics have in common is that they are all *historical*, in the strict sense that they all depend on the development of systems to record events and hence accumulate and transmit information to the future. No records, no history: so history is actually synonymous with the information age, since *pre-history* is that age in human development that precedes the availability of recording systems.

It follows that one may reasonably argue that humanity has been living in various kinds of information societies at least since the Bronze Age, the era that marks the invention of writing in Mesopotamia and other regions of the world (4th millennium BC). And yet, this is not what is typically meant by the information revolution. There may be many explanations, but one seems more convincing than any other: only very recently has human progress and welfare begun to depend mostly on the successful and efficient management of the life-cycle of information. Only recently have we entered into a *hyperhistorical predicament* (Floridi, 2012d).

The life-cycle of information typically includes the following phases: *occurrence* (discovering, designing, authoring, etc.), *recording*, *transmission* (networking, distributing, accessing, retrieving, etc.), *processing* and *management* (collecting, validating, merging, modifying, organizing, indexing, classifying, filtering, updating, sorting, storing, etc.), and *usage* (monitoring, modelling, analysing, explaining, planning, forecasting, decision-making, instructing, educating, learning, etc.). Figure 1 provides a very simplified illustration.

Now, imagine Figure 1 to be like a clock. The length of time that the evolution of information life-cycles has taken to bring about the information society should not be surprising. According to recent estimates, life on Earth will last for another billion years,

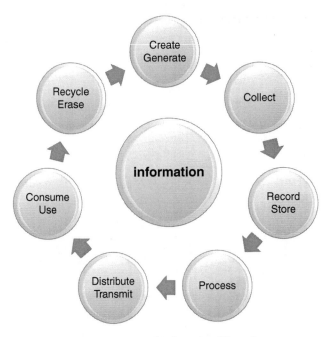

Figure 1. A typical information life-cycle

until it is destroyed by the increase in solar temperature. So imagine an historian writing in the near future, say in a million years. She may consider it normal, and perhaps even elegantly symmetrical, that it took roughly six millennia for the agricultural revolution to produce its full effect, from its beginning in the Neolithic (10th millennium BC) until the Bronze Age, and then another six millennia for the information revolution to bear its main fruit, from the Bronze Age until the end of the 2nd millennium AD. During this span of time, ICTs evolved from being mainly recording systems—writing and manuscript production—to being also communication systems—especially after Gutenberg and the invention of printing—to being also processing and producing systems, especially after Turing and the diffusion of computers. Thanks to this evolution, nowadays, the most advanced societies highly depend on information-based, intangible assets (knowledge-based economy), information-intensive services (especially business and property services, communications, finance, insurance, and entertainment), and information-oriented public sectors (especially education, public administration, and health care). For example, all members of the G7 group—namely Canada, France, Germany, Italy, Japan, United Kingdom, and the United States of America—qualify as information societies because, in each country, at least 70 per cent of the Gross Domestic Product (GDP) depends on intangible goods, which are information-related, not on material goods, which are the physical output of agricultural or manufacturing processes. Their functioning and growth requires and generates immense amounts of data, more data than humanity has ever seen in its entire history.

1.2 The zettabyte era

A few year ago, researchers at Berkeley's School of Information Management and Systems Lyman and Varian (2003) estimated that humanity had accumulated approximately 12 exabytes of data[4] in the course of its entire history, until the commodification of computers. However, they also calculated that print, film, magnetic, and optical storage media had already produced more than 5 exabytes of data just in 2002. This is equivalent to 37 000 new libraries the size of the Library of Congress. Given the size of the world population in 2002, it turned out that almost 800 megabytes (MB) of recorded data had been produced per person. It is like saying that at the beginning of the new millennium every newborn baby came into the world with a burden of 30 feet of books, the equivalent of 800 MB of data printed on paper. Of these data, 92 per cent were stored on magnetic media, mostly in hard disks, thus causing an unprecedented 'democratization' of information: more people own more data than ever before. In the last decade, such exponential escalation has been relentless. According to a more recent study, in 2011 we passed the zettabyte (1 000 exabytes) barrier:

In 2011, the amount of information created and replicated will surpass 1.8 zettabytes (1.8 trillion gigabytes)—growing by a factor of 9 in just five years. (Gantz and Reinsel, 2011)

This figure is now expected to grow fourfold approximately every three years. Every day, enough new data is being generated to fill all US libraries eight times over. As a result, there is much talk about 'big data' (Floridi, 2012c).

'Exaflood' is a neologism that has been coined to qualify this tsunami of bytes that is submerging the world. Of course, hundreds of millions of computing machines are constantly employed to keep us afloat and navigate through such an exaflood. All the previous numbers will keep growing steadily for the foreseeable future, not least because computers are among the greatest sources of further exabytes. Thanks to them, we are living in *the age of the zettabyte*. It is a self-reinforcing cycle and it would be unnatural not to feel overwhelmed. It is, or at least should be, a mixed feeling.

On the one hand, ICTs have brought concrete and imminent opportunities of enormous benefit to people's education, welfare, prosperity, and edification, as well as great economic and scientific advantages. Unsurprisingly, the US Department of Commerce and the National Science Foundation have identified Nanotechnology, Biotechnology, Information Technology and Cognitive Science (NBIC) as research areas of national priority. Note that the three—NBC—would be virtually impossible without the I. In a comparable move, the EU heads of States and governments acknowledged the immense impact of ICTs when they agreed to make the EU the most competitive and dynamic knowledge-driven economy by 2010. That was, of course, before the financial crisis.

On the other hand, ICTs also carry significant risks and generate dilemmas and profound questions about the nature of reality and of our knowledge of it, the

---

[4] One exabyte corresponds to 10^{18} bytes or a 50 000-year-long video of DVD quality.

development of information-intensive sciences (e-science), the organization of a fair society (consider the digital divide), our responsibilities and obligations to present and future generations, our understanding and management of a globalized world, and the scope of our potential interactions with the environment. The almost sudden burst of a global information society, after a few millennia of a relatively quieter gestation, has generated new and disruptive challenges, which were largely unforeseeable only a few decades ago. As a very simple illustration, consider identity theft, the use of information to impersonate someone else in order to steal money or get other benefits. According to the Federal Trade Commission, frauds involving identity theft in the USA accounted for approximately $52.6 billion of losses in 2002 alone, affecting almost 10 million Americans (I shall return to the topic of identity theft in Sections 12.5.4 and 12.6).

1.3 ICTs as re-ontologizing technologies

In order to grasp the new scenarios that we might witness in the near future, and hence the sort of ethical problems with which we might be expected to deal, it is useful to introduce two key concepts at the outset, those of 'infosphere' and of 're-ontologization'.

Infosphere is a neologism I coined some years ago (Floridi, 1999b) on the basis of 'biosphere', a term referring to that limited region on our planet that supports life. As will become clearer in the course of this book, it is a concept that is quickly evolving. *Minimally*, it denotes the whole informational environment constituted by all informational entities (thus including information agents as well), their properties, interactions, processes, and mutual relations. It is an environment comparable to, but different from, cyberspace, which is only one of its sub-regions, as it were, since it also includes offline and analogue spaces of information. *Maximally*, it is a concept that, given an informational ontology, can also be used as synonymous with reality, or Being. The difference between the two readings is a function of our understanding of information, as something that has only semantic properties (e.g. Wikipedia) or also ontic properties (information as data patterns, e.g. the magnetic structure of a digital support).[5]

Re-ontologizing is another neologism that I have introduced in Floridi (2007a) in order to refer to a very radical form of re-engineering, one that not only designs, constructs, or structures a system (e.g. a company, a machine, or some artefact) anew, but one that also fundamentally transforms its intrinsic nature, that is, its ontology or essence. In this sense, for example, nanotechnologies and biotechnologies are not merely re-engineering but actually re-ontologizing our world.

Using the two previous concepts, it becomes possible to formulate succinctly the following thesis: ICTs are re-ontologizing the very nature of (and hence what we mean by) the infosphere, and here lies the source of some of the most profound

---

[5] On the distinction between information as something (e.g. a structure), for something (e.g. an algorithm), and about something (e.g. a piece of news), see Floridi (2010c, 2011a).

transformations and challenging problems that we will experience in the close future, as far as technology is concerned.

The most obvious way in which ICTs are re-ontologizing the infosphere concerns the transition from analogue to digital data and then the ever-increasing growth of our informational space. Both phenomena are very familiar and require no explanation, but a brief comment may not go amiss. This radical re-ontologization of the infosphere is largely due to the fundamental convergence between digital resources and digital tools. The ontology of the information technologies available (e.g. software, algorithms, databases, communication channels, and protocols, etc.) is now the same as (and hence fully compatible with) the ontology of their objects, the raw data being manipulated. This was one of Turing's most consequential intuitions: in the re-ontologized infosphere, populated by ontologically equal entities and agents, where there is no ontological difference between *processors* and *processed*, interactions become equally digital. They are all interpretable as 'read/write' (i.e. access/alter) activities, with 'execute' the remaining type of process. The digital deals effortlessly and seamlessly with the digital. This potentially eliminates one of the long-standing bottlenecks in the infosphere and, as a result, there is a gradual erasure of *ontological friction*. I shall discuss the latter concept in more detail in Chapter 12. Here, suffice it to say that because of their 'data superconductivity', ICTs are well known for being among the most influential factors that facilitate the flow of information in the infosphere. We are all acquainted daily with aspects of a *frictionless infosphere*, such as *spamming* (because every email is virtually free) and *micrometering* (because every fraction of a penny may now count). Three other significant consequences include:

(1) a substantial erosion of the *right to ignore*: in an increasingly frictionless infosphere, it becomes progressively less credible to claim ignorance when confronted by easily predictable events (e.g. as George W. Bush did with respect to Hurricane Katrina's disastrous effects on New Orleans's flood barriers) and hardly ignorable facts (e.g. as Tessa Jowell, a British Labour MP, did with respect to her husband's finances). We all swim in such information, it is hard to argue we are not being touched by it. I shall return to this in Chapter 10. And therefore

(2) an exponential increase in *common knowledge*. This is a technical term from epistemic logic, which basically refers to cases in which everybody not only knows that *p* but also knows that everybody knows that everybody knows, . . . , that *p*. More on this in Chapter 12.

The impact of (1) and (2) is quickly increasing also because meta-information about how much information is, was or should have been available is becoming overabundant. From (1) and (2) it follows that:

(3) we are witnessing a steady increase in agents' *responsibilities*. As I shall argue in Chapter 8, ICTs are making humanity increasingly responsible, morally speaking, for the way the world is, will and should be. This is a bit paradoxical since,

as we shall see in Chapter 13, ICTs are also part of a wider phenomenon that is making the clear attribution of responsibility to specific individual agents more difficult and ambiguous.

1.4 The global infosphere, or how information is becoming our ecosystem

During the last decade or so, we have become accustomed to conceptualizing our life online as a mixture between an evolutionary adaptation of human agents to a digital environment, and a form of post-modern, neo-colonization of the latter by the former. This is probably a mistake. ICTs are as much re-ontologizing our world as they are creating new realities. The threshold between *here* (*analogue, carbon-based, offline*) and *there* (*digital, silicon-based, online*) is fast becoming blurred, but this is as much to the advantage of the latter as it is to the former. Adapting Horace's famous phrase, 'captive infosphere is conquering its victor', the digital-online is spilling over into the analogue-offline and merging with it. This recent phenomenon is variously known as 'Ubiquitous Computing', 'Ambient Intelligence', 'The Internet of Things', or 'Web-augmented Things'. I prefer to refer to it as the *onlife experience*. It is, or will soon be, the next stage in the development of the information age.[6] With a slogan: hyperhistory happens onlife.

The increasing re-ontologization of artefacts and of whole (social) environments suggests that it is becoming difficult to understand what life was like in pre-digital times, and, in the near future, the very distinction between online and offline will become blurred and then disappear. To someone who was born in 2000 the world will always have been wireless, for example. To her and any other member of what Janna Quitney Anderson calls Generation AO, the Always-On Generation, the peculiar clicking and whooshing sounds made by conventional modems while handshaking, also known as the whale song, will be as alien as the sounds made by a telegraph's Morse signals are to us. To put it dramatically, the infosphere is progressively absorbing any other ontological space. Let me explain.

In the (fast approaching) future, more and more objects will be *ITentities* able to learn, advise, and communicate with one other. A good example is provided by RFID (Radio Frequency IDentification) tags, which can store and remotely retrieve data from an object and give it a unique identity, such as a barcode. Tags can measure 0.4 mm^2 and are thinner than paper. Incorporate this tiny microchip in everything, including humans and animals, and you have created *ITentities*. This is not science fiction. According to a report by Market Research Company InStat, the worldwide production of RFID increased more than 25-fold between 2005 and 2010 to reach 33 billion. Imagine

---

[6] Coroama et al. (2004), Bohn et al. (2004), and Brey (2005) offer an ethical evaluation of privacy-related issues in Ambient Intelligence environments. For a technically informative and balanced assessment I would recommend Gow (2005).

networking these 33 billion *ITentities* together with all the hundreds of millions of other ICT devices of all kinds already available, and you will see that the infosphere is no longer 'there' but 'here', and it is here to stay. Your Nike shoes and iPod have been talking to each other for some time, with predictable (but amazingly unforeseen) problems in terms of privacy (Saponas et al., 2007). Your next fridge[7] could already inherit from the previous one your tastes and wishes, just as your new laptop can import your favourite settings from the old one; and it could interact with your new way of cooking and with the supermarket website, just as your laptop can talk to a printer or to another computer. We have all known this in theory for some time; the difference is that now it is actually happening in our kitchen.

Nowadays, we are still used to considering the space of information as something we log-in to and log-out from. Our naïve metaphysics is still modern or Newtonian: it is made of 'dead' cars, buildings, furniture, clothes, which are non-interactive, irresponsive, and incapable of communicating, learning, or memorizing. However, what we still experience as the world offline is bound to become a fully interactive and responsive environment of wireless, pervasive, distributed, *a2a* (anything to anything) information processes, that works *a4a* (anywhere for anytime), in real time. The day when we routinely google the location of physical objects ('where are the car keys?') is very close.[8]

As a consequence of such re-ontologization of our ordinary environment, I shall argue in Chapter 15 that we are already living in an infosphere that will become increasingly *synchronized* (time), *delocalized* (space), and *correlated* (interactions). Although this might be interpreted, optimistically, as the friendly face of globalization (see Chapters 14 and 15), we should not harbour illusions about how widespread and inclusive the evolution of the information society will be. Unless we manage to solve it, the digital divide will become a chasm, generating new forms of discrimination between those who can be denizens of the infosphere and those who cannot, between insiders and outsiders, between information rich and information poor. It will redesign the map of worldwide society, generating or widening generational, geographic, socio-economic, and cultural divides. Yet the gap will not be reducible to the distance between rich and poor countries, since it will cut across societies. Pre-historical cultures have virtually disappeared, with the exception of some small tribes in remote corners of the world. The new divide will be between historical and hyperhistorical ones. We might be preparing the ground for tomorrow's informational slums.

---

[7] See e.g. 'The Internet of Things: Smart Houses, Smart Traffic, Smart Health', *Science Daily*, 26 June 2012 <http://www.sciencedaily.com/releases/2012/06/120626065009.htm>.

[8] In 2008, Thomas Schmidt, Alex French, Cameron Hughes, and Angus Haines (four 12-year-old boys from Ashfold Primary School in Dorton, UK) were awarded the 'Home Invention of the Year' Prize for their Speed Searcher, a device for finding lost items. It attaches tags to valuables and enables a computer to pinpoint their location in the home. See '12 year old inventors use wireless to solve the problem of lost keys', *Public Technology*, 7 April 2008 <http://www.publictechnology.net/sector/12-year-old-inventors-use-wireless-solve-problem-lost-keys>.

1.5 The metaphysics of the infosphere

The previous transformations already invite us to understand the world as something 'a-live' (artificially live). Such animation of the world will, paradoxically, make our outlook closer to that of pre-technological cultures, which interpreted all aspects of nature as inhabited by teleological forces. The first thing one may do these days when looking at an IT screen is to tap it, instead of looking for a keyboard. Unfortunately, such 'animation' of artefacts seems to go hand in hand with irrational beliefs about the power of ICTs. There are always passengers at Heathrow Airport who believe that the IRIS system, which checks passengers' ID by scanning their irises, will somehow work even if they never registered for such service in the first place. Somehow ICTs are seen as omniscient and omnipotent gods.

The next step will be a re-conceptualization of our ontology in informational terms. It is happening before our very eyes. It will become normal to consider the world as part of the infosphere, not so much in the dystopian sense expressed by a *Matrix*-like scenario, where the 'real reality' is still as hard as the metal of the machines that inhabit it; but in the evolutionary, hybrid sense represented by an environment such as New Port City, the fictional, post-cybernetic metropolis of *Ghost in the Shell*. The infosphere will not be a virtual environment supported by a genuinely 'material' world behind; rather, it will be the world itself that will be increasingly interpreted and understood informationally, as part of the infosphere. At the end of this shift, the infosphere will have moved from being a way to refer to the space of information to being synonymous with Being itself.

We are modifying our everyday perspective on the ultimate nature of reality, from a materialist one, in which physical objects and processes play a key role, to an informational one. This shift means that objects and processes are *de-physicalized*, in the sense that they tend to be seen as support-independent (consider a music file). They are *typified*, in the sense that an instance of an object (my copy of a music file) is as good as its type (your music file of which my copy is an instance). And they are assumed to be by default perfectly *clonable*, in the sense that my copy and your original become indistinguishable and hence interchangeable. Less stress on the physical nature of objects and processes means that the right of usage is perceived to be at least as important as the right to ownership. From a hyperhistorical perspective, re-purposing, updating, or upgrading are not merely expressions of plagiarism or sloppy morality, but also ways of appropriating and appreciating the malleable nature of informational objects. Our Newtonian and Historical educational system still has to catch up with such transformation. Finally, the criterion for existence—what it means for something to exist—is no longer being actually immutable (the Greeks thought that only that which does not change can be said to exist fully), or being potentially subject to perception (modern philosophy insisted on something being perceivable empirically through the five senses in order to qualify as existing), but being potentially subject to interaction, even if intangible. To be is to be interactable, even if the interaction is only indirect or virtual. The following examples should help to make the previous points clearer and more concrete.

In recent years, many countries have followed the USA in counting acquisition of software not as a current business expense but as an investment, to be treated as any other capital input that is repeatedly used in production over time, like a factory.[9] Spending on software now regularly contributes to GDPs. So software is acknowledged to be a (digital) good, even if somewhat intangible. It should not be too difficult to accept that virtual assets too may represent important investments.

Computing resources themselves are usually provided by hardware, which then represents the major constraint for their flexible deployment. Yet we are fast moving towards a stage when cloud computing is 'softening' our hardware through 'virtualization', the process whereby one can deliver computing resources, usually built-in hardware—like a specific CPU, a storage facility or a network infrastructure—by means of software. For example, virtualization can be adopted in order to run multiple operating systems on a single physical computing machine so that, if more machines are needed, they can be created as a piece of software—i.e. as virtual machines (VMs)—and not purchased as physical hardware equipment. The difference between deploying a virtual and a physical machine is dramatic. Once the virtualization infrastructure is in place, the provider of virtualized hardware resources can satisfy users' requests in a matter of minutes and, potentially, to a very large scale. Likewise, terminating or halting such a provision is equally immediate. The VMs are simply shut down without leaving behind any hardware component that needs to be reallocated or dismantled physically. Clearly, this will further modify our conception of what a machine is, in favour of a utility-based approach. Dropbox, Google Documents, Apple's iCloud, and Microsoft SkyDrive have provided everyday experiences of cloud computing to millions of users for some time now. The quick disappearance of any kind of 'drive' (the old floppy disk drive, the more recent CD or DVD drive) in favour of 'ports' (USB, etc.) is a clear signal of the virtualization movement.

Next, consider the so-called 'virtual sweatshops' in China. In claustrophobic and overcrowded rooms, workers play online games, like *World of Warcraft* or *Lineage*, for up to twelve hours a day, to create virtual goods, such as characters, equipment, or in-game currency, which can then be sold to other players. At the time of writing, End User License Agreements (EULA, this is the contract that every user of commercial software accepts by installing it) of massively multiplayer online role-playing games (MMORPG) such as *World of Warcraft* still do not allow the sale of virtual assets. This would be like the EULA of MS-Office withholding from users the ownership of the digital documents created by means of the software. The situation will probably change, as more people invest hundreds and then thousands of hours building their avatars and assets. Future generations will inherit digital entities that they will want to

---

[9] 'Software Investment—Now They See It', *The Economist*, 16 February 2006 <http://www.economist.com/node/5523570>.

own. Indeed, although it was forbidden, there used to be thousands of virtual assets on sale on eBay. Sony, more aggressively, offers a 'Station Exchange', an official auction service that

provides players a secure method of buying and selling [in dollars—my specification] the right to use in game coins, items and characters in accordance with SOE's license agreement, rules and guidelines.[10]

Once ownership of virtual assets has been legally established, the next step is to check for the emergence of property litigations. This is already happening: in May 2006, a Pennsylvania lawyer sued the publisher of *Second Life* for allegedly having unfairly confiscated tens of thousands of dollars worth of his virtual land and other property. Insurances that provide protection against risks to avatars may follow, comparable to pet insurance one can buy at the local supermarket. Again, *World of Warcraft* provides an excellent example. With 11.1 million subscribers as of June 2011, *World of Warcraft* is currently the world's most-subscribed MMORPG.[11] It would rank 71st in the list of 221 countries and dependent territories ordered according to population. Its users, who (will) have spent billions of man-hours constructing, enriching, and refining their digital properties, will be more than willing to spend a few dollars to insure them.

The combination of virtualization of services and virtual assets offers an unprecedented opportunity. Nowadays it is still common and easy to insure a machine (e.g. a laptop or a mobile) on which the data are stored, but not the data it stores. This is because, although data may be invaluable and irreplaceable, they are also perfectly clonable at a negligible cost, contrary to physical objects, so it would be impossible for an insurer to ascertain their irrecoverable loss or corruption. However, cloud computing decouples the physical *possession* of data (by the provider) from their *ownership* (by the user), and once it is the provider that physically possesses the data and is responsible for their maintenance, the user/owner of such data should rightly expect to see them insured, for a premium of course, and to be compensated in case of damage, loss, or downtime. Users should be able to insure *their* data precisely because they do not physically possess them. So-called 'cyber insurance' has been around for more than a decade, it is the right thing to do, but it is only with cloud computing that it can become truly feasible. We are likely to witness a welcome shift from hardware to data in the insurance strategies used to hedge against the risk of irreversible losses or damages.

Despite some important exceptions (e.g. vases and metal tools in ancient civilizations, engravings, and then books after Gutenberg), it was the industrial revolution that really marked the passage from a nominalist world of unique objects to a Platonist world of types of objects, all perfectly reproducible as identical to one other, therefore epistemically indiscernible, and hence pragmatically dispensable because replaceable

---

[10] Wikipedia, 'Station Exchange' <http://en.wikipedia.org/wiki/Station_Exchange>.
[11] Source: Wikipedia entry 'World of Warcraft' <http://en.wikipedia.org/wiki/World_of_warcraft>.

without any loss in the scope of interactions that they allow. When our ancestors bought a horse, they bought *this* horse or *that* horse, not 'the' horse. Today, we find it obvious that two automobiles may be virtually identical and that we are invited to test-drive and buy the model rather than an individual 'incarnation' of it. We buy the type not the token. Indeed, we are fast moving towards a commodification of objects that considers repair as synonymous with replacement, even when it comes to entire buildings. This has led, by way of compensation, to a prioritization of informational *branding*—a process compared by Klein to the creation of 'cultural accessories and lifestyle philosophies' (2000, p. 16)—and of *re-appropriation*. The person who puts a sticker in the window of her car, which is otherwise perfectly identical to thousands of others, is fighting an anti-Platonic battle in support of a nominalist philosophy. The information revolution has further exacerbated this process. Once our window-shopping becomes Windows-shopping and no longer means walking down the street but browsing through the web, the processes of de-physicalization and typification of individuals as unique and irreplaceable entities start eroding our sense of personal identity as well. We begin to act and conceptualize ourselves as mass-produced, anonymous entities among other anonymous entities, exposed to billions of other similar informational organisms online. We conceive ourselves as bundles of types, from gender to religion, from family role to working position, from education to social class. So we construct, self-brand, and re-appropriate ourselves in the infosphere by using blogs and Facebook entries, homepages, YouTube videos, and Flickr albums, fashionable clothes, and choices of places we visit, types of holidays we take, and cars we drive, and so forth. It is perfectly reasonable that *Second Life* should be a paradise for fashion enthusiasts of all kinds. Not only does it provide a new and flexible platform for designers and creative artists, it is also the right context in which users (avatars) intensely feel the pressure to obtain visible signs of self-identity and unique personal tastes. After all, your free avatar looks like anybody else's. Likewise, there is no inconsistency between a society so concerned about privacy rights and the success of services such as Facebook. We use and expose information about ourselves to become less informationally anonymous and indiscernible. We wish to maintain a high level of informational privacy almost as if that were the only way of saving a precious capital that can then be publicly invested (squandered, pessimists would say) by us in order to construct ourselves as individuals easily discernible and uniquely re-identifiable by others (see Chapters 11 and 12).

At the roots of the processes I have just sketched in this section there seems to be a far deeper metaphysical drift caused by the information turn, what I have described in the past as *the fourth revolution*.

1.6 The information turn as the fourth revolution

Oversimplifying, science has two fundamental ways of changing our understanding. One may be called *extrovert*, or about the world, and the other *introvert*, or about ourselves. Three scientific revolutions have had great impact both extrovertly and introvertly. In changing our understanding of the external world, they also modified our conception of who we are. After Copernicus, the heliocentric cosmology displaced the Earth and hence humanity from the centre of the universe. Darwin showed that all species of life have evolved over time from common ancestors through natural selection, thus displacing humanity from the centre of the biological kingdom. And following Freud, we acknowledge nowadays that the mind is also unconscious and subject to the defence mechanism of repression, thus displacing it from the centre of pure rationality, a position that had been assumed as uncontroversial at least since Descartes. The reader who, like Popper and myself, would be reluctant to follow Freud in considering psychoanalysis a strictly scientific enterprise like astronomy or evolutionary theory, might yet be willing to concede that contemporary neuroscience is a likely candidate for such a revolutionary role. Either way, the result is that today we acknowledge that we are not immobile, at the centre of the universe (Copernican revolution), we are not unnaturally separate and diverse from the rest of the animal kingdom (Darwinian revolution), and we are very far from being stand-alone Cartesian minds entirely transparent to ourselves (Freudian or neuroscientific revolution).

One may easily question the value of this classic picture. After all, Freud himself was the first to interpret these three revolutions as part of a single process of reassessment of human nature (Freud, 1955; Weinert, 2009). His hermeneutic manoeuvre was, admittedly, rather self-serving. But it did strike a reasonable note. In a similar way, when we now perceive that something very significant and profound has happened to human life after the informational turn, I would argue that our intuition is once again perceptive, because we are experiencing what may be described as a fourth revolution in the process of dislocation and reassessment of humanity's fundamental nature and role in the universe. Since the fifties (and Turing may easily be elected as the representative figure of such revolution), computer science and ICTs have exercised both an extrovert and an introvert influence. They have not only provided unprecedented epistemic and engineering powers over natural and artificial realities; but by doing so they have also cast new light on who we are, how we are related to the world, and hence how we understand ourselves. Today, we are slowly accepting the idea that we are not Newtonian, standalone, and unique entities, but rather informationally embodied organisms (*inforgs*), mutually connected and embedded in an informational environment, the *infosphere*, which we share with both natural and artificial agents similar to us in many respects. Let me explain.

1.7 The evolution of inforgs

We have seen that we are probably the last generation to experience a clear difference between online and offline environments. Some people already live *onlife*. Some cultures are already hyperhistorical. A further transformation worth highlighting concerns the emergence of artificial and hybrid (multi)agents, i.e., partly artificial and partly human (consider, for example, a family as a single agent, equipped with digital cameras, laptops, tablets, smart phones, mobiles, wireless network, digital TVs, DVDs, CD players, etc.). These new agents already share the same ontology with their environment and can operate within it with much more freedom and control. We (shall) delegate or outsource, to artificial agents and companions, our memories, decisions, routine tasks, and other activities in ways that will be increasingly integrated with us and with our understanding of what it means to be an agent. Yet all this is rather well known, and it is not what I am referring to when I talk about *inforgs*. The fourth revolution and the evolution of inforgs concern a transformation in our philosophical anthropology. It should not be confused with the sci-fi vision of a 'cyborged' humanity, or a revised version of the extended mind thesis. Walking around with something like a Bluetooth wireless headset implanted in your ear does not seem the best way forward, not least because it contradicts the social message it is also meant to be sending: being on call 24/7 is a form of slavery, and anyone so busy and important should have a personal assistant instead. The truth is rather that being a sort of cyborg is not what people will embrace, but what they will try to avoid, unless it is inevitable. I am not referring to the widespread phenomenon of 'mental outsourcing' and integration with our daily technologies either. This is interesting, but it is a vision still based on a Cartesian mind at the centre of the world, overflowing into the world. Nor am I referring to a genetically modified humanity, in charge of its informational DNA and hence of its future embodiments. This post-humanism, once purged of its most fanciful and fictional claims, is something that we may see in the future, but it is not here yet, both technically (safely doable) and ethically (morally acceptable), so I shall not discuss it. As I anticipated in the previous pages, I have in mind a quieter, less sensational, and yet more crucial and profound change in our conception of what it means to be an agent. We have begun to see ourselves as *inforgs* not through some transformations in our bodies but, more seriously and realistically, through the re-ontologization of our environment and of ourselves. It is our world and our metaphysical interpretation of it that is changing.

By re-ontologizing the infosphere and ourselves in it, as I shall argue in Chapter 11, ICTs have brought to light the intrinsically informational nature of human identity. This is not equivalent to saying that people have digital alter egos, some Messrs Hydes represented by their @s, blogs, and https. This trivial point only encourages us to mistake ICTs for merely *enhancing* technologies. Our informational nature should not be confused with a 'data shadow' either, an otherwise useful term introduced by Westin (1968) to describe a digital profile generated from data concerning a user's habits online. The change is more radical. To understand it, consider the distinction between *enhancing* and *augmenting* appliances (I shall return to this distinction in Section 14.4). The switches and dials of the former are interfaces meant to plug the appliance into the user's body ergonomically. Drills and guns are perfect examples. It is the cyborg idea. The data and control panels of augmenting appliances are instead interfaces between different possible worlds: on the one hand, there is the human user's *Umwelt*,[12] and on the other hand, there are the dynamic, watery, soapy, hot, and dark world of the dishwasher; the equally watery, soapy, hot, and dark but also spinning world of the washing machine; or the still, aseptic, soapless, cold, and potentially luminous world of the refrigerator. These robots can be successful because they have their environments 'wrapped' and tailored around their capacities, not *vice versa*. Imagine someone trying to build a droid like *Star Wars*' C3PO capable of washing dishes in the sink in exactly the same way as a human agent would. Now, despite some superficial appearances, ICTs are neither enhancing nor augmenting in the sense just explained. They are re-ontologizing devices because they engineer environments that the user is then enabled to enter through (possibly friendly) gateways. It is a form of initiation. Looking at the history of the mouse, for example, one discovers that our technology has not only adapted to, but also educated, us as users. Douglas Engelbart once told me that he had even experimented with a mouse to be placed under the desk, to be operated with one's knee, in order to leave the user's hands free. After all, we were coming from a past in which typewriters could be used more successfully by relying on both hands. Luckily, the story of the mouse did not go the same way the story of the QWERTY keyboard went. Today, we just expect to be able to touch the screen directly. HCI (Human–Computer Interaction) is a symmetric relation. Or consider current attempts to eliminate screens in favour of bodily projections, so that you may dial a telephone number by using a virtual keyboard appearing on the palm of your hand. No matter how futuristic, this is not what I mean. Imagine instead the possibility of dialling a number by indicating it with one's fingers or merely vocalizing them.

To return to the initial distinction, whilst a dishwasher interface is a panel through which the machine enters into the user's world, a digital interface is a gate through which a user can be (tele)present in the infosphere (more on telepresence in Chapter 3). This simple but fundamental difference underlies the many spatial metaphors of 'cyberspace', 'virtual reality', 'being online', 'surfing the web', 'gateway', and so forth. It follows that we are witnessing an epochal, unprecedented migration of humanity from its Newtonian, physical space to the infosphere itself as its *Umwelt*, not least because the latter is absorbing the former. As a result, humans will be inforgs among other (possibly artificial) inforgs and agents operating in an environment that is friendlier to informational creatures. And as digital immigrants like us are replaced by digital natives like our children, the latter will come to appreciate that there is no ontological difference between infosphere and physical world, only a difference in levels of abstraction (see Chapter 3). When the migration is complete, we shall increasingly feel deprived, excluded, handicapped, or impoverished to the point of paralysis and psychological trauma whenever we are disconnected from the infosphere, like fish

---

[12] The outer world, or reality, as it affects the agent inhabiting it.

out of water. One day, being an inforg will be so natural that any disruption in our normal flow of information will make us sick.

CONCLUSION

For some time, the frontier of cyberspace has been the human–machine interface. For this reason, we have often regarded ourselves as lying outside cyberspace. In his famous test, Turing (1950) posited a keyboard/screen interface to blanket human and computer. Half a century later, that very interface has become part of our everyday reality. Helped perhaps by the ubiquitous television and the part it has played in informing and entertaining us, we now rely on interfaces as our second skins for communication, information, business, entertainment, socialization, and so forth. We have moved inside the infosphere, the all-pervading nature of which also depends on the extent to which we accept its interface as integral to our reality and transparent to us (in the sense of no longer perceived as present). What matters is not so much moving bits instead of atoms—this is an outdated, communication-based interpretation of the information society that owes too much to mass-media sociology—as the far more radical fact that our understanding and conceptualization of the very essence and fabric of reality is changing. Indeed, we have begun to accept the virtual as reality. So the information society is better seen as a neo-manufacturing society in which raw materials and energy have been superseded by data and information, the new digital gold and the real source of added value. Not just communication and transactions then, but the creation, design, and management of information are the keys to the proper understanding of our hyperhistorical predicament.

All this means that ICTs are creating the new informational environment in which future generations will live most of their time. On average, Britons, for example, already spend more time online than watching TV, while American adults already spend the equivalent of nearly five months a year inside the infosphere.[13] This population is quickly ageing. According to the Entertainment Software Association, for example, in 2011, the average game player was 37 years old and had been playing games for 12 years; the average age of the most frequent game purchaser was 41 years old; 42 per cent of all game players were women, with women over the age of 18 representing a significantly greater portion of the game-playing population (37 per cent) than boys aged 17 or younger (13 per cent); and 29 per cent of Americans over the age of 50 played video games, an increase from 9 per cent in 1999.[14]

---

[13] See House of Commons, Information Committee, 'Digital; Technology: Working for Parliament and the Public', First Report of Session 2001–2 <http://www.publications.parliament.uk/pa/cm200102/cmselect/cminform/1065/1065.pdf>; and Office of National Statistics, 'Focus On The Digital Age, 2007 Edition', 15 March 2007 <http://www.ons.gov.uk/ons/rel/social-trends-rd/focus-on-the-digital-age/2007-edition/index.html>.

[14] Entertainment Software Association, 'Industry Facts' <http://www.theesa.com/facts/index.asp>.

Previous revolutions in the creation of wealth (especially the agricultural and the industrial ones) led to macroscopic transformations in our social structures and architectural environments, often without much foresight, normally with deep metaphysical, epistemological, and ethical implications. The information revolution—understood both in terms of wealth creation (third revolution) and in terms of re-self-conceptualization (fourth revolution)—is no less dramatic. We shall be in serious trouble if we do not take seriously the fact that we are constructing the new physical and intellectual environments that will be inhabited by future generations. In view of this important change in the sort of ICT-mediated interactions that we will increasingly enjoy with other agents, whether biological or artificial, and in our self-understanding, it seems that a fruitful way of tackling the new ethical challenges posed by ICTs is from an environmental approach, one which does not privilege the natural or untouched, but treats as authentic and genuine all forms of existence and behaviour, even those based on artificial, synthetic, hybrid, and engineered artefacts. The task is to formulate an ethical framework that can treat the infosphere as a new environment worth the moral attention and care of the human inforgs inhabiting it. Such an ethical framework must be able to address and solve the unprecedented challenges arising in the new environment. It must be an *e-nvironmental ethics* for the whole infosphere. This sort of *synthetic* (both in the sense of holistic or inclusive, and in the sense of artificial) *environmentalism* will require a change in how we perceive ourselves and our roles with respect to reality, what we consider worth our respect and care, and how we might negotiate a new alliance between the natural and the artificial. It will require a serious reflection on the human project. These are the topics of the rest of the book. Unfortunately, I suspect it will take some time and a whole new kind of education and sensitivity to realize that the infosphere is a common space, which needs to be preserved to the advantage of all. My hope is that the following chapters will contribute to such a change in perspective.

2

What is information ethics?

> If we are to go forward, we must go back and rediscover these precious values—
> that all reality hinges on moral foundations and that all reality has spiritual control.
>
> Martin Luther King Jr., 'Rediscovering Lost Values' (1992), vol. 2, p. 255.

SUMMARY

Previously, in Chapter 1, I highlighted some crucial transformations brought about by ICTs in our lives. I summarized them under the general concepts of a 'fourth revolution' and the 'hyperhistorical predicament', in order to stress the fact that we are experiencing a transformation in our philosophical anthropology and metaphysical outlook, not just a change in our technologies and how far they empower us. Towards the end of the chapter, I suggested that information ethics should be able to clarify and solve the ethical challenges arising in the infosphere. Such a statement is more problematic than it might seem at first sight. As we shall see in some detail in this chapter, in recent years, 'information ethics' (IE) has come to mean different things to different researchers working in a variety of disciplines, including computer ethics, business ethics, medical ethics, computer science, the philosophy of information, social epistemology, ICTs studies, and library and information science. This is not surprising. Given the novelty of the field, the urgency of the problems it poses, and the multifarious nature of the concept of information itself and of its related phenomena, a Babel of interpretations was probably going to be inevitable. It is, however, unfortunate, for it has generated some confusion about the specific *nature*, *scope*, and *goals* of IE. Fortunately, the problem is not irremediable, for a unified approach can help to explain and relate the main senses in which IE has been discussed in the literature. This approach will be introduced in Section 2.1. Once outlined, I shall rely on it in order to reconstruct, partly historically and partly conceptually, three *microethical* approaches to IE, in Sections 2.2–2.3. These will then be critically assessed in Section 2.4. In Section 2.5, I shall indicate how these three approaches can be overcome by a fourth, which I shall qualify as *macroethical*. The conclusion outlines how the next chapter investigates IE as a macroethical theory.

2.1 Introduction: a unified model of information ethics

A unified model of information ethics may be built, schematically, by focusing our attention on a typical, human moral agent A. Throughout this book, I shall refer to such agent as *Alice*. ICTs affect Alice's moral life in many ways. They may be simplified as follows.

Suppose Alice, our abstract moral agent A, is interested in pursuing whatever she considers her best course of action, given her predicament. We shall assume that Alice's evaluations and interactions have *some* moral value, but no specific value needs to be introduced at this stage. Intuitively, Alice can avail herself of some information (information as a *resource*) to generate some other information (information as a *product*), and in so doing affect her informational environment (information as *target*). This simple model, summarized in Figure 2, will help us to get some initial orientation in the multiplicity of issues belonging to information ethics. I shall refer to it as the R(esource) P(roduct) T(arget) or RPT model.

The RPT model is useful to explain, among other things, why any technology that radically modifies the 'life of information' epistemologically or ontologically is bound to have profound moral implications for any moral agent. Moral life is a highly information-intensive activity and ICTs, by re-ontologizing the informational context in which moral issues arise, not only uncover new aspects of old problems, but invite us to reconsider some of the grounds on which our ethical positions are based (I shall return to this point in Chapters 3, 4, and 5, when discussing the foundations of computer ethics).[1] At the same time, the model rectifies an excessive emphasis occasionally placed on specific technologies. This happens most notably in *computer* ethics (Johnson and Miller, 2009), *Internet* ethics (Langford, 2000), *machine* ethics (Anderson and Anderson, 2011), *nano*ethics (Allhoff, 2007), and *robo*-ethics (Wallach and Allen,

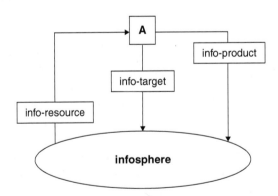

Figure 2. The 'external' R(esource) P(roduct) T(arget) model

---

[1] For a similar position in computer ethics see Maner (1996).

2009b). The RPT model calls our attention to the more fundamental phenomenon of information in all its varieties and long tradition. This was also Wiener's position:[2]

> To live effectively is to live with adequate information. Thus, communication and control [of information] belong to the essence of man's inner's life, even as they belong to his life in society. (Wiener, 1954, p. 18)

The various difficulties encountered in the conceptual foundations of information ethics are arguably connected to the fact that the latter has not yet been recognized as primarily an *environmental* ethics, the main concern of which is (or should be) the ecological management and well-being of the whole infosphere, as we shall see in Chapters 4 and 5. Instead, since the appearance of the first works in the eighties,[3] IE has been claimed to be the study of moral issues arising from one or another of the three distinct 'information vectors' in the RPT model. This, in turn, has paved the way for a fruitless compartmentalization and false dilemmas, with researchers either ignoring the wider scope of IE, or arguing as if only one 'vector' in the RPT model and its corresponding *microethics* (that is, a practical, field-dependent, applied, and professional ethics) provided *the* right approach to IE. The limits of such narrowly constructed interpretations of IE become evident once we look at each 'information vector' more closely, in what is a roughly chronological order of appearance of four stages.

2.2 First stage: IE as an ethics of informational resources

According to Froehlich (1997),[4] the expression 'information ethics' was introduced in the eighties, by authors such as Koenig et al. (1981) and Hauptman (1988), who then went on to establish the *Journal of Information Ethics* in 1992. It was used as a general label to discuss issues regarding information or data confidentiality, reliability, quality, and usage. Unsurprisingly, the disciplines involved were initially library and information science and business and management studies. They were only later joined by information technologies studies.

It is easy to see that this initial interest in information ethics was driven by concerns about information as a *resource* that should be managed efficiently, effectively, and fairly. Using the RPT model, this meant paying attention to the crucial role played by information as something extremely valuable for Alice's evaluations and actions, especially in moral contexts. Moral evaluations and actions have a large epistemic component, since Alice may be expected to proceed 'to the best of her information'; that is, we normally expect an agent to avail herself of whatever information she can muster in order to reach (better) conclusions about what she can or ought to do in

---

[2] The classic references here are Wiener (1950, 1954, 1961, 1964). Terry Bynum has convincingly argued that Wiener may be considered one of the founding fathers of information ethics (Bynum 1998, 2001a, 2010).

[3] An early review is provided by Smith (1996).

[4] For a reconstruction of the origins of IE see further Capurro (2006).

some given circumstances. Socrates had already argued that a moral agent is naturally interested in gaining as much valuable information as the circumstances require, and that a well-informed agent is more likely to do the right thing. The ensuing 'ethical intellectualism' analyses evil and morally wrong behaviour as the outcome of deficient information. We do evil because we do not know better, in the sense that the better the information management is the less the moral evil that is caused. Conversely, Alice's moral *responsibility* tends to be directly proportional to her degree of information: any decrease in the latter usually corresponds to a decrease in the former. This is the sense in which information occurs in the guise of judicial evidence. It is also the sense in which one speaks of an agent's informed decision, informed consent, or well-informed participation. In Christian ethics, even the worst sins may be forgiven in the light of the sinner's insufficient information, as a counterfactual evaluation is possible: had Alice been properly informed, she would have acted differently and hence would not have sinned. She might be forgiven because she is misinformed about what she is doing (*Luke* 23:34). In a secular context, Oedipus and Macbeth remind us how the inadvertent mismanagement of informational resources may have tragic consequences.

From an *information-as-resource* perspective, it seems that the machinery of moral thinking and behaviour needs information, and quite a lot of it, to function properly. However, even within the limited scope adopted by an analysis based solely on information as a resource, care should be exercised, lest all ethical discourse is reduced to the nuances of higher quantity, quality, intelligibility, and usability of informational resources. The more the better is not the only, nor always the best, rule of thumb. For the (sometimes explicit and conscious) withholding of information can often make a positive and significant difference. This is not to be confused with the head-in-the-sand problem, to which I will return in Chapter 10. It is rather a well-known, ethical strategy: Alice may need to lack (or intentionally preclude herself from accessing) some information in order to achieve morally desirable goals, such as protecting anonymity, enhancing fair treatment, or implementing unbiased evaluation. Famously, Rawls' 'veil of ignorance' exploits precisely this aspect of information-as-a-resource, in order to develop an impartial approach to justice (Rawls, 1999). Being informed is not always a blessing and might even be morally dangerous or wrong, distracting or crippling.

Whether the (quantitative and qualitative) presence or the (total) absence of information-as-a-resource is in question, it is obvious that there is a perfectly reasonable sense in which information ethics may be described as the study of the moral issues arising from 'the triple A': *availability*, *accessibility*, and *accuracy* of informational resources, independently of their format, type, and physical support. Rawls' position has been already mentioned. Since the eighties, other important issues have been unveiled and addressed by IE understood as an *ethics of informational resources*: the so-called *digital divide*, the problem of *infoglut*, and the analysis of the *reliability* and *trustworthiness* of information sources (Froehlich, 1997; Smith, 1997). Courses on IE, taught as part of information sciences degrees, tend to share this approach, as researchers in library and information sciences are particularly sensitive to such issues, also from a professional perspective (see e.g. Alfino and Pierce (1997); Mintz (1990); Stichler and Hauptman (1998)). One may recognize in this original approach to information ethics a position broadly defended by Van den Hoven (1995) and more recently by Mathiesen (2004), who criticizes Floridi and Sanders (1999) and is in turn criticized by Mather (2005). Whereas Van den Hoven purports to present this approach to IE as an enriching perspective, which contributes to the wider debate on a more broadly constructed conception of IE, Mathiesen appears to present her view, restricted to the informational needs and states of the individual moral agent, as the only correct interpretation of IE. Her position seems thus undermined by the problems affecting any univocal interpretation of IE, as argued by Mather.

2.3 Second stage: IE as an ethics of informational products

It seems that information ethics began to merge with computer ethics only in the nineties. Then, through the mature diffusion of personal computers and the Internet, the impact of ICTs became so widespread as to give rise to new issues not only in the management of information-as-a-resource by professional figures (librarians, journalists, scholars, scientists, IT specialists, and so forth) but also in the distributed and pervasive creation, consumption, sharing, and control of all kinds of information, by a very large and quickly increasing population of people online, commonly used to dealing with digital tools of all sorts (games, mobile phones, emails, the web, etc.). In other words, the Internet highlighted how IE could also be understood in a second, but closely related sense, in which information plays an important role as a *product* of Alice's moral evaluations and actions (Cavalier, 2005). To understand this transformation, let us consider the RPT model again.

Obviously, Alice is not only an information consumer but also an information producer, who may be subject to constraints while being able to take advantage of opportunities in the course of her activities. Both constraints and opportunities may call for an ethical analysis. Thus, IE, understood as an *ethics of informational products*, covers moral issues arising, for example, in the context of *accountability, liability, libel legislation, testimony, plagiarism, advertising, propaganda, misinformation, disinformation, deception*, and more generally of *pragmatic rules of communication à la* Grice. The debate on peer-to-peer (P2P) software provides a good example but, once again, this way of looking at information ethics is far from being a total novelty. Kant's classic analysis of the immorality of *lying* is one of the best-known case studies in the philosophical literature concerning this kind of information ethics. The boy crying wolf, Iago misleading Othello, or Cassandra and Laocoon, pointlessly warning the Trojans against the Greeks' wooden horse, remind us how the ineffective management of informational products may have tragic consequences. More generally, anyone exposed to mass media studies will have encountered these sorts of ethical issues.

It is hard to identify researchers who uniquely support this specific interpretation of IE, since works on IE as an ethics of informational *products* tend to be inclusive; that is, they tend to build on the first understanding of IE as an ethics of informational *resources* as well, and add to it a new layer of concerns (see e.g. Moore (2005)). However, the shift of the main focus, from the first to the second sense of IE (from resource to product), can be noted in some successful anthologies and textbooks, which were carefully revised when undergoing new editions. For example, Spinello (2003) explicitly highlights to a much greater extent the ethical issues arising in the hyperconnected society, when compared to the first edition (Spinello, 1997), thus emphasizing a sort of IE that is closer to the sense clarified in this section rather than in the previous one. Likewise, Severson (1997), after the typical introduction to ethical ideas, dedicates a long chapter to respect for intellectual property. Finally, it would be fair to say that the new perspective can be more often found shared, perhaps implicitly, by studies that are socio-legally oriented and in which IT-professional issues appear more prominently.

2.4 Third stage: IE as an ethics of the informational environment

The emergence of the information society has further expanded the scope of IE. The more people have become accustomed to live and work immersed within informational environments, the easier it has become to unveil new ethical issues involving informational realities. Returning to our initial model, independently of Alice's information input (informational resources) and output (informational products), in the nineties there appeared works highlighting a third sense in which information may be subject to ethical analysis, namely when Alice's moral evaluations and actions affect her informational environment. Think, for example, of Alice's respect for, or breach of, someone's information *privacy* or *confidentiality*. Although a rather old and classic problem, *hacking*, understood as the unauthorized access to a (usually computerized) information system, is another good example, because it shows the change in perspective quite clearly. In the eighties, it was not uncommon to mistake hacking for a problem to be discussed within the conceptual framework of an ethics of informational resources. This misclassification allowed the hacker to defend his position by arguing that no use (let alone misuse) of the accessed information had been made. Yet hacking, properly understood, is a form of breach of privacy. What is in question is not what Bob does with Alice's information, which has been accessed without authorization, but what it means for Alice's informational environment to be accessed by Bob without authorization. So the analysis of hacking belongs more accurately to an *ethics of the informational environment*. Other issues here include *security* (including issues related to information warfare, cyberwar, and terrorism), *vandalism* (from the burning of libraries and books to the dissemination of viruses), *piracy, open source software, freedom of expression, censorship, filtering,* and *contents control.* Mill's analysis 'Of the Liberty of Thought and Discussion' is a classic of IE interpreted as an ethics of the informational environment. Juliet, simulating her death, and Hamlet, re-enacting his father's homicide, show how the risky management of one's informational environment may have tragic consequences.

Works in this third stage in IE are characterized by environmental and global concerns. Rowlands (2000), for example, proposes an interesting approach to environmental ethics in terms of an ethics of naturalized *semantic* information. According to him,

There is value in the environment. This value consists in a certain sort of information, information that exists in the relation between affordances of the environment and their indices. This information exists independently of... sentient creatures.... The information is there. It is in the world. What makes this information valued, however, is the fact that it is valued by valuing creatures [because of evolutionary reasons], or that it would be valued by valuing creatures if there were any around. (p. 153)

Similar approaches foster the merging process of information ethics and computer ethics begun in the nineties (Woodbury, 2003), moving towards what Charles Ess has labelled ICE, *information and computer ethics* (Weckert and Adeney (1997) and Floridi (1995b, 1999a) were among the first works to look at IE as an environmental ethics). Concerning the topic of globalization of IE, Bynum and Rogerson (1996) soon became an important reference (but see as well Buchanan (1999) and Ess (2006)), together with the regular publication of *Ethics and Information Technology.*

2.5 Fourth stage: IE as a macroethics

So far we have seen that the RPT model, summarized in Figure 2, may help one get some initial orientation concerning the multiplicity of issues belonging to different interpretations of IE. Despite its advantages, however, the model can still be criticized for being inadequate, for at least two reasons.

First, the model is still too simplistic. Arguably, several important issues belong mainly, but not only, to the analysis of just one 'information vector'. The reader may have already thought of several examples that well illustrate the problem: someone's *testimony* (e.g. Iago's) is someone else's *trustworthy information* (i.e. Othello's); Alice's *responsibility* may be determined by the information that Alice holds ('apostle' means 'messenger' in Greek), but it may also concern the information that Alice issues (e.g. Judas' kiss); *censorship* affects Alice both as a user and as a producer of information; *misinformation* (i.e. the deliberate production and distribution of misleading information) is really an ethical problem that concerns all three 'informational vectors'; *freedom of speech* also affects the availability of *offensive content* (e.g. child pornography, violent content and socially, politically or religiously disrespectful statements) that might be morally questionable and should not be disseminated. Historically, all this means that the previous simplifications, associating decades with specific phases of the evolution of IE to specific approaches to IE, are just that, simplifications that should be taken with a lot of caution. The 'vectors' are normally much more twisted and entwined.

Second, the model is insufficiently inclusive. There are many important issues that cannot easily be placed on the map at all, for they really emerge from, or supervene on, the interactions among the 'information vectors'. Two significant examples may suffice: the 'panopticon' or 'big brother', that is, the problem of *monitoring and controlling* anything that might concern Alice; and the combined debate about information *intellectual property* (including copyright and patents legislation) and *fair use*, which affects both users and producers while shaping their informational environment.

Both criticisms are justified: the RPT model is indeed inadequate. Yet *why* it is inadequate is a different matter. The tripartite analysis just provided helps to structure both chronologically and analytically the development of IE and its interpretations. But it is unsatisfactory, despite its initial usefulness, precisely because any interpretation of information ethics based upon only one of the 'information vectors' is bound to be too reductive. As the examples mentioned above emphasize, supporters of narrowly constructed interpretations of information ethics as a *microethics* (that is, a one-vector-only ethics, to use our model) are faced with the problem of being unable to cope with a wide variety of many other relevant issues, which remain either uncovered as inexplicable, or shoehorned into an inadequate conceptual framework. In other words, the model shows that idiosyncratic versions of IE, which privilege only some limited aspects of the *information cycle*, are unsatisfactory. We should not use the model to attempt to pigeonhole problems neatly, which is impossible. We should rather exploit it as a useful first approximation to be superseded, in view of a more encompassing approach to IE as a *macroethics*; that is, a theoretical, field-independent, applicable ethics. Philosophers will recognize here the Hegelian *Aufhebung* or Wittgenstein's ladder, which can be used to reach a new starting point, but then can be discharged.

In order to climb up on, and then throw away, any narrowly constructed conception of information ethics, while keeping its advantages, a more encompassing approach to IE needs to

(1) bring together the three 'information vectors';
(2) consider the whole information-cycle; and
(3) analyse informationally all entities involved (including Alice and any other moral agents) and their changes, actions, and interactions, by treating them not apart from, but as part of the informational environment itself, or *infosphere*, to which they belong as information systems themselves.

Figure 3 illustrates the revised model.

Whereas steps (1) and (2) do not pose particular problems, and may be shared by any of the three approaches already seen, step (3) is crucial but involves an 'update' in the ontological conception of 'information' at stake. Instead of limiting the analysis to

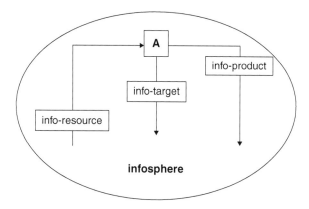

Figure 3. The 'internal' R(esource) P(roduct) T(arget) model

semantic contents—as any narrower interpretation of IE as a microethics inevitably tends to do—an ecological approach to information ethics also looks at information from an object-oriented perspective, and treats it as an entity as well. In other words, one moves from a (broadly constructed) epistemological conception of information ethics—in which information is roughly equivalent to news or semantic content—to one that is typically ontological, and treats information as equivalent to patterns or entities in the world. Thus, in the revised RPT model, the agent is embodied and embedded, as an information agent, in an equally informational environment.

The revision of our perspective, just introduced, requires what in Chapter 3 will be defined as a change in our *level of abstraction*. Here, the point may be illustrated with a simple analogy. Imagine looking at the whole universe from a chemical perspective. Every entity and process will satisfy a specific chemical description. Alice, for example, will be an open thermodynamic system consisting of approximately 70 per cent water and 30 per cent other elements (mainly carbon, then nitrogen, etc.). Now consider an informational perspective. The same Alice will be a cluster of data and processes; that is, not a chemical but an informational object. At this level of analysis, information systems as such, rather than just living systems in general, are raised to the role of agents and patients, that is, senders and receivers of actions, with environmental processes, changes, and interactions equally described informationally.

Understanding the *nature* of IE ontologically, rather than epistemologically, modifies the interpretation of its *scope* and *goals*. Not only can an ecological IE gain a global view of the whole life-cycle of information, thus overcoming the limits of other micro-ethical approaches, but it can also claim a role as a macroethics, that is, as an ethics that concerns the whole realm of reality, at an informational level of abstraction. This is what we shall see in Chapter 4. Indeed, the crucial point is to determine whether an informational level of abstraction is not only fruitful but preferable these days to a 'Newtonian' one of material things, brains, and bodies. Before entering into this discussion, however, it is clear that we need to understand the methodological tool itself, that is, the method of levels of abstraction. As mentioned above, this is required in order to shift our perspective on information from one that is exclusively epistemological and semantic (information about something or for something) to one which is *also* ontological (information as something; Section 16.2 shows what happens if this ontological perspective is missed). This is what we are going to see in the next chapter.

CONCLUSION

As a social organization and way of life, the information society has been made possible by a cluster of ICT-infrastructures. And as a full expression of *techne*, the information society has already posed fundamental ethical problems. Nowadays, a pressing task is to formulate an information ethics that can treat the world of data, information, and knowledge, with their relevant life-cycles, as a new environment, the *infosphere*, in which human beings, as informational organisms, may be flourishing. In this chapter, I have outlined a view of IE as a kind of macroethics, the kind that in Chapter 1 has been called *e-nvironmental ethics* or *synthetic environmentalism*. Such a view shifts our interpretation of information ethics from a microethical to a macroethical perspective by modifying our interpretation of information from an exclusively epistemological one to one that is also ontological. Such modification requires a change in our perspective. This perspective has been called a level of abstraction or simply LoA. The analysis of what a LoA is and how the method of LoAs works is the task of the next chapter. Once the methodological analysis is complete, I shall return to the investigation of IE as a macroethics. The reader already acquainted with the method of abstraction could easily skip to Section 3.4 to read about the application of the method to telepresence and two ethical problems that exemplify the application of the method itself.

3

The method of abstraction

> In the development of our understanding of complex phenomena, the most powerful tool available to the human intellect is abstraction.
>
> C. A. R. Hoare, *Notes on Data Structuring* (1972), p. 83.

SUMMARY
Previously, in Chapter 2, I suggested that the development of information ethics as a macroethics requires a change in our view of the world, in our ontological perspective. Such a change is made possible by the *method of (levels of) abstraction* (LoA). In this chapter, I present a simplified version of it, which will be used in the rest of the book. The reader interested in the theory behind the method of abstraction may wish to consult Chapter 3 of volume one (Floridi, 2011a), but neither this chapter nor the rest of this second volume presupposes any acquaintance with that material.[1]

The chapter is divided into two parts. The first part provides a brief introduction to the method of abstraction. In Section 3.1, I shall illustrate what a LoA is intuitively. In Section 3.2, I shall offer a definition of the basic concepts fundamental to the method. Although the definitions require some rigour, all the main concepts are introduced without assuming any previous knowledge. The definitions are illustrated by some simple examples, which are designed to familiarize the reader with the method. A comparison between levels of abstractions and interfaces is also developed in order to make the method more accessible and easily applicable in the following analyses. In Section 3.3, I shall conclude the first part by showing how the choice of a LoA commits a theory ontologically.

The second part provides a detailed application of the method to the analysis of telepresence (hereinafter simply *presence*, whenever the term causes no confusion). This is not a mere sandbox, since, by analysing a concrete case study methodologically, in this part of the chapter I shall seek to accomplish two other tasks. First, I shall articulate and defend a new theory of presence, which will be needed in the rest of the book, whenever agents present in the infosphere are in question. Second, I shall introduce the discussion

---

[1] Dijkstra (1968) and Parnas (1972) are two classic articles introducing the concept of LoA; Medvidovic et al. (1996) provides a review; and Hoare (1972) offers a more philosophical approach where the connection between Russell's theory of types and the concept of typed variable is made fully explicit.

of two important ethical issues, virtual pornography and informational privacy; I shall return to the latter in Chapter 12. More specifically, in Section 3.4 I shall introduce a case study. In Section 3.4.1, I shall reconstruct the standard model of *presence as epistemic failure*. In Section 3.4.2, I shall criticize it for being inadequate and replace it by a new model of *presence as successful observation*, based on the LoA method. I shall first apply it to virtual pornography in Section 3.4.3, criticize and refine it in Section 3.4.4, further test it against the analysis of informational privacy, and then show it to be effective. In the conclusion, I shall briefly summarize the results obtained and introduce the next chapter.

3.1 Introduction: on the very idea of levels of abstraction

Suppose Bob meets Alice at a party in Oxford. He would like to meet her again and, at the end of the evening, he asks her for her address. Her reply is: '56B Whitehaven Mansions, Charterhouse Square'. She does not specify the town, so Bob takes it to be an Oxford address, and asks no further questions. Bob has just assumed that Alice's address is a specific *observable* of a specific *level of abstraction* (LoA). Call the latter OXFORD. The following day, Bob checks the address using an online Oxford map. The interface (LoA) returns no data matching that observable, that is, the OXFORD LoA does not provide access to that observable address. Frustrated, Bob changes LoA, and uses ENGLAND instead; same level of granularity (addresses), but much wider scope of the LoA/interface. Nothing. He further expands his LoA to UK, but still fails to recover Alice's address. In desperation, he googles it. Now the LoA has completely changed, not only in scope, but also in granularity and type. All manner of information online related to that address will be returned. The first entry makes Bob feel like a fool: '56B Whitehaven Mansions, Charterhouse Square' is indeed a place in Smithfield, London W1. But it is the home and work address of a retired Belgian police officer, Monsieur Hercule Poirot. Alice used the LoA of a novel to misinform him.

The idea of a 'level of abstraction' plays an absolutely crucial role in how we handle any information process, and so in how we negotiate our interactions with the world, and therefore in how we develop our philosophy of information, including our information ethics. This is so even when a specific LoA is wrong or left implicit, as in Alice's reply to Bob. Using a different example, whether Bob observes the presence of oxygen at the party depends on the LoA at which he is operating; to abstract it—to assume that oxygen is not what matters in that context—is not to overlook its vital importance, but merely to acknowledge its lack of immediate relevance to the current discourse, which could always be extended to include oxygen, should that become an interesting observable. So what is a LoA exactly?

The method of abstraction comes from modelling in science, where the variables in the model correspond to observables in reality, all others being abstracted. The terminology has been influenced by an area of computer science, called Formal

Methods, in which discrete mathematics is used to specify and analyse the behaviour of information systems. Despite that heritage, the idea is not at all technical and, for the purposes of this book, no mathematics is required. Let us begin with another everyday example.

Suppose we join Alice, Bob, and Carol at a party. They are in the middle of a conversation. Alice is a collector and potential buyer; Bob tinkers in his spare time; and Carol is an economist. We do not know the subject of their conversation, but we are able to hear this much:

- Alice observes that it (whatever 'it' is) has an anti-theft device installed, is kept garaged when not in use, and has had only a single owner;
- Bob observes that its engine is not the original one, that its body has been recently re-painted but that all leather parts are very worn;
- Carol observes that the old engine consumed too much, that it has a stable market value, but that its spare parts are expensive.

The participants view the system under discussion according to their own interests, which orient the choice of their conceptual interfaces or, more precisely, of their own levels of abstraction. We may guess that they are probably talking about a car, or perhaps a motorcycle, but it could be an airplane, since any of these three systems would satisfy the descriptions provided by A, B, and C above. Whatever the reference is, it provides the source of information under discussion. We shall call it the *system*. A LoA consists of a collection of observables, each with a well-defined possible set of values or outcomes. For the sake of simplicity, let us assume that Alice's LoA matches that of an owner, Bob's that of a mechanic, and Carol's that of an insurer. Each LoA (imagine a computer interface) makes possible a determinate analysis of the system. We shall call the result or output of such analysis a *model* of the system. Evidently, a system may be described at a range of LoAs and so can have a range of models. We are now ready for a more formal definition.

3.2 The definition of a level of abstraction

The term 'variable' is commonly used throughout science for a symbol that acts as a place-holder for an unknown or changeable referent. A 'typed variable' is understood as a variable qualified to hold only a declared kind of data. For example, if Bob asks Alice for her telephone number, whatever the latter is (variables), he expects natural numbers to be the TYPE of the variables she will provide. Since the system investigated may be entirely abstract or fictional—recall the example of Monsieur Poirot's address—the term 'observable' should not be confused here with 'empirically perceivable'. Historically, it might be an unfortunate terminological choice, but, theoretically, an *observable* is just an *interpreted typed variable*; that is, a typed variable together with a statement of what feature of the system under consideration it represents, for example a set of data could have NATURAL NUMBERS as a type and *telephone number* as a feature of the system. A LoA is (usually) a finite but non-empty set of observables, which are expected to be the building blocks in a theory characterized by their very choice. An 'interface' (called a 'gradient of abstractions') consists of a collection of LoAs and is used in analysing a system from varying points of view or at varying LoAs.

We saw that models are the outcome of the analysis of a system developed at some LoA(s). The *method of abstraction* consists in formalizing the model, often implicitly, and only in *qualitative* rather than quantitative terms, by using the concepts just introduced (and others relating to system behaviour which we do not need here). In the previous example, Alice's LoA might consist of observables for SECURITY, METHOD OF STORAGE, and OWNER HISTORY; Bob's might consist of observables for ENGINE CONDITION, EXTERNAL BODY CONDITION, and INTERNAL CONDITION; and Carol's might consist of observables for RUNNING COST, MARKET VALUE, and MAINTENANCE COST. For the purposes of discussion, the interface might consist of the set of all three LoAs. In this case, the LoAs happen to be disjoint, but in general they do not have to be. LoAs can be nested, disjoint, or overlapping and may be, but do not have to be, hierarchically related or ordered in some scale of priority, or support some syntactic compositionality (the molecular is composed of atomic components). A particularly important case is that in which one LoA includes another. Suppose, for example, that Dave joins the discussion and analyses the system using a LoA that includes those of Alice and Bob. Dave's LoA might match that of a buyer. Then Dave's LoA is said to be more concrete, finely grained, or lower than Alice's or Bob's, each of which is said to be more abstract, more coarsely grained, or higher; for both Alice's and Bob's LoAs abstract some observables that are available at Dave's. Basically, Dave can obtain all the information about the system that Alice and Bob might have, for example the name of the previous owner, and that it is rather expensive to maintain, and so he can obtain some information that is, in principle, unavailable to one or the other of them, since Alice does not know about running costs and Bob has no clue about the ownership history.

A LoA qualifies the level at which a system is considered. In the following chapters, I shall rely on the method of abstraction in order to refer to the LoA at which the properties of the system under analysis can sensibly be discussed. In general, it seems that many uninteresting disagreements might be clarified, if the various interlocutors could make their LoAs explicit and precise. Yet a crucial clarification is in order. It must be stressed that a clear indication of the LoA at which a system is being analysed allows *pluralism* without endorsing *relativism*. Elsewhere I have called such middle-ground *relationism*. Since I shall discuss epistemic and ethical relativism quite often in the rest of this book, this might be the right place to offer a clarification. When I criticize a position as *relativistic*, or when I object to *relativism*, I do not mean to equate such positions to non-absolutist, as if there were only two alternatives, e.g. as if either moral values were absolute or relative, or truths were either absolute or relative. The method of abstraction enables one to avoid exactly such a false dichotomy, by showing that a subjectivist position, for example, need not be relativistic, but only relational. To use a simple example: Alice may be tall when compared to Bob, but not when compared to someone in the basketball team. It does not mean that her measure changes, but only that she is or is not tall depending on the frame of reference, that is, on the LoA. I equate relativism, when I criticize it, to an 'anything goes' position. Now, it is a mistake to think that 'anything goes' as long as one makes explicit the LoA, because LoAs are mutually comparable and assessable. Consider again the example of Alice's telephone number. There might be some significant differences in the way in which Alice communicates it to Bob. She might add a plus and the relevant country code at the beginning, thus modifying the overall TYPE of the information provided. She might omit the plus, the country code, and the city code, if it is a local mobile phone. So there is quite a lot of 'relationism' ('it depends on . . . ') but no 'relativism': it would be silly to conclude that any LoA would do. A string of letters would not work, nor would a mix of letters and numbers, or numbers and non-alphanumeric symbols, or an endless string. Using a different example, when we are asked to provide the number of our credit card, the type is (a finite number of) natural numbers. This is why an interface can easily constrain the sort of input required. In general, only some LoAs are possible and, among those, some are better than others. Crucially, the assessment and corresponding preference is usually dictated by the purpose driving the original request of information. Introducing an explicit reference to the LoA clarifies that:

(1) the model of a system is a function of the available observables;
(2) different interfaces may be correctly ranked depending on how well they satisfy modelling specifications (e.g. informativeness, coherence, elegance, explanatory power, consistency with the data, etc.) and the purpose orienting the choice of the LoA (LoAs are teleologically oriented); and
(3) different analyses can be correctly compared provided that they share the same LoA.

Let us now agree that a system is characterized at a given LoA by the properties it satisfies at that LoA (Cassirer, 1953). We are interested in systems that change, which means that some of those properties change value. The evolution of a changing system is captured at a given LoA and at any instant by the values of its observables (the attributes of the system). Thus, a system can be thought of as having states, determined by the value of the properties that hold at any instant of its evolution, for then any change in the system corresponds to a state change and *vice versa*. Generalizing, this enables one to view any system as having states and transitions. The lower the LoA, the more detailed the observed changes and the greater the number of state components required to capture the change. Each change corresponds to a transition from one state to another. A transition may be non-deterministic. Indeed, it will typically be the case that the LoA under consideration abstracts the observables required to make the transition deterministic. As a result, the transition might lead from a given initial state to one of several possible subsequent states.

We have now moved from a static to a dynamic observation of a system, analysed as a transition system. We shall see in Chapter 7 that the notion of 'transition system' provides a convenient means to support the identification of the necessary and sufficient criteria for agency, being general enough to embrace the usual notions like automaton and process. In scientific investigations, it is frequently used to model interactive phenomena. Here we need only the idea; for a formal treatment of much more than is required in this context, the reader might wish to consult (Arnold and Plaice, 1994).

A *transition system* comprises a (non-empty) set S of states and a family of operations, called the *transitions* on S. Each transition may take input and may yield output, but, at any rate, it takes the system from one state to another and in that way forms a relation on S. If the transition does take input or yield output, then it models an interaction between the system and its environment and so is called an *external* transition; otherwise the transition lies beyond the influence of the environment (at the given LoA) and is called *internal*. It is to be emphasized that inputs and outputs are, like states, observed at a given LoA. Thus, the transitions that model a system are dependent on the chosen LoA. At a lower LoA, an internal transition may become external; at a higher LoA an external transition may become internal.

Returning to our example, the system being discussed by Alice might be further qualified by state components for LOCATION, WHETHER IN-USE, WHETHER TURNED-ON, WHETHER THE ANTI-THEFT DEVICE IS ENGAGED, HISTORY OF OWNERS, and ENERGY OUTPUT. The operation of garaging the system might take as input a driver, have the effect of placing the system in the garage with the engine off and the anti-theft device engaged, leaving the history of owners unchanged, and outputting a specific amount of energy. The 'in-use' state component could non-deterministically take either value, depending on the particular instantiation of the transition. Perhaps the system is not in use, being garaged for the night; or perhaps the driver is listening to a programme broadcast on its radio in the quiet solitude of the garage. The precise definition depends on the LoA. Alternatively, if speed were observed but time, accelerator position, and petrol consumption abstracted, then accelerating to 60 miles per hour would appear as an internal transition.

3.3 Abstraction and ontological commitment

We can now use the method of abstraction and the concept of LoA to make explicit the ontological commitment of a theory, in the following way.

A theory comprises at least a LoA and a model. The LoA allows the theory to analyse a given system and to elaborate a model that identifies some properties of the system at the chosen LoA (see Figure 4).

The ontological commitment of a theory can be clearly understood by distinguishing between a *committing* and a *committed* component within the scheme. A theory commits itself ontologically by opting for a specific LoA. Compare this to the case in which one has chosen a specific kind of car (say a Volkswagen Polo) but has not bought

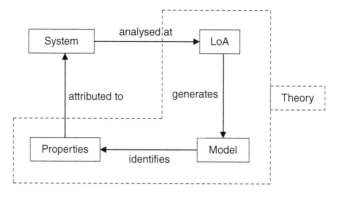

Figure 4. The scheme of a theory

one yet. On the other hand, a theory is ontologically committed in full by its model, which is therefore the bearer of the specific commitment. The analogy here is with the specific car one has actually bought (that blue, four-wheeled, etc. specific vehicle in the car park that one owns). To summarize, by adopting a LoA, a theory commits itself to the existence of some specific types of observables characterizing the system and constituting the LoA (by deciding to buy a Volkswagen Polo one shows one's commitment to the existence of that kind of car), while, by adopting the ensuing models, the theory commits itself to the corresponding tokens (by buying that particular vehicle, which is a physical token of the type Volkswagen Polo, one commits oneself to that token, e.g. one has to insure it). Figure 5 summarizes this distinction.

By making explicit the ontological commitment of a theory, it is clear that the method of abstraction plays an absolutely crucial role in ethics. For example, different theories may adopt androcentric, anthropocentric, biocentric, or ontocentric LoAs, even if this is often left implicit. IE is committed to a LoA that interprets reality—that is, any system—informationally. The resulting model consists of information systems and processes.

We shall see that an informational LoA has many advantages over a biological one, adopted by other forms of environmental ethics. Here, it can be stressed that, when any

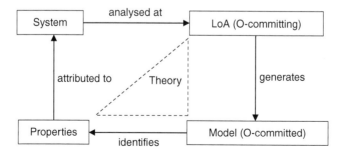

Figure 5. The SLMS scheme with ontological commitment

36 THE ETHICS OF INFORMATION

other level of analysis becomes irrelevant, IE's LoA can still provide some minimal normative guidance. That is, when even land ethics fails to take into account the moral value of 'what (the theory is committed to treat as what) there is', I shall argue that IE still has the conceptual resources to assess the moral situation and indicate a course of action.

A further advantage of an informational LoA is that it allows the adoption of a unified model for the analysis of the three information vectors and their environment in the RPT model we saw in Chapter 2. In particular, this means gaining a more precise and accurate understanding of what can count as a moral agent and a moral patient, as we shall see in Chapters 6 and 7.

The introduction of the method of abstraction and the basic concepts concerning the levels of abstraction is now complete. We are ready to move to the second part of this chapter, dedicated to the application of the method itself to a specific case.

3.4 An application of the method of abstraction: telepresence

Telepresence is a philosopher's gold mine. It is such a rich concept and experience, a phenomenon so intuitive and yet so difficult to capture in all its nuances and implications that its potential as a source of philosophical questions and insights seems inexhaustible.[2]

Some of the classic issues in philosophy could easily be re-conceptualized as problems concerning (tele)presence.[3] Examples include action at distance; the semantics of possible worlds understood as the availability and accessibility of spaces different from the actual; the tension between appearance and reality (where is the agent, really?) and the issuing sceptical challenges (is the agent's brain inside or outside a vat on Alpha Centauri?); testimony as 'knowledge at distance' in time as well as in space; the nature of individual identity in different contexts; the mind–body problem; consciousness as awareness of 'there-being'. Heidegger without the semantics of presence would be inconceivable,[4] and Christian theology has been struggling for centuries with the idea of omnipresence as one of the most significant of God's attributes.[5]

The previous list could easily be expanded but this is not the task of the rest of this chapter. Instead, in the following pages I wish to show how the method of abstraction

---

[2] Ijsselsteijn and Harper (2001) still provide a good introduction to telepresence that prepares the ground for the philosophical debate.

[3] Goldberg (2000) is a collection of essays concerning several philosophical themes related to telepresence.

[4] For an ecological Heideggerian–Gibsonian approach to telepresence, see e.g. Zahorik and Jenison (1998).

[5] The debate on divine presence from a telepresence research perspective is reviewed in Biocca (2001), who argues against several conceptual confusions in Sheridan (1999)—who builds on earlier work by Zahorik and Jenison (1998)—and in Mantovani and Riva (2001). On the same debate see further Lauria (2001).

may be usefully applied to such an important phenomenon and crucial concept. In so doing, we will explore the conceptual foundation of telepresence theory and investigate two ethical implications of telepresence brought about by ICTs.

3.4.1 Presence as epistemic failure

Presence is notoriously a polymorphic phenomenon and a polysemantic concept.[6] However, after almost twenty-five years of research—Minsky (1980) is usually acknowledged as having pioneered presence studies—some convergence on a general conceptual framework has begun to emerge.[7] In current studies, presence is often understood as a *type of experience* of 'being there', one loosely involving some technological mediation and often depending on virtual environments. An authoritative and influential source like the International Society for Presence Research (ISPR), for example, endorses the following analysis (asterisks and numbering in the original):

[1] Presence (a shortened version of the term 'telepresence') is a psychological state or subjective perception in which even though part or all of an individual's current experience is generated by and/or filtered through human-made technology, part or all of the individual's perception fails to accurately acknowledge the role of the technology in the experience. Except in the most extreme cases, the individual can indicate correctly that s/he is using the technology, but at *some level* and to *some degree*, her/his perceptions overlook that knowledge and objects, events, entities, and environments are perceived as if the technology was not involved in the experience. Experience is defined as a person's observation of and/or interaction with objects, entities, and/or events in her/his environment; perception, the result of perceiving, is defined as a meaningful interpretation of experience.[8]

This standard view of presence has been popular at least since the work of Lombard and Ditton (1997). It consists of three fundamental steps:

(1) presence is reduced to a *type of perception*, visual perception for instance. An example could then be seeing some geographical shapes and colours;

(2) the type of perception in (1) is then specified, cognitively, as a special *kind of experience*, namely a psychological, subjective, meaningful interpretation of the experienced; for example, experiencing the above-mentioned colours and shapes as a specific type of environment, e.g. a valley on Mars;

(3) the special kind of experience in (2) is further qualified, semantically, as a *perception of contents* that *fails*, at least partially, momentarily or occasionally, to be a perception of its machine-mediated nature as well; in our example, this means having the impression of being on Mars, failing to realize that it is actually a computer-mediated environment.

---

[6] Schuemie et al. (2001) and Ijsselsteijn et al. (2000) provide two surveys of several ways in which presence has been interpreted and analysed, but see further Lombard and Ditton (1997).

[7] See e.g. Sacau et al. (2003) and Ijsselsteijn et al. (2000).

[8] International Society for Presence Research (2000). *The Concept of Presence: Explication Statement* <http://ispr.info/about-presence-2/about-presence>.

Since these three steps are primarily epistemic, one may refer to (1)–(3) as a model of presence as *epistemic failure* (the EF model). The roots of the EF model are *philosophically Cartesian* and *culturally mass-mediatic*, in the following sense.

The philosophically Cartesian nature of EF can be evinced from the priority assigned to the understanding of presence in epistemic terms. In the quotation above from the International Society for Presence Research website, for example, even the reference to interaction is actually a reference to the *perception* of interaction.[9] When Descartes speculates in the *Meditations* on the possibility of living in a dream or, as we would say nowadays, in a Matrix-like simulation, somehow artificially generated by a malicious yet omnipotent demon, the emphasis is precisely on the completely realistic perception of the environment, despite the possibility of an unperceived mediation that makes the perception itself possible, even though the environment, and thus our presence within it, is entirely fictional. Descartes construes the sceptical challenge in terms of a fundamental tension between the actual experience of something—e.g. Descartes being in his room, in front of the fire, looking at his hands—and the possibility of its (i.e. of the perception's) unreliability as a source of access to, or presence in (front of), the real nature of the experienced environment.

The EF model is eminently modern, strictly related as it is to that priority of epistemology over ontology that characterizes philosophy after the scientific revolution, from Descartes to Kant. The mass-mediatic character of EF (see especially Lombard and Ditton (1997)) is a reasonable consequence of its Cartesian roots. For modernity—known for the primacy it attributes to knowledge and epistemology—makes increasing room for (one may argue that it was bound to lead to) a culture in which the production (fiction) and representation (communication) of realities become socially and psychologically predominant. Simplifying: having placed knowledge at centre stage for some centuries, Western thought made the next move almost inevitable, namely promoting the products of knowledge, and hence the world of information (the semantic infosphere), as the primary environment inhabited by the human mind. Correspondingly, the understanding of presence mutates from

(a) the mere possibility of an epistemic failure to perceive the difference between what is and what is not real, leading to the sceptical challenge; to
(b) the actual engagement with realities that are known to be artificial or synthetic because demonstrably constructed through the (mass-) mediation of increasingly powerful technologies, which replace Cartesian dreams and demons as the condition of possibility of the experience.

---

[9] A representative case of a Cartesian approach is Biocca (2001), who defends an approach based on the philosophy of mind and the classic mind–body dualism of Cartesian origins. Note that, although Biocca seems justified in criticizing some metaphysical approaches, this is not a reason to consider Cartesianism as the only available alternative.

In EF, the logical possibility of failure—e.g. one may be dreaming—becomes the wilful failure to perceive the technology that may be making one dream.

To summarize, the EF model promotes an understanding of presence as the Cartesian failure to recognize the technologically (mass-)mediated nature of the experiences enjoyed by an epistemic agent. As a consequence, EF enables one to catalogue as presence a variety of radically different phenomena otherwise largely unrelated, including oral and textual representations, immersions in VR scenarios, radio narratives, online games, television and cinema, tele-robotics, etc.; witness the very wide scope of MIT's journal *Presence: Teleoperators and Virtual Environments*, the main peer-reviewed journal in the field.

It is unclear whether the very wide scope of EF is actually an advantage—providing a conceptual reduction of a broad spectrum of phenomena to a single, unifying frame of interpretation—or arguably the sign of some serious misunderstanding. Several reasons may incline one to take the latter view.

EF embeds an unresolved tension between the subjective, introspective, single-agent understanding of presence—which the model inherits from a Cartesian approach—and the social, public, intra-subjective, and multi-agent understanding of presence proper to a mass-mediatic and new social-media approach. Is telepresence a personal and private experience or is it something made possible only by social interaction? Is solipsistic telepresence an oxymoron? Consider just ordinary presence, not telepresence: was Robinson Crusoe present (did he feel present) on the island before meeting Friday? Of course, there is no straightforward answer to this type of question, because, trivially and somewhat boringly, it all depends on what one means by '(tele)presence'. However, the fact that similar questions are reasonably prompted by the EF model and yet appear so poorly posed is evidence that there might be something wrong with EF itself. The model starts looking like a position that allows misconceived questions to be asked, the sorts of question that make research go amiss. This suspicion paves the way to another, more substantial criticism. EF manages to be, in different ways, both too exclusive and too inclusive, resulting in being literally eccentric with respect to its correct focus.

On the one hand, by adopting an anthropocentric perspective—typically Cartesian—the model considers any investigation of cases of presence *of* (not just *through*, or *by means of*) artificial agents beyond its scope, and this is because, at least at this stage in the evolution of artificial intelligence (AI), no machine is capable of subjective experience of any sort, let alone one of a Cartesian nature. Yet telepresent robotics is not just about devices remotely controlled by human operators, it is also and significantly about devices that are able to be present remotely by telecontrolling other devices, while keeping human agents entirely out of the loop, or at most on the loop but not in it, as mere external observers. Along the same lines, even if more hypothetically, it is hard to see how EF can analyse the concept of presence when the agent involved is a hybrid inforg, that is, an agent who may enjoy some technologically mediated experiences of presence while, at the same time, perceiving them precisely as mediated (e.g. one eye sees only the dark night, its stars and moon, the other enjoys an active infrared night vision and hence a detailed, monochrome scene through a display device). On the other hand, the EF model grants full citizenship in the realm of telepresence studies to experiences such as reading a novel or watching a movie, an oddity that causes a loss of specificity and an irrecoverable metaphorization of the concept of 'presence at distance'.

Such a metaphorical way of approaching presence can be related to a further difficulty. EF provides a *merely negative understanding* (more on this presently) of presence—as failure to perceive the technologically-mediated nature of the experience—and this is bound to be unsatisfactory. The approach by negation (*per via negativa*) means that one attempts to define or conceptually capture a *definiendum* by saying what the *definiendum* is not. It may work with dichotomies and Boolean concepts: for example, if one understands what 'left' means one may also understand the meaning of 'right' negatively, as 'not-left'. It is a standard method in mathematics, where the method of false position or *regula falsi* helps one to estimate the roots of a nonlinear equation $f(x) = 0$ by an iterative process of 'false' approximations. But 'failure to perceive' fails itself to be either a Boolean description or a precise concept that can be further refined by iteration. It is comparable to defining a zebra as not quite a horse but close: it includes far too many things (might it be like a centaur? A mule? A camel?) and, although correct, it begs the question, since we might as well speak of a zebra as not a donkey but almost. That we speak of a zebra in terms of not being a horse—that we conceptualize presence as epistemic failure—only shows that we do already possess and implicitly rely upon some fairly detailed idea of what we wish to define—the zebra in front of us or at least in our memory, the actual experience of being telepresent—but that we surrender to the difficulty of providing a tight conceptual analysis. Instead, we opt for what is in fact a merely generic indication, a 'you know what I mean'. Such hand-waving brings us to a further problem.

EF allows odd cases of *nested telepresence*. Consider the *Odyssey*. A large part of Odysseus' adventures is recounted by Odysseus himself after having landed on Scheria, Phaeacians' island. One of these adventures is the encounter and blinding of Polyphemus on the Cyclops' island. According to EF, the reader, by being in Homer's narrative space, is also telepresent in Scheria where, by listening to Odysseus, she is also telepresent on the Cyclops' island. Only a semantic space can allow this nesting. But then only a metaphorical sense of 'telepresence' may be at work here, since this nesting has nothing to do with the ordinary set-theoretic sense in which, by being telepresent in a given space S_1, say a hotel room, one is also telepresent in the space S_2 that includes S_1, say in the hotel where the room is. In the latter case, the co-(tele)presence is a logical necessity. In the former case, it can only be a matter of possible mental experience, since the Cyclops' island is not 'inside' Scheria.

We have now reached the last problem. EF cannot clearly define *absence*. This is not a philosophical gimmick. Any conceptual analysis of telepresence should also be able to discriminate between, and possibly explain, cases of unachieved telepresence, of failure or interruption of telepresence, of faulty or insufficient telepresence, in short, of 'not being there'. So here lies another clear sign that the EF model is unsatisfactory. Fortunately, it is also the condition of possibility for a better approach, the end of the *pars destruens* and the beginning of the *pars construens*. Consider a counterfactual analysis: had the agent *not* failed to perceive the technologically mediated nature of her experience, she would not have been telepresent. This is the inevitable logical consequence of EF, but it is also a *reductio ad absurdum*. For surely the doctor teleoperating on a patient is still present, independently of her perception (or lack thereof) of the technological mediation. Surely the soldier is still telepresent on the minefield through a robot, despite all the possible perception of the artificial nature of the experience. The same holds true for the pilot controlling a drone, the IT technician fixing a customer's laptop remotely, and so forth. The fact is that epistemic failure is not the right criterion to identify cases of telepresence. The good news is that, precisely by focusing on absence, we can gain a better perspective on presence and hence acquire a vantage point to frame some relevant ethical issues. This is where the application of the method of levels of abstraction finally makes its entry as an essential tool.

3.4.2 Presence as successful observation

Concentrating on absence has the immediate advantage of clarifying that speaking of presence in a vacuum of references makes little sense. Something is present or absent only for an observer and only at a given LoA.[10] We saw that a LoA is a specific set of typed variables. Intuitively, it is representable as a (dynamic) interface. Through a LoA, the observer accesses the environment, so a LoA could be, for example, the five senses unaided, a microscope, a CCTV camera, or a Geiger counter. Consider now a security system with a motion detector (Figure 6). In the past, motion detectors triggered an alarm whenever a movement was detected within the range of the sensor, including the swinging of a tree branch (object *a* in Figure 6). The old LoA_1 consisted of a single

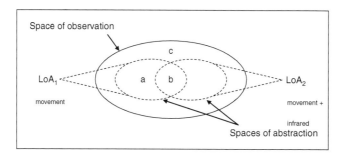

Figure 6. An example of levels of abstraction

---

[10] Ijsselsteijn (2002) seems to defend a similar perspective.

typed variable, which may be labelled MOVEMENT. Nowadays, when a passive infrared (PIR) motion detector registers some movement, it also monitors the presence of an infrared signal, so the entity detected has to be something that also emits infrared radiation—usually perceived as heat—before the sensor activates the alarm. The new LoA$_2$ consists of two typed variables: MOVEMENT and INFRARED RADIATION. Clearly, a cat (object *b* in Figure 6) walking in the garden is present for both LoAs, but for the new LoA$_2$, which is more finely grained, the branch of the tree swinging is absent. Likewise, a stone in the garden (object *c* in Figure 6) is absent for both the new and the old LoA, since it satisfies no typed variable of either one.

What the two sensors detect (the word is used here in a purely engineering sense of extracting data from a signal) is movement, a change in the environment, some form of action (e.g. walking) or interaction (e.g. interrupting the flow of a signal), or a transition in the system. More generally, this is one of the two senses in which something is present or absent in a space of observation, that is, as a *dynamic source of action/interaction* or *change*. The other sense is as a *static property-bearer*. The immobile branch of the tree is absent both for the old-fashioned sensor and for the new PIR sensor. It is still absent for the latter, even if it moves, because it fails to satisfy another typed variable, the infrared one. The cat, on the contrary, is constantly (i.e. non-intermittently) present for an infrared sensor, even if it does not move, because it is a heat-generator. Clearly, the method of LoA is an efficient way of making explicit and managing the ontological commitment of a theory of presence. This is crucial. Mantovani and Riva (2001), for example, acknowledges that

> the meaning of presence depends on the concept we have of reality (from the ontology which we more or less explicitly adopt) and that different ontological positions generate different definitions of presence, telepresence and virtual presence. (p. 541)

It seems that what is needed is a method of levels of abstraction. The method clarifies that *to be present is to be the value of a typed variable of a LoA* (to paraphrase Quine). To be absent is, of course, to fail to be any such value. This view is consistent with the general thesis, defended in Mantovani and Riva (1999), that presence is an ontology-dependent concept. The social construction of presence, further supported by Mantovani and Riva (1999), may be interpreted as a specific case of the broader view articulated in this chapter.

As we have just seen, depending on the class of typed variables in question, there might be three ways of being (simply, not yet tele-) present/absent at a given LoA:

(1) as a source of action/interaction;
(2) as a property-bearer;
(3) as both (1) and (2).

Without clause (2), one would be unable to define forms of 'passive' presence. Thus, a model according to which 'presence is tantamount to successfully supported action in the environment' (Zahorik and Jenison, 1998) would fail to acknowledge the fact that some *x* might be present even without any observable (let alone successful) interaction between *x* and *x*'s environment. Of course, a solution would be to modify our understanding of 'interaction' and 'environment', but this seems rather *ad hoc*. A more fruitful alternative is to accept that any analysis of presence requires the identification of a space of observation and a level of abstraction. Unperceivable subatomic particles are known to be present from their actions and our interactions. The sofa in the room is present because of its perceivable qualities. The flame of a candle in the room is present because of both. Absence may be equally gradual.

If we now extend the previous analysis to telepresence, the easiest thing is to refer to the new model as being based on *successful observation* (SO), thus:

(SO) an *x* observable at a given LoA in a local space of observation (LSO) is also telepresent in a remote space of observation (RSO) if and only if that *x* is also observable in RSO at a given LoA.

Note that LSO and RSO need to be different (LSO ≠ RSO), whereas the two LoAs may but do not have to be identical (see Figure 7).

The new model shifts the perspective from an internal and subjective assessment of a particular experience—presence as epistemic failure—to an external and objective evaluation—presence as successful observation—which requires an explicit definition of the LoAs adopted in the process of analysis. This has at least four major advantages.

The first and most obvious is that, contrary to the EF model, the SO model provides a clear criterion of discrimination between what does and what does not count as telepresence. It thus regiments very clearly the applicability of the concept, which now excludes intentional experiences that may be technologically enabled but are in themselves merely psychological—such as reading, listening to the radio or watching a movie—but includes standard cases of presence, such as operating in virtual environments (from immersive virtual realities and MMORPG to old-fashioned, text-based virtual worlds such as MUDs, MOOs, IRC, and chats), remotely controlling other artificial agents, being a member of a digital community, and playing online. Since there is no presence in a remote space unless the entity in question is observable there at some given LoA, one cannot be present on Scheria, but can be present on an island in *Second Life*.

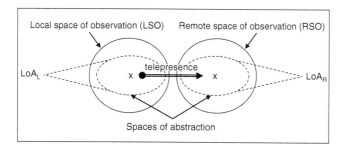

Figure 7. A model of telepresence

A further advantage of the new model is that all this is good news for mass-media and literature studies as well. For that peculiar experience of 'as if I were there' caused by many forms of communication will never be properly studied as long as it is catalogued under the wrong heading of telepresence. It requires the development of its own set of conceptual tools. There are, of course, borderline cases, and the new model contributes to explaining them. Watching *All My Children* on TV does not make the audience telepresent, either as a property-bearer or as a source of interaction. However, participating by *tele*-phone in a *tele*-vision programme does indeed satisfy the criterion laid down by the SO model, and quite rightly so, for the audience is now capable of some minimal interaction at a distance with the remote environment. Indeed, the example shows the need for a deeper understanding of the nature of environments conducive to telepresence. It takes two to interact. Of course, digital ICTs are far more open to the possibility of telepresence than classic mass media, but telepresence is possible even through the latter. The difference lies precisely in the re-ontologizing power of ICTs, which not only 'augments' the agent's capacities epistemically, but allows the construction of new spaces where the agent can be telepresent interactively. It won't be long before we are able to experience something like the 'wall-to-wall circuit' interactive TV described by Bradbury in *Fahrenheit 451*:

> She [Helen] didn't look up from her script again. 'Well, this is a play that comes on the wall-to-wall circuit in ten minutes. They mailed me my part this morning.... They write the script with one part missing. It's a new idea. The home-maker, that's me, is the missing part. When it comes time for the missing lines, they all look at me out of the three walls and I say the lines....' 'What's the play about?' 'I just told you. There are these people named Bob and Ruth and Helen.'[11]

Reality television shows can still get much worse.

A third, important advantage of SO is that it enables one to acknowledge a spectrum of ways of being present, from the *weak presence* of something barely detectable as a mere property-bearer (more on this below) to the *strong presence* of an agent endowed not only with observable properties, but also with the capacity of acting and interacting—the agent can be the receiver of an action and respond to it accordingly—with the environment, both pragmatically, by doing or changing things, and epistemically, by observing things locally. Presence is no longer a Boolean concept—as in the EF model—and the SO model justifies talks of augmented telepresence, or attempts at making telepresence resilient, and so forth.

The last advantage to be stressed finally leads us to the discussion of two ethical issues, in the next section.

3.4.3 Virtual pornography

Clearly 'being there remotely' as a mere property-bearer is far less useful and interesting than being telepresent also as an agent capable of some successful action and interaction

---

[11] Bradbury, R., *Fahrenheit 451* (New York, NY: Simon & Schuster, 1995), p. 17–18.

in the remote space. In both cases, however, telepresence, as defined by the new SO model, brings to light the need to analyse ethical problems that, on the one hand, inevitably escape the old EF model (recall the definition of telepresence as a sort of personal experience), and, on the other hand, do not seem mere updated versions of the standard problems occurring in everyday life.

The SO model makes explicit that we are confronted by a new ethical context in which teleagents and telepatients interact in technologically sustained environments. Of course, their actions have moral values and consequences, but our degree of understanding is still limited. Needless to say, the slightly sci-fi scenario should not mislead. Millions of people spend an enormous amount of time online, being present in remote spaces in which they both learn how to behave and show how they behave (Bracken and Skalski, 2010). Interestingly, the fact that many of the entities with which human teleagents and telepatients come into contact may be entirely artificial, enriches and sharpens our ethical discourse in general. The new scenarios require an upgrade of old conceptual tools and the creation of new ones. For environmentalism acquires a new meaning when one's environment is a remote virtual space, and the sort of things with which one interacts may have a digital, and not a biological, nature. A concrete example will help to illustrate the point.

According to SO and contrary to EF, classic pornography, in the form of texts, pictures, or movies, does not generate any form of presence. In this respect, there is no difference between De Sade's *Justine* and Voltaire's *Candide*: the reader is still left out of the remote space of observation. It follows that whatever might be morally significant with old-fashioned pornography cannot be grounded on an analysis of telepresence. The dialectics of exposure (see later) seems much more pertinent. This still holds true when the nature of media changes from analogue to digital: YouPorn.com, even if it provides some choices and options, fails to represent a case of presence, according to the new model.

Things are slightly more twisted, however, with new forms of ICTs-based 'pornography' (the quotation marks are required precisely by the novelty), which implement various degrees of interaction, without any form of physical intercourse: dedicated telephone services, chat-rooms, online services, and other multi-user environments, usually employed for role-playing games, or virtual reality scenarios inhabited by avatars. Here pornography (which is a semantic concept) and promiscuity and prostitution (which are pragmatic concepts) merge. In similar cases, the agent is indeed present remotely, at least in the sense supported by SO, in semantic spaces that also allow some degree of interaction. However, despite the obvious connection with more ordinary forms of pornography—consider for example pornographic comics or pornographic animated cartoons—one important difference is that the other teleagents with 'whom' the human agent *interacts* may be entirely synthetic. *S1m0ne*, the film directed by Andrew Niccol about a digitally created actress 'who' becomes a star, offers a great (and chaste) thought experiment. More realistically and less morally, erotic chatterbots nowadays are not science fiction, as a quick search on Google easily proves; some of the success of virtual worlds online is certainly linked to their interactive pornography. This is not entirely a novelty. People have been gallant and tried to date pieces of software for decades now, including ELIZA. The result is that, in such cases, arguments against pornography based on the crimes, immoralities, degradation, exploitation, and health hazards that may affect the people involved—an argument often rehearsed in the context of pornographic videos with real human actors—may become ineffective. Likewise, any Kantian argument to the effect that no human being should be used as a mere means to an end would need some sophistic, and not sophisticated, tuning to be applicable. Clearly, telepresence in informational environments inhabited by agents of unclear nature is forcing us to revise our well-entrenched, ethical assumptions. A patient-oriented approach (see next chapter) might be the best way forward. This too is far from being an original position. John Duke Coleridge had already grasped it in 1868, when he formulated the classic definition of criminal obscenity still used in British Law, in terms of what 'tends to deprave and corrupt' the *receiver* (user, or patient) of the message, not the *sender* (the producer, or agent):

[A]n article shall be deemed to be obscene if its effect... is, if taken as a whole, such as to tend to deprave and corrupt persons who are likely, having regard to all relevant circumstances, to read, see, or hear the matter contained or embodied in it.[12]

It is only if one adopts this patient-oriented LoA that the fact that the entities with which the agent is interacting are virtual is irrelevant, whereas the fact that an increasing degree of interaction might become available makes the tendency of virtual pornography to 'deprave and corrupt' the user ever more likely and serious. I shall return to this crucial shift in favour of a patient-oriented approach in the following chapters.

3.4.4 *An objection against presence as successful observation*

Checking the limits of the old EF model, one may be tempted to raise similar objections against the new SO model. True, SO does provide a definition of presence and a criterion of discrimination between presence and absence. But SO might still be eccentric in a very significant way, since one of the most important types of phenomena, commonly interpreted as presence, refers to the availability of tele-perceptual technologies such as radars, satellites, webcams, sonars, and CCTV cameras. It seems that, without being either a property-bearer or a source of interaction in a RSO, an entity, even an artificial one, might still be present in a RSO remotely, for example by means of a monitoring appliance. Yet SO fails to accommodate such types of presence, which might be qualified as *telepistemic*. It follows that SO needs to be revised, if not abandoned.

---

[12] Obscene Publications Act 1857, 1959, and 1964 (only the latter two are still in force in England and Wales) <http://www.legislation.gov.uk/ukpga/1964/74/>.

The previous objection is correct in drawing the inference, but mistaken in suggesting the need for a solution. What needs to be modified is our understanding of *telepistemics* itself. For what looks like telepresence is, in fact, something slightly different, and understanding the difference casts an interesting light on several issues.

Suppose you are in a room. You are just present in that room. Pull down the wall between that room and the next, and you will not say that you are now telepresent in the next room; you are merely present in a larger room. In chess, when a pawn reaches the opposite side of the board, it can be promoted to any piece except a king. Suppose the pawn is promoted to a queen. Suddenly most of the board becomes a local space, distant by only one move. Many telepistemic technologies are 'tele-promoting' in this sense. The queen is not telepresent in a remote space; it is the space of the pawn that has been enlarged. Take a digital camera. Start monitoring what is happening in your room. Again, you are not telepresent in your room, or at least not according to SO (at least because in this case we have LSO = RSO); the burden of proof that you are is on EF's shoulders. Now imagine making the digital camera one inch longer, and then another inch, and so forth, or just making your camera increasingly more powerful. Gradually, the camera allows you to monitor things that are increasingly further away from your local space. At what point are you telepresent? At ten metres? Fifty? A hundred? When only a cable connects you and the appliance? Or a radio signal? The answer is never, according to the SO model. Making a remote space epistemically available locally is different from being present in that remote space as an entity. It is like pulling down the wall between two rooms. This is why there is no point in using a portable baby-monitor unit with a range of several miles: the monitor guarantees the user only telepistemic access to the remote space but no actual interactive telepresence at distance. If something happens, it is only the more frustrating to know that nothing can be done in time, given the long distance. Indeed, it can be tragic, as I shall argue in Chapter 10.

The problem with telepistemics consists in a fallacious confusion between

(1) the successful observation *of x* not only locally, in LSO, but also remotely, in RSO; and
(2) the successful observation *by x*—which is in LSO—of some *y* that is present in RSO.

The former is a case of ontic telepresence, the latter is a case of epistemic access at distance. The two phenomena are separate and should not be confused. Compare this to the illusion of movement caused by web-browsing: one feels as if one were being uploaded in different spaces or visiting sites on the web, when in reality one is downloading those spaces into one's own. The web makes you a chess queen: everything is only one click (move) away.

Should we then abandon any talk of presence in all those cases of technologically mediated telepistemics? The answer is no. Telepistemics may still be a case of presence; it is just that the previous confusion impedes one seeing precisely who or what is telepresent where. It is not the observer *x* in LSO accessing the entity *y* in RSO that is present in RSO, but exactly the opposite: by being accessed telepistemically, the *y* in RSO is now also present in the observer's LSO, typically as a mere property-bearer. Using the previous analogy, once the wall is pulled down, you are not remotely present in the other room; it is the chair that was in the other room that is now locally present in your space.

In order to fix the distinction it may be useful to speak of *forward* and *backward presence* (see Figure 8).

The distinction between *forward* and *backward presence* fits SO perfectly well. Recall that something may be telepresent as a mere property-bearer. Take the map of a portion of the small crater encircling the Mars exploration rover called *Opportunity*. At some point, it showed the location of crystalline haematite. This is a case of *backward presence*. It made the Martian haematite present in our space, through a stratification of communication and spatial layers. At the same time, (not we but) the scientists who controlled *Opportunity* were also *forward present* on Mars, as interactive agents.

Telepistemic technologies have evolved dramatically since Galileo discovered the four satellites of Jupiter with his own telescope. Nevertheless, by looking at pictures on the screen of a computer we (you and I, not the NASA scientists mentioned above) are no more present on Mars now than Galileo was on those satellites.

Mere word playing, one may reply. Forward and backward presence is just like active and passive sentences: there is no substantial difference between 'Alice drives the car' and 'the car is driven by Alice'. You are where I am; I am where you are. The distinction is linguistically possible but conceptually useless.

This impatient complaint is understandable but unjustified. These are different cases, they should not be confused, and the distinction between forward and backward presence is no mere hair-splitting. First of all, it helps to clarify that 'local' and 'remote' are indexical concepts. Like other words and concepts, such as 'I', 'now', 'here', 'my', and so forth, they acquire a different meaning depending on the position of the observer. The observer's LSO is not the observed's LSO, obviously. It follows that the SO model is correct in stressing the need for an explicit statement of where the

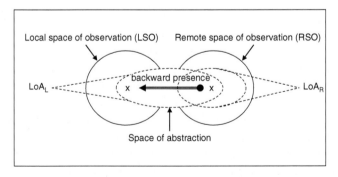

Figure 8. A model of telepistemics as backward presence

observer is before talking of telepresence. There is no obviously privileged space to be defined as *local*. Call this an anti-localist thesis.

Second, presence at its best is usually full, that is, both forward and backward presence. Alice is present in RSO as an interactive agent (full forward presence) and she also observes herself locally as being remotely present (backward presence). It is this feedback function that allows Alice to control her interactions with the remote environment.

Third, but not less important, being able to understand the difference between the two types of presence means equipping ourselves with a powerful conceptual tool that can help us to frame some ethical problems far more accurately. Some distinctions can be subtle, but sometimes this means that they are sharper and cut more surgically. We have seen that interactive pornography is a case of what we can now call *forward presence*. Let us have a look at informational privacy as a case of *backward presence*.

3.4.5 Informational privacy: from intrusion to abduction

The literature in computer ethics on informational privacy is vast and constantly growing (for a review and further references see Tavani (2011, ch. 5)). So, it is not my intention here even to sketch the current debate. The outlines of the problem are well known and I shall return to it in Chapter 12 in order to provide a solution from an IE perspective. What I wish to point out here is the need to acquire the correct overall perspective when approaching it, and hence the usefulness of adopting the method of abstraction.

Privacy is often discussed topologically, i.e. in terms of space. Yet, depending on how one understands presence, two fundamental scenarios become available.

If telepistemics is defined in terms of *remote presence of the observer*—what has been called above *forward presence*—it is natural to slip into a view that equates the observed's privacy with the protection of her (local or remote) space, whether physical, psychological, or informational. One may then be further inclined to apply familiar concepts of space ownership: a right to informational privacy may be an exclusive right to own and use one's own information or information about oneself, for example. Since information does not need to be removed to be stolen—as happens with a car, for example—this further reinforces the view that privacy is ownership of an informational space, which is not subtracted from the owner when exploited by someone else. The result is a metaphorical conceptualization of informational privacy breach as trespassing. Illegal, or simply unauthorized, access to a website or a database, or common cases of digital surveillance, are portrayed as intrusions in someone else's space or place. It is indicative that the standard line of defence by the intruder—'the gate was open' or 'there was no gate'—is not rejected as irrelevant, but rather as pertinent yet mistaken, with a 'yes, but' kind of rebuttal.

As we shall see in more detail in Chapter 12, the problem with this approach is that privacy is often exercised in public spaces; that is, in spaces which are not only socially and physically public—a street, a car park, a restaurant—but also informationally public—anyone can see the newspaper one buys, the bus one takes, the t-shirt one wears, the drink one is ordering. The tension is obvious. According to an old estimate, now outdated, in 2003 there were more than 1.5 million cameras monitoring public places in Britain, with the result that the average Briton was recorded by CCTV cameras 300 times a day.[13] Today, they seem to have grown to 4.5 million.[14] How could the telepresence of Alice observed through a CCTV system operated by a bank on a street, for example, be a breach of her privacy, given that she is accessing a space which is public in all possible senses anyway? How do the shop records of her purchase of a t-shirt breach her privacy, if what she wears is visible to all? Attempts at solving these apparent inconsistencies result in strange geometries of overlapping spaces and exercises in conceptual contortionism. The problem is at the origin: an analysis of telepistemics and hence informational privacy in terms of *forward* presence is simply not very helpful. If we do not rectify it, arguing against the growing number of CCTV cameras will be more difficult.

Consider now the conclusion reached in the previous section. Once telepistemics is understood as a way of making the observed locally present—what has been defined above as *backward presence*—a privacy breach is more easily comparable to a case of metaphorical abduction: the observed is moved to an observer's local space—space which is remote for the observed, recall that LSO ≠ RSO—unwillingly and possibly unknowingly. Of course, what is abducted is only some information; hence no actual removal of the material entity is in question—recall the example of the car above—but rather a cloning of the relevant piece of information. Yet the cloned information is not a space that belongs to the observed and which has been trespassed; it is rather part of the observed herself, or better something that (at least partly) constitutes the observed for what she is, given an informational LoA. It is a *Doppelgänger*, as Richard Avedon described it once, when speaking of his photograph of Henry Kissinger: 'Is it just a shadow representation of a man? Or is it closer to a doppelgänger, a likeness with its own life, an inexact twin whose afterlife may overcome and replace the original?' From this perspective, privacy becomes a defence of personal identity and uniqueness. The inconsistency concerning private vs. public spaces no longer arises: the observed wishes to preserve her integrity as an informational entity even when she is in an entirely public place. After all, kidnapping is a crime independently of where it is committed, whether in public or not. What one buys, wears or does in public belongs to a sphere that is no one's property in particular, but monitoring and recording it subtracts from this public sphere a portion of the information that constitutes the observed analysed as an information system at the adequate LoA, and makes it part of a space that belongs to, and is controlled by, the observer, to which the observed herself often has no other

---

[13] 'Survey: The Internet Society: An Interview', *The Economist*, 23 January 2003 <http://www.economist.com/node/1555772>.

[14] 'Is CCTV creeping too far?', *BBC News*, 11 January 2011 <http://www.bbc.co.uk/news/magazine-12224075>.

access, and in a way that may be completely invisible to the observed herself. Alice is often unaware that part of her information is being abducted.

Let me offer another example. When we hear someone speaking loudly on a mobile phone near us, perhaps in the constrained space of a train, we are often annoyed. We do not wish to listen to her business, but cannot help it. Paradoxically, we know that that person is breaching our privacy, yet the old model of presence would not enable us to say why. We could not stop being present and hence regain our privacy just by becoming fully aware of the technologically mediated nature of the experience. Our privacy is certainly not being breached because she is entering into our informational space: we, after all, are the ones who do the listening. It is because she is abducting us into her informational space, forcing us to be telepresent in her space against our wills. Our privacy is affected because this is a case of *imposed backward presence*.

3.4.6 The method of abstraction and telepresence

We have now completed the case study. The new model of presence based on successful observation suggests that to be (tele)present is to be the value of a typed variable of a LoA. The new model SO is no longer Cartesian, for it does not privilege the subjective, internal, mental perception (or lack thereof) of a technologically mediated displacement. The new model is not even mass-mediatic, for it does not refer to social or shared spaces of communication or fiction. SO is a model developed on principles using an anti-psychologistic, non-Cartesian, LoA methodology, which is minimalist in its ontological assumptions and firmly based in the philosophy of information. According to SO, what matters, in the analysis of presence is the occurrence and flow of information at a given level of abstraction. An entity is *forward present* in a remote space if it is successfully observed at a chosen LoA remotely, in RSO, either as a property-bearer or as a source of change or action/interaction, that is, as some kind of information system. An entity is *backward present* in a local space if it is successfully observed at a chosen LoA locally, in LSO, at least as a property-bearer, that is, again, as some kind of information system. Adopting the LoA methodology allows one to specify the ontological commitments in the assessment of presence, while avoiding any (intrinsically unreliable and inevitably opaque) psychologism or qualitative phenomenological description based on the agent's reports of subjective experiences.

CONCLUSION

Quine once remarked that:

The very notion of an object at all, concrete or abstract, is a human contribution, a feature of our inherited apparatus for organizing the amorphous welter of neural input.
...
Science ventures its tentative answers in man-made concepts, perforce, couched in man-made language, but we can ask no better. The very notion of object, or of one and many, is indeed as

parochially human as the parts of speech; to ask what reality is really like, however, apart from human categories, is self-stultifying. It is like asking how long the Nile really is, apart from parochial matters of miles or meters. (Quine, 1992, pp. 6, 9)

This chapter may be read as an attempt to clarify what 'human categories' in the first quotation above means, in terms of a less parochial method of levels of abstraction. The reader still unconvinced of the importance of being clear about one's LoA, or doubtful about the connection between the problem highlighted by Quine above and the solution provided by the method of LoA, may find the following reference helpful. In November 1999, NASA lost the $125m Mars Climate Orbiter (MCO) because the Lockheed Martin engineering team used English (also known as Imperial) units of measurement, while the agency's team used the metric system. Not exactly Quine's example of the length of the Nile, but still two different LoAs. As a result, the MCO crashed into Mars.

Regarding the method, in the first part of the chapter I showed its principal features and crucial value for any analysis of information processes. The method clarifies implicit assumptions, facilitates comparisons, enhances rigour, and hence promotes the resolution of possible conceptual confusions. If carefully applied, the method confers remarkable advantages in terms of consistency and clarity. Too often, philosophical debates seem to be caused by a misconception of the LoA at which the questions should be addressed. This is not to say that the method represents a panacea. Disagreement is often not based on confusion. Indeed, informed and reasonable disagreement is precisely what characterizes philosophical questions, which remain intrinsically open to debate. But the chances of resolving or overcoming it, or at least of identifying a disagreement as irreducible, may be enhanced if one is first of all careful about specifying what sort of observables are at stake and what goals are orienting their choice, and therefore what questions it is meaningful to ask in the first place.

In order to clarify the method of levels of abstraction, in the second part of the chapter I relied on an illustrative application of it to the analysis of telepresence, with the further goal of discussing two problems: virtual pornography and informational privacy. I argued that a new model, based on successful observation, can improve our understanding of telepresence and help us to approach important ethical issues, such as virtual pornography and informational privacy, from the right perspective. Whether the new model can withstand criticism and prove to be fruitful is a question open to further research, but one thing remains unchallenged: the more telepresence becomes an ordinary phenomenon, involving an increasing number of people, the more important it will become to understand its nature and its ethical implications in ways that may be utterly unprecedented and unexpected. ICTs increase the ontic and epistemic power of human agents enormously. Furthermore, vast moral issues are associated with these expanded capacities. More understanding seems the only key to their proper management. After all, the queen has responsibilities unknown to the pawn, as we shall see in the next chapter.

4

Information ethics as e-nvironmental ethics

> Throughout legal history, each successive extension of rights to some new entity has been, therefore, a bit unthinkable.... The fact is, that each time there is a movement to confer rights onto some new 'entity', the proposal is bound to sound odd, or frightening or laughable.... I am quite seriously proposing that we give legal rights to forests, oceans, rivers, and other so-called 'natural objects' in the environment—indeed, to the natural environment as a whole.
>
> C. D. Stone, *Should Trees have Standing? Law, Morality and the Environment* (2010), p. 3.

SUMMARY
Previously, in Chapter 3, I introduced the method of levels of abstraction as an essential tool in philosophical analysis and conceptual design. I argued in its favour and showed its fruitfulness by applying it to the investigation of telepresence and then to two moral issues: virtual pornography and informational privacy. I also began to argue that we might need a new theoretical approach in order to deal with moral problems caused by ICTs. This chapter relies on the aforementioned method in order to outline a specific interpretation of IE as an e-nvironmental ethics, or synthetic ethics, or ethics of the infosphere, for inforgs like us. Here is a quick overview. In Section 4.1, I shall introduce the foundationalist problem in computer ethics (CE). In Section 4.2, I shall argue that the ethical questions prompted by ICTs put classic macroethics, such as deontologism and consequentialism, under pressure to deliver satisfactory answers. In Section 4.3, I shall propose a model of IE as a macroethics that can deal with ICT-related ethical problems better than other macroethics, so, in Section 4.4, I shall argue that IE can provide the theoretical foundation for CE. In Section 4.5, I shall outline the patient-oriented, ontocentric nature of IE as a macroethics. In Section 4.6, I shall make explicit the four ethical principles supported by IE as a macroethics. In Section 4.7, I shall defend the view that IE counts as a macroethics rather than a microethics and compare it to other approaches. In Section 4.8, I shall further clarify the nature of IE as a macroethics by analysing three applications of IE. As usual, in the conclusion I shall summarize the results obtained and introduce the next chapter.

As should already be clear from this summary, this is a programmatic, meta-theoretical chapter, meaning that the reader will encounter more clarifying statements than supporting arguments in their favour. The main goal of the chapter is to make things explicit and possibly clearer, so that they might become convincing in the following chapters. I hope the reader interested only in the latter will bear with me, while I anticipate and explain the agenda and prepare the ground for what follows.

4.1 Introduction: the foundationalist problem

We saw in Chapters 1 and 2 that the information revolution has brought about some serious ethical challenges. For some decades now, they have been investigated by a branch of applied ethics known as computer ethics (CE). Yet lobbying, financial support and the undeniable importance of the very urgent issues discussed by CE have not yet succeeded in raising it to the status of a philosophically respectable area of research. This was true when I first wrote that sentence and articulated most of the contents of this chapter in Floridi (1999). Almost fifteen years later, it is unfortunately still the case that most mainstream philosophers, if they take any notice of CE, look down on it as a practical subject, a professional ethics unworthy of their analysis and intellectual efforts, to be investigated and taught by the Engineering Department. They treat it like carpentry ethics, to use a Platonic metaphor.

The inescapable interdisciplinarity of CE has certainly done the greatest possible harm to the prospects for recognition of its philosophical significance. Everyone's concern is often nobody's business, and CE is at too much of a crossroads of technical matters, moral and legal issues, social as well as political problems, and conceptual analyses to be anyone's own game. The flourishing of literature, conferences, and courses on CE has not yet made it move from its philosophically fringe position. Philosophers' notorious conservatism may also have been a hindrance. After all, Aristotle, Mill, or Kant never said a word about it, and professional philosophers, who know their syllabus, do not often hold very broad or flexible views about which new philosophical questions may qualify as philosophers' own special problems. You will not find CE positions advertised as an AOS (area of specialization) in the American Philosophical Association's *Jobs For Philosophers*. Yet these and other external factors, such as the novelty of its questions and the conspicuously applied nature of its answers, should not conceal the fact that the essential difficulty about CE's philosophical status lies elsewhere, and more internally, since this is a methodological problem that concerns its conceptual foundation as an ethical theory.

A quick look at some of the most popular textbooks[1] indicates that CE shares with other philosophical disciplines in the analytic tradition broadly conceived three important but rather too general features:

---

[1] See e.g. Himma and Tavani (2008); Johnson and Miller (2009); Spinello (2011); Tavani (2011); Weckert (2007); and Van den Hoven and Weckert (2008).

(1) it is logically argumentative, with a bias for analogical reasoning;
(2) it is empirically grounded, with a bias for scenario analysis; and
(3) it endorses a problem-solving approach.

Besides (1)–(3), CE also presents a more peculiar feature, which has acted as a major driving force, namely:

(4) it is intrinsically decision-making oriented.

These four features can be read in a roughly inverted order of importance. Why CE shares them, and whether it ought to, are questions sufficiently obvious to deserve no detailed comment here. Technological changes have outpaced ethical developments, bringing about unanticipated problems that Moor (1985) famously and influentially defined in terms of a 'policy vacuum' filled by CE. Thus, the latter has initially surfaced from practical concerns arising in the information society. Rational decisions have to be made; technical, educational, and ethical problems must be solved; legislation needs to be adopted; and a combination of empirical evidence and logical arguments seems to provide the most obvious and promising means to achieve such pressing goals. A rather more interesting point is that (1)–(4) constitute the theoretical justification for CE's inductive methodology, which leads us to a last feature:

(5) CE is based on case studies.

During the last three decades, CE has consistently adopted a bottom-up procedure, carrying out extended and intensive analyses of individual cases, amounting very often to real-world issues rather than thought experiments. The goal is to reach decisions based on principled choices and defensible ethical principles and, hence, to provide more generalized conclusions in terms of conceptual evaluations, moral insights, normative guidelines, educational programmes, or legal advice, which might then apply to whole classes of comparable cases.

On the grounds of such extensive evidence and analysis, defenders of the novelty and originality of a CE-approach to moral issues have developed two positions. One holds, perhaps too generally, that (1)–(5) are sufficient to qualify CE as a well-grounded philosophical discipline. The other holds, more specifically and somewhat more forcefully, the following three theses. First, the impact of ICTs, their scale, and complexity have created a whole new range of problems (computer crime, software theft, hacking, viruses, privacy, over-reliance on smart machines, new workplace issues, intellectual and social discrimination, etc., see Chapter 2) which, in turn, have given rise to a grey area of moral issues, not all of which are just ICTs' versions of old ones. Second, the new and old ethical problems on which CE works within the context of feature (5) above—privacy, accuracy, intellectual property, and access used to be a classic set of examples in the eighties,[2] but as

---

[2] This was known as the PAPA group. In the mid-eighties, Mason (1986) discussed some dilemmas thought to be unique to ICTs and identifies at least four main ethical issues for ICTs' professionals: privacy, accuracy, ownership, and access to information, summarized by the acronym PAPA. The essay was influential in the subsequent literature.

we have seen in Chapters 1 and 2, the list is now much longer and more complex—have been so transformed by ICTs that they have acquired an altered form and new meanings. Third, in both cases (the first two theses above), we are confronted by the emergence of an innovative ethical approach, namely CE, which is at the same time original and of an unquestionable philosophical value.

Unfortunately, neither position carries much weight. The more general position just fails to be convincing, whereas the more restricted thesis is, more interestingly, the actual source of the foundationalist crisis that afflicts CE. Let me elaborate.

I shall later defend the view that IE, understood as the philosophical foundation of CE, does have something distinctive and substantial to say on moral problems and, hence, can contribute a new and interesting perspective to the ethical discourse. At the moment, it is sufficient to realize that features (1)–(3) fail to make CE any different from, let alone better than, other ethical theories already available, most notably consequentialism and deontologism, while feature (4) may work equally well against CE's philosophical ambitions, for it leads to the carpentry problem. As for feature (5), it takes only a moment of reflection to realize that, together with (4), it is one of the factors that contribute to, rather than solve, the foundational problem, for the following reason. If new moral problems generated by ICTs and the information revolution have any theoretical value, either by themselves or because they are embedded in original contexts, they usually provide only further evidence for the discussion of well-established ethical doctrines. Thus, CE problems may work as counterexamples, show the limits, or stretch the conceptual resources of already available *macroethics*, but can never give rise to a substantially new ethical perspective, unless they are the source of some very radical re-interpretation.

As I mentioned in Chapter 2, ICTs are re-ontologizing the context in which ethical issues arise, and in so doing they not only transform old problems, but also invite us to explore afresh the foundations on which our ethical positions are based. Missing the latter perspective, even people who support the importance of the work done in CE are led to adopt a dismissive attitude towards its philosophical significance, and argue that there is no special category of CE, but just ordinary ethical situations in which various kinds of ICTs and digital technologies are involved. On these grounds, they conclude that CE is at most a *microethics*. This is the view that has been influentially defended for many years by Deborah Johnson (compare the four editions of Johnson and Miller (2009)), for example. Johnson seems to show some sympathy for a moderately Kantian perspective, but does not take an explicit position. Her main thesis is that ethical issues surrounding ICTs are not wholly new, and that it is unnecessary to create a new ethical theory or system to deal with them. They have some unique features, but we can rely on our traditional moral principles and theories. An even more radical position is taken by Langford (1995), who disregards a philosophical approach to CE as entirely dispensable: 'this book is not a work of theoretical analysis and discussion. Practical Computer Ethics is not for academic philosophers' (p. 1).

This view seems to be shared by many ethicists as well. If interest in CE is more justified than interest in carpentry ethics, this is only because, in the information society, ICTs rather than nails, hammers, and timber permeate and influence almost every aspect of our lives. So, the argument continues, we need a conceptual interface to apply ethical theories to new scenarios. If 'there [were] a world market for maybe five computers' (Thomas Watson, president of IBM, 1946) or 'there [were] no reason anyone would want a computer in their home' (Ken Olsen, founder of DEC, 1977), or at least all the ICTs available in the world were kept under very tight control, there would be neither CE, nor any need for it.

Behind CE's foundationalist problem, there lies a lack of a strong ethical programme. Although everyone seems to agree that CE deals with innovative ethical issues arising in ICT contexts within (5), instead of reflecting on their roots and investigating, as thoroughly as possible, what new theoretical insights they could offer, we are urged by features (3) and (4) to rush in and look immediately for feasible solutions and implementable decisions. It is the impact agenda, the applied drive, and the practical business that seem to dominate our investigations.

The result is inevitably disappointing: features (3) and (4) load feature (5) with an unduly *action-oriented* (see Section 4.3 for a clarification of such label) meaning and CE problems are taken to entail the fact that CE is exclusively concerned with the moral value of human actions mediated by ICTs. Understood as a mere decision-making and action-oriented theory, CE appears only as a practical subject, which can hardly add anything to already well-developed ethical theories.

This was the state of play at the end of the nineties (Floridi, 1999a). It is largely the state in which CE still finds itself presently. More developments and greater diffusion of ICTs have brought about more ethical challenges, a deeper sense of the significance of the novelties we are experiencing, but also some hardening in the various positions, with conservative defenders of 'business as usual' even more sanguine about the need for some immediate, practical solutions, and revolutionary supporters of 'never seen before' ever more inclined to speculate freely on science-fiction fantasies such as alleged singularity turning points and bizarre scenarios of lives currently spent in future-generated simulations. We are caught between a short-sighted and a starry-eyed alternative.

When ethical analysis is required, moral problems in CE, with their theoretical implications, are still invariably approached against the background of a deontologist, contractualist, sometime virtue-ethicist, or, more often, consequentialist position. Predictably, CE itself is either disregarded, as a mere practical field of no philosophical interest, or colonized, as a special domain of the application of action-oriented ethics in search of intellectual adventures.[3] Conceptually, it was, and still is, a most unsatisfactory situation, for two related clusters of reasons.

---

[3] See e.g. Ermann and Shauf (2003), especially the first part, entitled 'Ethical Contexts', for a philosophical perspective.

4.2 Classic macroethics and ICTs ethical problems

On the more negative side, the nature of the ethical problems brought about by ICTs seems to strain the conceptual resources of *action-oriented* theories (more on this qualification in Section 4.3) more seriously than is usually suspected. We have already seen two examples in the previous chapter. Two possible forms of distortion, sometimes caused by the application of inappropriate action-oriented analyses, are:

(1) the projection of human agency, intelligence, freedom, and intentionality (desires, fears, expectations, hopes, etc.) onto ICTs; and
(2) the tendency to delegate responsibility to ICTs as an increasingly authoritative intermediary agent (it is not unusual to hear people dismiss an error as only the fault of a computer).

In both cases, we witness the erosion of a sense of moral responsibility for their actions on the part of human agents. Without a *patient-oriented* approach (see later), computer ethics may end up anthropomorphizing ICTs. The fact that such limits have not been fully and explicitly investigated in CE literature, despite their decisive importance, is a clear mark of the sense of inferiority shown by CE towards philosophically better-established theories. However, when consistently applied, consequentialism, contractualism, and deontologism (for a discussion of an approach based on virtue ethics, see Chapter 8) seem unable to accommodate ICTs ethical problems easily and, in the end, may well be inadequate, for at least the following reasons.

First, we might expect that the empirical, decision-making orientation of ICTs ethical problems would tend to make deontologism, with its inflexible universal maxims and duty-based ethics, a much less likely candidate than either contractualism or consequentialism. The strength of the conflicting interactions between different rights, duties, and moral values, emerging from the case studies carried out so far, seems to undermine the viability of a purely deontological approach to CE. Think, for example, of society's right to security vs. cryptography, of privacy vs. public control of information, of freedom of expression vs. offensive information, and mix all these ingredients within a globalized information society where different cultures rub against each other daily. More specifically, Kant's moral imperatives appear to be challenged by at least two difficulties. Neither the law of impartiality (the Golden Rule) nor the law of universality (behave as a universal legislator) are sufficient to approach:

(1) CE problems not involving human beings. We tend to reject the idea that there might be victimless crimes (and so does IE, see Section 9.1), e.g. computer crimes against banks, or that vandalism may not be morally blameworthy. (I shall come back to this problem in Section 4.8.1.) Yet it is unclear how a deontological approach can cope with this kind of problem without some serious adaptations, since both Kantian imperatives apply only to anthropocentric contexts; and

(2) CE problems with a ludic nature. For example, Alice and Bob may perceive their online activities as games or intellectual challenges, and their actions as role-playing rather than computer crimes. Because of the remoteness of the processes, the immaterial nature of the environment, and the virtual interaction with faceless individuals, many corners of the infosphere can easily be perceived as magical, dream-like, or fictional environments, and anything but real, so agents may wrongly assume that their actions are as virtual and insignificant as the killing of enemies in a computer game. The consequence is that not only do the agents not feel responsible for their actions—no one has ever been blamed or charged with murder for having killed some monsters in a video game—they may be perfectly willing to accept the universal maxim, and to extend the rules of the game to all agents. The hacker can be a perfect Kantian because universality without any concern for the actual consequences of an action is ethically powerless in a moral game.

Second, the previous problems help to explain why, in practice, most of the theoretical literature on CE tends to adopt some *pragmatic* version of the MINMAX principle and the Golden Rule ('minimize harms, maximize benefits' and 'do unto others as you would have them do unto you') and is often more or less knowingly consequentialist and, sometimes, contractualist in orientation. Things, however, are not much more promising if we look at these two approaches, for they too end up strained by the nature of the problems in question. Let me list eight issues to articulate the point.

(1) The virtual nature of the actions in question often makes it possible for them to remain completely undetected and to leave no really perceptible effects behind.
(2) Even when (1) does not apply, ICTs distance agents from, and hence diminish their sense of direct responsibility for, their ICT-mediated, controlled, or generated actions. Besides, the increasing separation of actions and their effects, both in terms of the potential anonymity of the agents and in terms of conceptual distance, makes 'moral sanctions' (in Mill's sense) ever less perceptible by the agents the more indirect, distant, and obscure the consequences of their actions become.
(3) In connection with (1)–(2), there is a corresponding de-personalization and an increasing sense of the practical anonymity of actions/effects, in a context where an individual agent's behaviour is often rightly perceived as only a marginal and microscopic component of wider and more complex courses of action (I shall return to the topic of 'distributed morality' in Chapter 13). The diffusion of responsibility may bring with it a diminished ethical sensitivity in the agent and a corresponding lack of perceived accountability.
(4) In connection with (1)–(3), the high level of control and compartmentalization of actions tends to restrict them and their evaluation to specific areas of potential misbehaviour.
(5) In connection with (1)–(4), the ostensibly negative anthropology resulting from CE case-studies shows that human nature, when left to itself, is much more

Hobbesian and Darwinian than consequentialism may be ready to admit, and hence able to cope with. The increasing number and variety of ICT crimes committed by perfectly respectable and honest people shows the full limits of an action-oriented approach to CE. Agents who behave immorally when dealing with ICTs may not perceive, or perceive in a distorted way, the nature of their actions because they have been educated to conceive as potentially immoral only human interactions in real life, or actions involving physical and tangible objects. A cursory analysis of the justifications that hackers usually offer for their actions, for example, is sufficient to clarify that they often do not understand the real implications of their behaviour, independently of their technical competence. We have already seen that this problem affects a deontological approach as well (the ludic problem above).

(6) Even when (1)–(5) do not apply, the great complexity of the constantly changing infosphere often makes any reasonable calculation or forecasting of the long-term, aggregate value of the global consequences of individual actions impossible.

(7) Quite apart from (1)–(6), individuals and their rights acquire an increasing importance within the information society, not just as agents, but also as potential targets (patients, or receivers) of automatically tailored actions, yet individuals' rights are something that consequentialism finds difficult to accommodate.

(8) In connection with all the previous points but (5), the potentially asymmetric nature of 'virtual' actions gives rise to a 'state of nature' in which individuals are very far from having even a vaguely comparable strength, either technical or technological, and therefore the 'strongest' can behave perfectly rationally, 'opt out' of the social contract, and be successful. For example, a very appropriate game-theoretic approach to CE problems would show that, since there are never equal conditions, the 'game' is heavily biased towards the hacker; suffice it to say that—according to the *White Paper on Computer Crime Statistics* published by the International Computer Security Association Kabay (1998)—most computer crimes remain undetected by their victims, they are not just unpunished, and of the attacks which are detected, few are reported.

The previous problems can be extended to contractualism as well, if we treat the latter as a version of consequentialism based on a negative anthropology and a conception of the nature of actions as always rationally motivated only by self-interest (I shall briefly comment on a deontological form of contractualism in the next section).

If deontologism, consequentialism, and contractualism are not ready-to-use programmes, which need to be only slightly recompiled to become applicable in the context of CE in order to tackle the moral issues raised by ICTs and deliver the expected results, on the more positive side we may wish to re-consider the *action-oriented* nature of CE itself. To show this, I need to introduce first a simple model of macroethics and finally explain what I mean by 'action-oriented' and 'patient-oriented'.

4.3 An informational model of macroethics

When reduced to its minimal logical structure, any action, whether morally loaded or not, is a binary relation between an *agent* and a *patient* (see Chapter 6). The interpretation of what can then be inferred from the occurrence of *prima facie* moral actions, in terms of what is the primary object of the ethical discourse, is a matter of philosophical controversy.

Virtue ethics, and ancient philosophy more generally, concentrates its attention on the moral nature and development of the individual agent who is the *source* of the action. It can therefore be properly described as primarily an *agent-oriented* ethics. Think of the agent as a sender of an action/message. Since the agent is usually assumed to be a single human being, virtue ethics is intrinsically anthropocentric and individualistic. It is also mostly interested in what happens to the agent when the latter is also the patient or *receiver* of the secondary effects of moral action. Thus, in a textbook example, virtue ethics asks: if Alice behaves courageously, how is this going to affect her character? Agent and patient of an action need not be different, consider the case of a suicide. Most notably, virtue ethics concentrates on cases in which what matters most is the fact that the agent or source of the moral action is also the most significant patient or receiver of that action. As we shall see in Chapter 8, nothing prevents virtue ethics from being applicable to other kinds of agents, like political parties, companies, teams, or institutions. Indeed, this is a reasonable way of reading Plato's *Republic*, even if this is not the usual way in which virtue ethics is developed today. This is so partly because of a historical limitation, which has Greek roots in the individualist conception of the agent in question and the metaphysical interpretation of his or her functional development, and partly because of a contemporary empiricist bias, which consists in an anti-realist conception of non-individual entities—paradoxically, one may live in a materialist culture based on ICTs and yet be reluctant to treat informational entities as real objects—and in a pre-theoretical refusal to conceive of moral values also as holistic properties of complex systems. We shall see in the following chapters that the removal of such limitations has interesting consequences for the foundation of CE.

Developed in a world profoundly different from the small, non-Christian Athens, utilitarianism, or more generally consequentialism, contractualism, and deontologism are the three most well-known theories that concentrate on the moral nature and value of the *actions* performed by the agent. They are therefore *relational* and *action-oriented* theories, intrinsically social in nature. They obviously anchor the stability of the moral value of human actions very differently: consequentialism and contractualism *a parte post*, at least theoretically *after* the process that outputs the actions and through the assessment of their consequences in terms of global and personal welfare; deontologism *a parte ante*, *before* the process and through universal principles and the individual's sense of duty. Yet the principal target of their analysis remains unchanged, for they both tend to treat the *relata*, i.e. the agent and the patient, Alice and Peter, as secondary in importance, and may sometimes end up losing sight of their destiny, as notorious examples such as those of the duty not to lie to the Nazi and the trolley problem clarify. From their relational perspective, what Alice becomes or does in her autonomy, and quite irrespective of external factors, as may be the case in virtue ethics, now has less importance than the more significant interactions between her and the surrounding society, or even the simple possibility of such interactions (the Kantian universal maxim). These macroethics may be based on a central concept of self-interest (consequentialism and contractualism), but their analyses focus primarily on the nature of action and choice, understood as the function from human interests to moral values and, thus, shift the attention from a purely *agent-oriented* to a substantially interactive, *action-oriented* approach. What matters is neither Alice nor Peter, but what goes on between them. Thanks to this shift in perspective, the philosophy of history, understood as the ethical interpretation of the collection of all significant actions liable to moral evaluation, acquires more importance than pedagogy, understood as the development and evaluation of an individual's cultivation. Having thus made the conception of human nature more peripheral to the ethical discourse than humanity's deeds, action-oriented and relational theories can finally ease and promote the enlargement of the concept of a morally responsible agent as a free and rational centre of rights and duties, and slowly come to include in the ethical discourse not only the Athenian gentleman but also women, people of other cultures, minority groups, and members of all social classes, and so forth; in short, any free and rational human agent. Since agent-oriented, intra-subjective theories and action-oriented, infra-subjective theories are all inevitably anthropocentric, we may follow common practice and define them as 'standard' or 'classic', without necessarily associating any positive evaluation with either of these two adjectives.

Standard ethics take only a relative interest in the *patient*, the third element in a moral relation, which is on the receiving end of the action and endures its effects. A partial exception is represented by a Kantian version of contractualism *à la* Rawls, which stresses the crucial importance of the impartial nature of moral concern, thanks to the hypothetical scenario in which rational agents are asked to determine the nature of society in a complete state of ignorance of what their positions would be in it, thus transforming the agent into the potential patient of the action.

Ontological power brings with it new moral responsibilities. We can respect only what we no longer fear, yet knowledge is a process of increasing emancipation from reality. In a world in which humanity can influence, control, or manipulate practically every aspect of reality, philosophical attention is finally drawn to the importance of moral concerns that are not immediately agent- or action-oriented and anthropocentric. Medical ethics, bioethics, and environmental ethics are among the best-known examples of this non-standard or non-classic approach. They seek to develop a *patient-oriented* ethics in which the receiver of the moral action may be not only a human being, but also any form of life. Indeed, land ethics extends the concept of patient to any component of the environment (Stone, 2010) thus coming close to the approach defended by IE, as we shall see presently and in Section 5.7. Capturing what is a pre-theoretical but very widespread intuition shared by most people, a *patient-oriented* ethics holds the broad view that any form of life has some essential proprieties or moral interests that deserve and demand to be respected, at least initially, minimally, and overridably. They argue that the nature and well-being of the patient of any action constitute (at least partly) its moral standing, and that the latter makes important claims on the interacting agent that, in principle and when possible, ought to contribute to the guidance of the agent's ethical decisions and the constraint of the agent's moral behaviour. Thus, compared to classic ethics, bioethics, medical ethics, and environmental ethics turn out to be theories of nature and space. Their analyses start from the moral properties and values of what there is or is 'present' and are no longer grounded in history and time, that is, in human actions, motives, and consequences. In other words, whereas classic ethics is a philosophy of the wrongdoer, non-classic ethics is a philosophy of the victim. This is because according to classic analyses, any action may be inexorably stained with evil, either because of what it is not—as from a consequentialist perspective, where every action is always improvable, so any action can be only relatively good at most—or because of what it could be—as from a deontologist perspective, where the same action in itself leads either to morally disapprovable or just amoral behaviour if it does not spring from a sense of duty and does not conform to the universal maxims. Contrariwise, non-standard ethics may be described as an ethics of the listening (of the patient) rather than as an ethics of the seeing (by the agent): such an approach to ethics places the receiver of the action, the patient, at the centre of ethical discourse and moves its sender, the agent, to its periphery. In so doing, it helps to widen further our anthropocentric view of who or what may qualify in principle as a locus of moral concern.

On the one hand, standard ethics are inevitably egocentric and logocentric, where all theorizing concerns a conscious and self-assessing agent whose behaviour must be supposed sufficiently free, reasonable, and informed for an ethical evaluation to be plausible on the basis of her responsibility. On the other hand, non-classic ethics, being biocentric and patient-oriented, is allocentric—i.e. centred on, and interested in, the entity itself that receives the action, rather than in its relation or relevance to the agent. They can thus be morally altruistic and can now admit any form of life within the ethical sphere, including all vulnerable human beings, not just foetuses, newborn babies, and senile persons, but also physically or mentally ill, disabled, or disadvantaged people. This is an option that lies beyond the immediate scope of any classic ethics, from Athens to Königsberg.

4.4 From computer ethics to information ethics

The development of ethical theories, sketched in the previous two sections, offers a useful heuristic to begin interpreting the nature of ICTs ethical problems at a different level of abstraction. If one tries to pinpoint what common feature so many case-based studies in CE share, it seems reasonable to conclude that this is a widespread interest in the fate and well-being of the receiver of the action, the patient, and that the latter is often an information system or environment. Broadly speaking, despite its immediate decision-making approach and its obvious social concerns, CE is not so much interested in the moral value of the actions in question, or in the agents' virtues or vices. Instead, CE develops its analyses, and attempts to indicate the best course of action, as a consequence of the steady and careful attention paid to what happens to (regions of) the infosphere, its features and components, and its inhabitants. Right and wrong in CE may not just qualify actions in themselves, they may essentially refer to what is eventually better or worse for the informational environment in question. Therefore, far from being a standard, action-oriented ethics, as it may deceptively seem at first sight, CE is primarily an ethics of *Being* rather than of *conduct* or *becoming*, and hence qualifies as non-standard. So CE is better placed, both theoretically and historically, in the same non-standard camp where we find other contemporary applied ethics such as medical ethics, bioethics, and environmental ethics. The fundamental difference, which sets it apart from all other members of the same class of non-standard theories, is that CE raises information as such, rather than just life in general, to the role of the true and universal patient of any action. As we saw in Chapter 2, without information there is no moral action, but information now moves from being a necessary epistemological prerequisite for any morally responsible action to being its primary object ontologically. This means that, by adopting a level of abstraction at which the ontology becomes informational, CE presents itself as a patient-oriented ethics that is ontocentric rather than biocentric.

The crucial importance of this change in perspective cannot be overestimated. Typical non-standard ethics can reach a high level of universalization of its ethical discourse thanks to its biocentric nature. However, this also means that even bioethics and environmental ethics fail to achieve a level of complete universality and impartiality, because they are still biased against what is inanimate, lifeless, intangible, abstract, engineered, artificial, synthetic, hybrid, or merely possible. Even land ethics is biased against technology and artefacts, for example. From its perspective, only what is intuitively alive deserves to be considered as a proper centre of moral claims, no matter how minimal, so a whole universe escapes its attention. Now, as I shall argue in Chapter 6, this is precisely the fundamental limit overcome by an informational approach, which further lowers the condition that needs to be satisfied, in order to qualify as a potential centre of a moral concern, to the minimal common factor shared by any entity, namely its information state. Since any form of Being is, in any case, also a coherent body of information, to say that CE is infocentric is tantamount to interpreting it correctly as an ontocentric patient-oriented theory.

4.5 Information ethics as a patient-oriented and ontocentric theory

From an IE perspective, the ethical discourse now comes to concern information as such; that is, not just all persons, their cultivation, well-being, and social interactions, and not just animals, plants, and their proper natural life either, but also anything that exists, from paintings and books to stars and stones; anything that may or will exist, like future generations; and anything that was but is no more, like our ancestors. Unlike other non-standard ethics, IE is more impartial and universal—or one may say less ethically biased—because it brings to ultimate completion the process of enlarging the concept of what may count as a centre of moral claims, which now includes every instance of information, no matter whether physically implemented or not. Such an all-embracing approach is made possible by the fact that IE adopts a LoA at which Being and the infosphere are co-referential. This enables one to endorse the following principles and concepts. I list them here with only brief comment, since they will be a matter of further elaboration throughout the rest of this book.

4.5.1 Uniformity of Being

An entity is a consistent packet of information, that is, an item that contains no contradiction in itself and can be named or denoted in an information process. A contradiction, when directly and positively *used* (i.e. not used at a meta-theoretical level, or just *mentioned*), is an instance of total negation of information, i.e. a mark left where all information has been completely erased, a scratch in the fabric of the infosphere. Since an information process positively involving a contradiction ends up being itself a source of contradiction, it is also a case of total negation, an information black hole, as it were. It follows that there are no information processes that fruitfully involve contradictions (obviously this is not to say that there are no contradictory information processes), that an information process can involve anything which is in itself logically possible, and that IE treats every logically possible entity as an informational entity. The infosphere is the totality of Being, hence the environment constituted by the totality of informational entities, including all agents, along with their processes, proprieties, and mutual relations.

4.5.2 Uniformity of nothingness, non-Being, or metaphysical entropy

Non-Being is the absence or negation of any information. Since the nineties, I have called this 'entropy'. This terminological choice is no longer rectifiable, but it was very unfortunate. It has led to endless misconceptions and misunderstandings because it is quite normal to read 'entropy' either thermodynamically or information-theoretically. Indeed, the metaphysical concept of 'entropy' that I have been using now for more than a decade and that pervades this book is strictly related to the previous two other concepts, but should not be equated to them. The point has turned out to be so demanding for so many readers that I am certainly the one to be blamed for the misleading communication. It is too late to undo the past. So let me add a clarification for the future.

Broadly speaking, entropy is a quantity specifying the amount of disorder, degradation, or randomness in a system bearing energy or information. More specifically, in *thermodynamics*, entropy is a parameter representing the state of randomness, disorder, or 'mixed-upness' of a physical system at the atomic, ionic, and molecular level: the greater the disorder, the higher the entropy. In a closed system undergoing change, entropy is a measure of the amount of thermal energy unavailable for conversion into mechanical work: the greater the entropy, the smaller the quantity of energy available. When your glass of sparkling water becomes flat, the entropy has increased. The same glass of water with an ice cube in it has less entropy than the same glass of water after the ice cube has melted. According to the second law of thermodynamics, during any process, the change in entropy of a system and its surroundings is either zero or positive, so the entropy of the universe as a whole inevitably tends towards a maximum.

Information in Shannon's sense is also known as *entropy*. More precisely, in information theory, assuming the ideal case of a noiseless channel of communication, entropy is a measure of three equivalent quantities:

(a) the average amount of information per symbol produced by the sender, or
(b) the corresponding average amount of data deficit (Shannon's uncertainty) that the receiver has before the inspection of the output of the sender, or
(c) the corresponding informational potentiality of the same source, that is, its *informational entropy*.

The informational and the thermodynamic concept of entropy are related through the concepts of probability and randomness. Entropy is a measure of the amount of 'mixed-upness' in processes and systems bearing energy or information. It can also be seen as an indicator of reversibility: if there is no change in entropy then the process is reversible. A highly structured, perfectly organized message contains a lower degree of entropy or randomness, less information in Shannon's sense, and hence it causes a smaller data deficit, which can be close to zero. By contrast, the higher the potential randomness of the symbols in the alphabet, the more bits of information can be produced by the device. Entropy assumes its maximum value in the extreme case of uniform distribution, which is to say that a biased coin has less entropy than a fair coin. As I have just mentioned, in thermodynamics, the greater the entropy, the less available the energy. This means that high entropy corresponds to high energy deficit. Similarly, so does entropy in information theory: higher values of entropy correspond to higher quantities of data deficit.

Both in thermodynamics and in information theory, entropy is a syntactic and quantitative concept: neither information nor entropy refers to the actual meaning, content, interpretation (semantics), or to the existence and nature (ontology) of the system. In IE, we still treat the two concepts of information and entropy as being related, but we are concerned with the semantic and ontological nature of information. For example, as the infosphere becomes increasingly meaningful and rich in content, the amount of information increases and (what one may call, for the sake of clarity) *metaphysical entropy* decreases; or as entities wear out and finally disappear, metaphysical entropy increases and the amount of information decreases. Thus, in IE, entropy is not a merely syntactic concept, but, as the opposite of semantic and ontic information, it indicates the decrease or decay of information leading to absence of form, pattern, differentiation, or content in the infosphere. It is therefore most emphatically *not* the physicists' or engineers' concept of entropy. Metaphysical entropy refers to any kind of *destruction* or *corruption* of entities understood as informational objects (mind, not just of semantic information, or messages), that is, any form of impoverishment of *Being*. *Destruction* is to be understood as the complete annihilation of the entity in question, which ceases to exist; compare this to the process of 'erasing' an entity irrevocably. *Corruption* is to be understood as a form of pollution or depletion of some of the properties of the entity, which ceases to exist as that entity and begins to exist as a different entity minus the properties that have been corrupted or eliminated. This may be compared to a process that degrades the integrity of the entity in question. So entropy, which has many meanings, is here comparable to the metaphysical concept of *nothingness*, to phrase it more metaphysically or theologically. The reference here is to the classic conception of evil as *privatio boni*, the thesis according to which only good is substantial, and evil is a 'privation of good' (I shall return to this point in Chapters 6 and 9). It is, after all, how Mephistopheles introduces himself:

> Ich bin der Geist, der stets verneint!
> Und das mit Recht; denn alles, was entsteht,
> Ist wert, daß es zugrunde geht;
> Drum besser wär's, daß nichts entstünde.
> So ist denn alles, was ihr Sünde,
> Zerstörung, kurz, das Böse nennt,
> Mein eigentliches Element.
>
> I am the Spirit, that ever denies!
> And rightly so; for everything that comes into being,
> deserves to perish;
> since it were better if nothing had come forth.
> Thus is everything that you call Sin,
> Destruction, in short, Evil,
> my proper Element.
> (Goethe, *Faust*, I, 1338–44, translation mine)

Henceforth, whenever I speak of *entropy*, the reader may safely assume that I mean *metaphysical entropy* (the Mephistophelean kind) as just defined.

4.5.3 *Uniformity of change*

At an informational LoA, all operations, changes, actions, and events can be treated as information processes, not in a procedural sense (e.g. as part of a program that performs some task), but as state transitions. Hence, here, the dynamics of reality are analysed in terms of information flows.

Any information process necessarily generates a trail of information. The absence of an information process is also an information process. This is an extension of the general principle underlying any static encoding of information to information dynamics, and it will prove important later in order to make sense of the distinction in ethics between action and omission.

An agent is any entity, to be defined more precisely in Chapter 7, capable of producing informational phenomena that can affect the infosphere. At its minimal level, agency is the mere presence of an implemented informational entity in terms that remind one of Heidegger's *Dasein*—the mere being-there or presence—of an informational entity in the infosphere. Let me anticipate (see Chapter 7) that, of course, not all informational entities are agents (compare to abstract informational entities); that many agents may often fail to be in a position to affect the infosphere significantly, beyond their mere presence (think of a grain of sand in the desert, as opposed to the last grain flowing through an hourglass determining the explosion of a bomb); and that not all agents are *accountable* (e.g. an earthquake). Furthermore, only some accountable agents are *responsible* (e.g. Alice), that is, able to be aware of the situation and capable of planning, withholding, and implementing their interactions with the infosphere with some degree of freedom and according to their evaluations.

4.5.4 The ontological equality principle

When the ethical discourse attempts to motivate, persuade, or even coerce Alice to act morally, an anthropocentric and self-interested justification of goodness may well be inevitable (see Section 6.2 for a similar distinction between 'intellectual' vs. 'strategic' reasons and Section 14.4 for an analysis of normative pressure points in a system). One may argue,[4] for example, that in the USA the death penalty is increasingly controversial not because people have come to grips with its immorality, but because they have come to realize that it is uneconomical, since in the American system it is far more expensive to execute someone than to jail that person for life. Independently of whether the argument is correct, it seems clear that any increase in the unpopularity of the death penalty is certainly welcome; no matter how disappointing the explanation may be, ethically speaking. Likewise, one could motivate agents to act morally towards the whole infosphere *instrumentally* (perhaps using egoistic or economic arguments) without having to cease being concerned with ethics. However, when the primary aim of an ethical investigation is to understand what is right and wrong and, hence, what the best course of action would be, irrespective of a specific agent's behaviour and motivation, it becomes possible to adopt a more *philosophical* viewpoint. In this respect, we shall see in Chapter 6 that IE holds that every informational entity,

---

[4] 'Revenge Begins to Seem Less Sweet', *The Economist*, 30 August 2007 <www.economist.com/node/9719806>.

insofar as it is an expression of Being, has a dignity constituted by its mode of existence and essence, defined here as the collection of all the elementary proprieties that constitute it for what it is. This dignity *prima facie* deserves to be respected and hence may place moral claims on any interacting agent. It ought to contribute towards constraining and guiding her ethical decisions and behaviour, even if only initially and in an overridable way. This ontological equality principle (see the corresponding notion of ontic trust in Chapter 15) means that any form of reality—that is, given the informational structuralist approach defended by IE, any instance of information—simply for the fact of being what it is, enjoys an initial, overridable, minimal right to exist and develop in a way appropriate to its nature. The conscious recognition of the ontological equality principle presupposes, *a parte ante*, a disinterested judgement of the moral situation from an absolute perspective, i.e. a perspective which is as patient-oriented as possible. Indeed, moral behaviour is less likely without this epistemic virtue. At most, we can only act to the best of our knowledge concerning the options available, the likely consequences and implications of the action undertaken, and so forth, yet this is hardly sufficient to ensure that our actions will be morally right, if our knowledge is either limited or biased towards the agent and what is best only for her and does not include a wider degree of attentiveness to the patient as well. Thus, a form of moral luck arises when an interested and agent-oriented judgement leads to a course of action which turns out to be respectful of the rights of the patients as well, though only by chance. The application of the ontological equality principle is achieved, *a parte post*, whenever actions are impartial, universal, and caring (I shall return to this point in Section 15.6). This means that IE endorses the Golden Rule, and its subsequent refinements such as the Kantian moral imperative or Rawls' choice in a state of ignorance, as part of the main explicit principle of its ethical analysis, even if in informational terms. We can do justice to any form of reality and deal fairly with it only if the principles we follow and the actions we perform:

- are independent of the position we enjoy in the moral situation, as patient or agent. We would make the same choices and behave in the same way even if we were at the receiving end of the action (impartiality);
- can in theory regulate the behaviour of any other agent placed in any other similar moral situation. Anyone else would make the same choices and behave in the same way in a similar situation (universality);
- look after the well-being of both the agent and the patient (care-fulness). Our choices and behaviour are equally agent-oriented and patient-oriented, insofar as the interests of both agent and patient are taken into account.

Biocentric ethics ground its analyses of the moral standing of bio-entities on the intrinsic worthiness of life and the intrinsically negative value of suffering. IE argues that there is something more elementary and fundamental than life and pain, namely Being, understood at an informational level of abstraction as equivalent to the infosphere, and entropy, understood metaphysically as any decrease or impoverishment of Being/infosphere.

The description of classes of informational entities in terms of their specific essence is a task to be left to a plurality of ontologies.[5] When the informational entities in question are human beings, for example, we refer to the analysis of human rights. Unfortunately, this clear limit in our knowledge is of the greatest importance, for it reminds us that, like many other macroethics, IE relies on human understanding for the implementation of the right action. As in the case of consequentialism, IE relies on moral education and the transmission of whatever past generations have been able to grasp about the nature of the world, and hence its intrinsic value, thus adopting a rule-ethics rather than an act-ethics approach. Yet it must also acknowledge the fact that even a good will acts in the dim light of uncertainty and that, as human beings, we shall always lack full ethical competence. This is why our first duty is epistemic: whenever possible, we must try to understand before acting. This also explains why moral education consists primarily in negative principles and a fundamental training not to interfere with the world, to abstain from engaging in positive actions and from tampering with reality. In most cases, we simply do not know where a *prima facie* positive interaction with the infosphere would lead us, or what negative outcome actions may have, even when they are well meant. I shall return to the risky nature of such actions in the following chapters. The time has now come to turn to the general principles that, according to IE, should guide, modify, and constrain information processes and, hence, also contribute to the foundation of the moral codes by which we could live.

4.6 The normative aspect of information ethics: four ethical principles

What is good for an informational entity and the infosphere in general? This is the moral question asked by IE. We have seen that the answer is provided by a minimalist approach to what Being deserves: any informational entity is recognized to be the centre of some basic ethical claims that deserve recognition and should help to regulate the implementation of any information process involving it. Approval or disapproval of any information process is then based on how the latter affects the essence of the informational entities it involves and, more generally, the well-being of the whole infosphere, i.e. on how successful or unsuccessful it is in respecting the claims attributable to the informational entities involved and, hence, in enriching or impoverishing the infosphere. More analytically, IE determines what is morally right or wrong, what ought and ought not be done, and so what the duties of a moral agent are, by means of

---

[5] In Floridi (1999a), I offered a tentative list of some interesting informational properties characterizing the infosphere. I am no longer convinced it was a felicitous attempt, though I still believe that it should be possible to develop an ethical ontology that indicates what is worth preserving and taking care of in reality. I have more reservations now about the feasibility of that project in a short space and within this context.

Table 1. The four ethical principles of IE

0 entropy ought not to be caused in the infosphere (null law)
1 entropy ought to be prevented in the infosphere
2 entropy ought to be removed from the infosphere
3 the flourishing of informational entities as well as of the whole infosphere ought to be promoted by preserving, cultivating, and enriching their well-being

four basic ethical principles (Table 1). Unsurprisingly, they closely resemble similar principles in medical ethics with which they share the same patient-oriented approach. I shall formulate them here in a patient-oriented version, but an agent-oriented one is easily achievable in terms of 'dos' and 'don'ts'.

The principles are listed in order of increasing moral value. They clarify, in very broad terms, what it means to live as a responsible and caring agent in the infosphere.

On the one hand, a process is increasingly disapprovable and its agent-source is increasingly blameworthy, the lower is the number index of the specific law that it fails to satisfy. Let us agree to define any process that is disapprovable in the sense just specified as a case of impoverishment of the infosphere. Moral mistakes may occur and entropy may increase because of a wrong evaluation of the impact of one's actions—especially when local goodness, i.e. the improvement of a region of the infosphere, is favoured to the overall disadvantage of the whole infosphere—because of conflicting or competing projects—even when the latter are aiming at the satisfaction of IE moral laws—or more simply because of the wicked nature of the agent—this possibility is granted by IE's negative anthropology, more on this in Chapters 6 and 9.

On the other hand, a process is already approvable and its agent-source praiseworthy, if it satisfies the *combination* of the null law with at least one other law, not the *sum* of the resulting effects. Note that, according to this definition, an action is unconditionally approvable only if it never generates any metaphysical entropy in the course of its implementation, that no positive law has a morally higher status ($0 \wedge 1 = 0 \wedge 2 = 0 \wedge 3$), and that the best moral action is the action that succeeds in satisfying all four laws at the same time. Most of the actions we judge morally good do not satisfy such a strict criterion, for they achieve only a balanced positive moral value; that is, although their performance causes a specific quantity of entropy, we acknowledge that the infosphere is in a better state after their occurrence. Finally, a process that satisfies only the null law—the level of entropy in the infosphere remains unchanged after its occurrence—either has no moral value, that is, it is morally irrelevant, insignificant, or negligible (see Chapter 13), or it is equally disapprovable and approvable, though in different respects. This last point requires some clarification.

Although it is logically conceivable, it seems that, strictly speaking, there can be no actual information process that is disapprovable and approvable in exactly the same measure, that is, such that its output leaves the infosphere in exactly the same entropic state in which it was before. Consequentialist analyses, for example, do not really take into account the possibility that Alice may escape any moral evaluation by perfectly balancing the amount of happiness and unhappiness generated by her actions. The possibility of a zero-sum moral game does not seem very interesting. However, it is also the case that, strictly speaking, there can be very few, if any, information processes that are morally insignificant. More likely, any process will always make a difference, either positive or negative, and therefore will always be subject to moral appraisal. This, however, would not only be counterintuitive, but it is not even the view defended by IE. We ordinarily treat most of the processes/actions that take place in life as amoral, i.e. lying beyond the scope of the ethical discourse, and we do so for good reasons. First, because we usually adopt a less strict criterion and accept some latitude in the levels of entropy before and after the occurrence of the process. Second, because we are acquainted with macroscopic forms of impoverishment of the infosphere (killing, stealing, lying, torturing, betraying, causing injustice, discriminating, etc.) such that a lot of minor fluctuations in the level of global entropy become irrelevant. Finally, and more importantly, because many processes do not easily modify the global level of entropy even when they are positively immoral. People who argue for the *fragility* of goodness sometimes do so on the mistaken assumption represented by its *non-monotonic* nature while forgetting its *resilience*. The point is sufficiently important to deserve to be discussed in a separate section.

4.6.1 *The non-monotonic nature of goodness and its resilience*

Consider an action—for example torturing a child—that is utterly morally wrong. This means that it generates a net increase in the level of entropy in the infosphere, and for IE, as well as for our pre-theoretical intuitions, this fact is irrevocable in itself and unredeemable by later events. Furthermore, in such a case, there is no way of re-engineering the process so that it loses its negative moral value. Drawing on the conceptual vocabulary of logic, this stability can be defined as the *monotonicity of evil*. The difficulty encountered by any pure form of consequentialism is that, since human rights and values (such as integrity) are, in principle, always overridable depending on the overall happiness generated *a posteriori* by an action's consequences, consequentialism must treat evil as non-monotonic: in theory, it is always possible to collect and trace a sufficient amount of happiness back to an utterly wicked action and thus force a modification in the latter's evaluation. Now, the advantage of IE is that, like our moral intuition, it attributes a non-monotonic nature only to goodness: unlike evil, goodness can, in principle, turn out to be less morally good and sometimes even morally wrong unintentionally, depending on how things develop, that is, on what new state the infosphere enters into, as a consequence of the process in question. This is one way of understanding the fragility of goodness: perhaps there is no action that could count as absolutely good at all times and in all places, so do what you wish and evil will remain evil, but make a mistake and what was initially morally good may be corrupted or turned into evil.[6] However, to describe goodness as fragile owing to its non-monotonicity would be a mistake because non-monotonicity is only one of the relevant features to be taken into account. If utter evil is monotonic, *prima facie* goodness, such as disinterested love or friendship, has the property of being *resilient*, both in the sense of *fault-tolerance*:

- to some degree, goodness has the ability to keep the level of entropy within the infosphere steady, despite the occurrence of a number of negative processes affecting it;

and in the sense of *error-recovery*:

- to some extent, goodness has the ability to resume or restore the previous entropic state of the infosphere, erasing or compensating any new entropy that may have been generated by processes affecting it.

Resilience—what we often find described by terms such as tolerance, forbearance, forgiveness, reconciliation, or simply other people's positive behaviour—makes goodness much more robust than its non-monotonic nature may lead one to conclude at first sight. It explains the presence and possibility of entropic balance in the infosphere, which in turn clarifies why so many actions often lie beyond our ethical concern: they simply fail to modify the goodness/evil balance of the infosphere significantly.

For further clarification, consider the following: unlike what Kant holds, moral actions are risky because only a fraction of their value can depend on our good will. We recognize this when we acknowledge that a bad action is forgivable but not excusable, while only a failed good action is excusable, and therefore that it is moral to do *x* only when *x* would be *prima facie* a good action, but immoral to do *x* when *x* is *prima facie* a bad action. Evil is monotonic, so one should not intentionally bet on one's own good luck. This holds true even when some morally risky actions (processes, behaviours)—such as driving too fast in a city centre—come close to the threshold between what is morally insignificant and what is morally wrong (e.g. a person may be injured because of such dangerous driving, thus making speeding a morally wrong action). Such morally risky actions can usually keep on the amoral side thanks to their (more or less lucky) reliance on the fault tolerance and error-recovering properties of the region of the infosphere they involve. In our example, for instance, this would include, among

---

[6] I am grateful to OUP's anonymous reviewer for reminding me that, usually,

the point [about the fragility of goodness] is not that we tend to get good mixed with evil in everything we do, but that there are vastly many more ways to get things wrong than there are to get things right. For example, in Aristotle courage is having just the right amount of concern for safety, neither too much nor too little, implying a willingness to take risks always and only for the right sorts of things in the right ways and the right times, and this is just one part of flourishing, which requires having all the virtues (exactly the right amount of concern for every good) and having the right amount of those goods as a result (which requires good fortune and not just virtue). There are a number of ways to get things wrong, but only one way (or relatively few ways) to get it just right.

I find the two senses—mixed-upness and balance—strictly related, but this is a matter of scholarly debate rather than philosophical argument.

other factors, the careful attitude of other drivers and pedestrians. Although it would not be morally right to rely on it, as Paul of Tarsus beautifully reminds us in 1 *Corinthians* 13 (more on this in Chapter 13), the strength of goodness should not be undervalued: it takes a fatal process to generate some permanent entropy.

4.7 Information ethics as a macroethics

The reader will recall that our investigation into the nature of IE has been prompted by the question of whether CE can fruitfully dialogue with other macroethical positions at the same conceptual level, having something important to contribute to the overall ethical discourse that may perhaps escape their conceptual frameworks. In search of an answer, we have first freed CE from its conceptual dependence on other macroethics and then disposed of the mistaken interpretation of CE as a standard, action-oriented theory. IE, the philosophical foundational counterpart of CE, has emerged as a non-standard, patient-oriented, ontocentric theory. The previous sections sketched its outline. The following chapters will seek to flesh out such a sketch. Our next task here is to evaluate whether this interpretation of IE is sufficient to vindicate the initial claim that the philosophical foundation of CE qualifies as a macroethics. Has IE anything to contribute to what is already provided by other standard and non-standard macroethics? What kind of new insights may IE offer to improve our understanding of what is morally right and wrong? The defence of the value of IE as a macroethics can be articulated in three stages, the last of which will require a new section of its own.

4.7.1 IE is a complete macroethics

From a meta-ethical perspective, IE is a naturalist and realist macroethics: the ontological features and well-being of the infosphere provide a patient-oriented approach for judgements of right and wrong and generate equally patient-oriented reasons for action. They are action-pulling, while the moral system, based on the nature and enhancement of the infosphere and the corresponding moral claims, is universally binding, i.e. binding on all agents, in all places, at all times. IE is not an ethics of virtue, happiness, or duty, but of 'respect' for the patient and 'care' exercised by the agent. According to IE, sometimes the right question to be asked is not 'what ought *I* to be?' nor 'what ought *I* to do?', but '*what* ought to be respected or improved?', for it is the 'what's' well-being that may matter most. The agent is invited to displace herself, to concentrate her attention on the nature and future of the action-receiver, rather than on its relation or relevance to herself, and hence to develop an allocentric attitude, i.e. a profound interest in, and respect for, the infosphere and its values for their own sake, together with a complete openness and receptivity toward it. The concept of care, as employed by IE, is the secular equivalent of the Pauline concept of ἀγάπη ('loving treatment with affectionate regard') or *caritas* ('dearness, love founded on esteem'). After the death of god,[7]

---

[7] On the death of god (small g intentional) and the philosophy of information, see Floridi (2011a, ch. 1).

the spiritual value of Being has become much more questionable. It certainly does not impose itself as obvious to the agent's attention anymore, so it is the agent who needs to be sensitized. An agent cares for the patient of her actions when her behaviour enhances the possibilities that the patient may come to achieve whatever is good for it. Though an action, which is universal and impartial, may be morally appropriate, it becomes morally good only when it is driven by care for the patient's sake. This is moral altruism for IE.

The reader will probably agree that IE is a controversial theory, even if it is to be hoped that it will be less so by the end of this book. Yet, one needs to realize that it is controversial as a macroethics, since most of the problems that may afflict it are problems concerning the whole class of macrotheories. To put it briefly, any substantial problem IE encounters is unlikely to be just IE's problem, whereas whatever solutions and insights IE provides are its own original contributions. For example, IE takes Being/infosphere as intrinsically good, and non-Being/entropy as evil, so that moral prescriptivity becomes, at least also, an intrinsic property of information: some features of the infosphere are descriptive and action-guiding and generate reasons for action independently of any motives or desires that agents may actually have. Of course, this is an original position, but it remains a rather controversial one. However, other theories are also based on first principles, such as $\epsilon\dot{v}\delta\alpha\iota\mu o\nu\acute{\iota}\alpha$, happiness, duty, or life, which are equally open to discussion: what is morally good in itself? Why is x rather than y to be considered morally good in itself? And so forth. Two arguments offered in support of IE are its explanatory power and its degree of universality (see next paragraph). That IE's position may still be subject to criticism at this level reinforces the view that IE does represent a new perspective, which involves the whole ethical discourse. This is all that matters in this chapter.

In the following chapters, I shall argue that IE provides a valuable perspective from which to approach, with insight and adequate discernment, not only moral problems in its own special field but also the whole range of conceptual and moral phenomena that form the ethical discourse. Contrary to other macroethics, IE has its own domain of special application, but what was a weakness now becomes a strength: action-oriented and anthropocentric or patient-oriented and biocentric theories seem to be inadequate to tackle moral problems emerging in our information society because the latter are peculiarly ontocentric and patient-oriented in nature. On the other hand, although I remarked before that non-standard ethics move the ethical focus from history and time to nature and space, it would be a mistake to think that, similarly, IE manages only to shift our focus a step further. On the contrary, by enlarging the perspective of the ethical discourse to the whole infosphere and anything in it, IE clearly comes to include both history and nature, both time and space within the scope of its analysis. This has a remarkable consequence in terms of the kind of relation that occurs between IE and other macroethics, since IE may rightly claim the whole domain of ethics as its area of interest. To see that this is the case, let me briefly compare IE with other ethical approaches in the next four sections. I shall then analyse three applications of IE in Section 4.9.

4.7.2 IE and other non-standard ethics

The general advantage of IE over other non-standard ethics consists in the fact that IE provides a more comprehensive philosophy of nature and history and thus can absorb their positive contributions without sharing the same limits. As for any more specific comparison, three points may be explicitly mentioned here.

First, IE does not attribute to information the same absolute value that biocentric theories attribute to life. This allows a more intuitive organization of the infosphere into a scale of classes of informational entities according to their potential capacities to implement processes that may improve regions of influence in the infosphere. All entities have some minimal moral value, but they do not share the same degree of dignity. Intuitively, from the point of view of the infosphere and its potential flourishing and enrichment, responsible agents, such as human beings, full-AI agents, extraterrestrial minds, angels, gods, and God, have greater dignity and are the most valuable informational entities, deserving the highest degree of respect, because they are the only ones capable of both knowing the infosphere and taking care of it according to the conscious implementation of their self-determined projects by increasing or decreasing the anti-entropy level of their actions. After all, the 'divinity' of God consists also in his omnipotence.

Second, since IE does not limit its own area of interest to the biophysical environment, the applicability of its ethical principles is in fact field-independent and universal, because the infosphere includes also any other environment.

Finally and most importantly, IE does not tend to be purely conservative, like other *green* ethics. On the contrary, it is a *blue ethics* (the expression comes from blue-print), like virtue ethics, that is, an ethics of projects and meaningful construction in a very strong sense. As we shall see in Chapter 8, IE advocates a constructionist approach as the right basis on which to think, remodel, and constructively improve the world (the infosphere), and implement new realities. Information processes are goal-driven but their goals are internally directed toward the development of the infosphere, so they cannot be completely heteronomous. To phrase it more metaphysically, the infosphere has its own development as its own goal, so there is no 'external space' determining its teleology from outside its teleology, i.e. heteronomously. The best thing that can happen to the infosphere is for it to be the target of a process of enrichment, extension, and improvement without any ontological loss. In the vocabulary of IE, this means without any increase, and possibly a decrease in metaphysical entropy. It follows that the most approvable courses of action always have a caring and constructionist nature. The moral agent is an agent that looks after the infosphere and brings about positive improvements in it, so as to leave the infosphere in a better state than it was in before the agent's intervention. Given its constructionist nature, it is easy to see that IE may approach questions concerning abortion, eugenics, human cloning, or bioengineering more open-mindedly than other biocentric ethics. With all due attention and care, we should be ready to innovate.

4.7.3 IE and virtue ethics

The well-being of an entity as well as of the whole infosphere consists in the preservation and cultivation of its properties. So, IE can dialogue with virtue ethics on the basis of its patient-oriented and non-functionalist standpoint: the well-being and flourishing of an informational entity—what an informational entity should be and become—can be determined by the good qualities in, or that may pertain to, that informational entity as a specific instance of Being. The similarity between virtue ethics and IE is that both treat the human being as an entity under construction, a work in progress in charge of itself. The difference between the two approaches lies in their ontologies and in the much broader conception of what may count as a good entity endorsed by IE. If anything, this seems to be a feature that works in favour of an IE approach, as we shall see in Chapter 8.

4.7.4 IE and deontologism

It is possible to develop a deontological version of IE. An IE moral imperative could be, for example: 'act so that you never treat the infosphere as a means but also always as an end at the same time'. Even this modified maxim, however, already shows that IE's advantage over deontologism lies, again, in its much wider concept of what qualifies as a centre of ethical claims. We have already seen that this was one of the reasons why ethical theories have enlarged their perspective beyond the Kantian approach. Like deontologism, IE treats evil as monotonic: nothing justifies the infringement of the first moral law. An increase in entropy may often be inevitable, but is never morally justified, let alone approved. In this sense, IE counts as what Max Weber called an *ethics of conviction*. However, unlike deontologism, IE does not adopt an agent-oriented perspective—Alice's reliance on her sense of duty—for determining whether an action deserves to qualify as moral. For IE, an action qualifies as moral only from the patient's perspective. It is only the ontology of the victim that can really define an action as 'right', not the deontology of the wrongdoer or of the impartial judge. So a natural tendency to care for the well-being of the infosphere and a spontaneous desire to make it progress can be highly commendable virtues.

4.7.5 IE and consequentialism

What has been said about deontologism holds true for a consequentialist version of IE as well. Broadly speaking, both macroethics share the view that a morally good action is an action that improves the environment in which it takes place. Hence, as far as its four ethical principles are concerned, IE qualifies, like consequentialism, as what Max Weber calls an *ethics of responsibility*. Adopting the vocabulary of consequentialism, we may say that the restraint of entropy and the active protection and enhancement of (parts of) the infosphere are conducive to maximal utility. We can even rephrase the utilitarian principle and say that 'actions are right in proportion that they tend to increase (ontological) information and decrease (metaphysical) entropy'. However, the difference between IE and consequentialism remains significant because of at least four problems.

(1) *The monotonic problem*. This has been already discussed above. We have just seen that, as far as rights and moral evil are concerned, IE adopts a position closer to deontologism.

(2) *The mathematical problem*. If any quantification and calculation is possible at all in the determination of a moral life, then IE is clearly in a much better position than consequentialism. Consequentialism already treats individuals as units of equal value and relies on a calculus of aggregate happiness, which in the end is too simplistic, utterly unsatisfactory because it is computationally intractable, and amounts to little more than a metaphorical device, despite its crucial importance within the theory. On the contrary, if required, IE may resort to a highly developed mathematical field (information theory) and *try to adapt* to its own needs a very refined methodology, statistical methods, and important theorems, in terms of Sigma logarithms and balanced statistics. I have become increasingly sceptical about any possibility that quantities and algorithmic procedures might play more than a conceptual role in solving moral problems, since the passage from a quantitative and syntactic context to a qualitative, semantic, and ontic one seems to be impossible. However, if a consequentialist should seriously think otherwise, it can easily be shown that IE's approach is more powerful (Floridi and Sanders, 1999a). That a mathematical theory of information may not be sufficient to introduce mathematical certainty or computability into our moral reasoning is not a crucial problem for IE—which has nowhere been described as an algorithmic approach—but may work as a *reductio ad absurdum* for any naïve form of quantitative consequentialism.

(3) *The supererogatory problem*. There is no limit to how much better a course of action could be, or to the amount and variety of good actions that an agent may perform but does not. As a result, since goodness is a relative concept—relative to the amount of happiness brought about by the consequences of an action—consequentialism may simply be too demanding, place excessive expectations on the agent, and run into the supererogatory problem, asking the agent, who wishes to behave morally, to perform actions that are above and beyond the call of duty or even beyond the scope of her good will. In IE, this does not happen because the morality of a process is assessed on the basis of the state of the infosphere only, i.e. relationally, rather than relative to other processes. So while consequentialism is in principle satisfied only with the best action in absolute terms, in principle, IE praises any single action which improves the infosphere according to the principles specified in Section 4.6, as a morally approvable action, independently of the alternatives. According to IE, the state of the world is always morally disapprovable—there is always some entropic process that is impoverishing the infosphere—so that any process that improves it is already a morally good process. This is the advantage of a minimalist approach, which is more flexible and capable of appreciating thousands of little good actions, over a maximalist approach, which is capable of praising only

the single, best action. In a society that has become used to metering cents and seconds of used-time, the minute attention given to even small marginal values by IE can be appreciated as a much more successful alternative (see Chapter 13).

(4) *The comparative problem*. Consequentialism must accept that, since all actions are evaluated in terms of their consequences and all consequences are comparable according to a single quantitative scale, lives may in turn be judged morally better or worse merely for contingent reasons: an agent may simply be born in a context or find herself in circumstances where her actions can achieve more good than those of other agents. This is another sense in which we may speak of moral luck. Furthermore, this is not a problem faced by IE. Of course, IE shares the very reasonable point that different agents can implement the four ethical principles more or less successfully and with different degrees of efficacy, depending on their existential conditions. However, unlike consequentialism, which endorses a global conception of happiness, IE assesses the value of a process locally, in relation to the outcome it can achieve in the specific region of the environment it affects. This means that IE does not place different processes in competition with each other and, so, does not have to rank what two agents in different situations have, may have, or could have done. This is different from the problem of assessing what has been done and what the same agent in the same situation or under comparable circumstances could have done. Situations and circumstances count both for the kinds of process implementable and for the level of implementation, but they are irrelevant when comparing different courses of action. Thus, maintaining one's dignity in a Nazi prison-camp is simply no better or worse, morally speaking, than giving a lift to an unknown person on a rainy day, not just because the two experiences are worlds apart, but because both agents have done their best to improve the infosphere, and this is all that matters in order to consider their actions morally approvable. If comparable at all, they are so only in the vague and non-gradable sense in which the goodness of a good knife is comparable to the goodness of a good pencil. Consequentialism is not equally flexible.

4.8 Case analysis: three examples

The thesis to be articulated now is that not only can IE dialogue with other macroethics, it can also contribute a new[8] and important ethical perspective: a process or an action may be morally right or wrong irrespective of its consequences, motives, universality, or virtuous nature, but because it affects positively or negatively its patient (understood informationally) in particular and the infosphere in general. The

---

[8] To be scholarly precise, I take it to be more like a revival of a Platonist and Spinozian approach.

whole book might be read as an explanation and defence of this thesis: without IE's contribution, our understanding of moral facts in general, not just of CE problems in particular, would not be fully satisfactory. In order to begin supporting it, let me analyse three indicative examples: vandalism, bioengineering, and death. They are all negative in nature, but this is just for the sake of simplicity.

In Chapter 3, we have already encountered two cases, virtual pornography and informational privacy. Both examples illustrated how a patient-oriented approach might be helpful to understand what goes wrong in these cases and what would be the right course of action. IE seems to be able to cast some new light on CE problems, but one may still object that IE cannot successfully treat other types of moral problems not based on ICTs. For instance, one may fairly wonder how something that is not a sentient being or does not even exist may still have a moral standing, no matter how minimal, and, hence, impose any significant claim on an interactive agent so as to influence and shape her behaviour. This doubt may seem reasonable until we realize that it is in clear contrast to a rather common view of what is morally right or wrong and that this is precisely the problem solved by IE, as I shall argue in the analysis of the following three cases.

4.8.1 *Vandalism*

Imagine a boy playing in a dumping-ground. Nobody ever comes to the place. Nobody ever uses anything in it, nor will anyone ever wish to do so. There are many old cars abandoned there. The boy entertains himself by breaking their windscreens and lights, skilfully throwing stones at them. He enjoys himself enormously, yet most of us would be inclined to suggest that he should entertain himself differently, that he ought not to play such a destructive game, and that his behaviour is not just morally neutral, but is positively disapprovable, though of course very mildly so when compared to more serious mischiefs. In fact, we express our contempt by defining his course of action as a case of 'vandalism', a word loaded with an explicitly negative moral judgement. So let us suppose, for the sake of argument, that his course of action is indeed morally disapprovable. Which macroethics can help us to understand our sense of dissatisfaction with the boy's behaviour?

Any biocentric ethics is irrelevant, and broad environmental issues are out of the question as well, since, by definition, breaking the car windscreens does not modify the condition of the dumping-ground.

More radically, consequentialism, in its turn, finds it difficult to explain why the boy's behaviour is not actually approvable, since, after all, it is increasing the level of happiness in the world. Certainly, the boy could be asked to employ his time differently, but then we would be only saying that, as much as his vandalism is morally appreciable, there is something better he could be doing. We would be running into the supererogatory problem without having explained why we feel that his game is a form of vandalism and hence blameworthy. The alternative view, that his behaviour is causing our unhappiness, just begs the question: for the sake of the argument, we must be treated as mere external observers of his childish game.

Deontologism soon runs out of answers too. Its ends–means maxim is inapplicable, for the boy is playing alone and no human interaction is in view. Deontologism's imperative to behave as a universal legislator may be a bit more promising, but we need to remember that it often generates only drastic reactions and thus more problems than solutions: the boy can bite the bullet and make a rule of his misbehaviour. In this case, though, the problem is even more interesting. For Kant apparently never thought that people could decide to behave as universal legislators without taking either the role or the task seriously, but just for fun, setting up mad rules as reckless players. The *homo ludens* can be Kantian in a very dangerous way, as Stanley Kubrick's *Dr Strangelove* illustrates. The boy may agree with Kant and act as a universal legislator, as happens in every game: he is not the only one allowed to break the cars' windscreens in the dumping-ground, and indeed anyone else is welcome to take part in the game. With its stress on the universal extension of a particular behaviour, deontologism may well increase the gravity of the problem. Just think what would happen if the boy were the president of a military power playing a war game in the desert, searching for weapons of mass destruction.

Virtue ethics is the only macroethics that comes close to offering a convincing explanation, though in the end it too fails. From its perspective, the boy's destructive game is morally disapprovable not in itself, but because of the effects it may have on his character and future disposition. However, by so arguing, virtue ethics is quietly begging the question: it is because we find it disapprovable that we infer that the boy's vandalism will lead to negative consequences for his own development. Nobody grants that breaking windscreens necessarily leads to a bad character; life is too short to care and, moreover, a boy who has never broken a car windscreen might not become a better person after all, but a repressed maniac, who knows? Where did David practise before killing Goliath? Besides, the context is clearly described as ludic, and one needs to be a real wet blanket to reproach a boy who is enjoying himself enormously and causing no apparent harm, just because there is a chance that his playful behaviour may perhaps, one day, slightly contribute to the possible development of a moral attitude that is not praiseworthy. We come then to IE, and we know immediately why the boy's behaviour is a case of blameworthy vandalism: he is not respecting the objects for what they are, and his game is only increasing the level of entropy in the dumping-ground, pointlessly. It is his lack of care and respect, the absence of consideration of the objects' nature, that we find morally blameable. He ought to stop destroying bits of the infosphere and show more respect for what is naturally different from himself and yet similar as an informational entity to himself. He ought to employ his time more 'constructively'.

4.8.2 Bioengineering

Suppose one day we genetically engineer and clone non-sentient cows. They are alive but, by definition, they lack any sort of feelings. They are biological masses, capable of growth when properly fed, but their eyes, ears, or any other senses are incapable of any sensation of pain or pleasure. We no longer kill them, we simply carve into their living flesh whenever part of their body is needed. The question here is not whether it would be moral to create such monsters, for we may simply assume that they are available, but rather, what macroethics would be able to explain our sense of moral repugnance for the way we treat them? Most people would consider it morally wrong, not just because of our responsibility as creators, not just because of the kind of moral person we would become if we were to adopt such behaviour, not because of the negative effects, which are none, and not because of the Kantian maxims, neither of which would apply, but because of the bio-system in front of us and its values. Even if the senseless cow is just a biological mass, no longer feeling anything, this does not mean that any of our actions towards it would be morally neutral. IE can still argue, for example, that the cow is still a body whose integrity and unity demand respect. Affecting the essence of the body would still be wrong, even if the body is no longer sentient. Indeed, since the original status of the body was that of a sentient being, we ought to do our best to reinstate its former conditions for its own sake and well-being. Mind that I am not suggesting that we should not bioengineer some ways of producing meat, given the environmental damage produced by cattle, nor am I suggesting that, on the whole, it may not be better to produce non-sentient cows that would cause less environmental damage. The example is an attempt to illustrate how we could explain our sense of moral repugnance in creating a monster. Let me introduce a second example to clarify the point further. There seems to be nothing morally wrong in cloning one's lungs, or producing some extra litres of one's blood, which could turn out to be useful in the future, because when used they will be serving their purpose. But we find the idea of cloning a whole non-sentient twin, which we could then keep alive artificially and exploit as a source of organs, whenever necessary, morally repugnant. I would argue that we are right, and that this is so because to take an arm away from our twin would mean to affect its integrity adversely and transform it into something that it was not meant to be, a mutilated body. We would be showing no care whatsoever, and our actions would not be implemented for the sake of the patient.

4.8.3 Death

Standard ethics do not treat dead people; at most they try to teach the living how to face death. Non-standard biocentric ethics treat only the dying. Only IE has something to say about the actual corpse and its moral claims. This last example comes from the *Iliad*. Achilles has killed Hector. For many days, he has, in his fury, repeatedly dragged Hector's body behind his chariot around the tomb of his friend Patroclus. He has decided to take his full revenge for Patroclus' death by not accepting any ransom in exchange for Hector's body. Hector will have no burial and must be eaten by the dogs. Achilles' misbehaviour seems obvious, but there is more than one way of explaining why it is morally blameworthy. Other non-standard ethics can say nothing relevant and a deontological approach is not very useful. Just before dying, Hector asked Achilles to be kind and to accept his parents' offers in return for his body, yet Achilles rejected his prayers and was ready to face the consequences. He is not afraid of universalizing his behaviour. Although Priam tries to reason him into returning Hector's body using a deontological argument ('Think of your father, O Achilles like unto the gods, who is such even as I am, on the sad threshold of old age....'), Achilles has already been informed by his mother about the gods' will and is ready to change his course of action anyway. Actually, he finds Priam's line of reasoning rather annoying. The consequentialist, of course, can lead us to consider the pain that Achilles' behaviour has caused to Priam, Andromache, and all the other Trojans. A supporter of virtue ethics may easily argue that what is morally wrong is Achilles' attitude, for he is disrespectful towards the dead, his family, the gods, and the social customs regulating human relations even during war time. Yet Achilles changes his mind only because the gods intervene. Indeed, the speech made by Apollo in the last book of the *Iliad*, the speech that convinces the gods that it is time to force Achilles to modify his behaviour and return Hector's body, is perhaps best read from an IE perspective as a defence of the view that even a dead body, a mere lifeless object, can be outraged and deserves to be morally respected (*Iliad*, 24.40–50):

> Achilles has lost all pity! No shame in the man,
> shame that does great harm or drives men on to good.
> No doubt some mortal has suffered a dearer loss than this,
> a brother born in the same womb, or even a son...
> he grieves, he weeps, but then his tears are through.
> The Fates have given mortals hearts that can endure.
> [see above the argument against the simplistic fragility of goodness]
> But this Achilles—first he slaughters Hector,
> he rips away the noble prince's life
> then lashes him to his chariot, drags him round
> his beloved comrade's tomb. But why, I ask you?
> What good will it do him? What honour will he gain?
> Let that man beware, or great and glorious as he is,
> we mighty gods will wheel on him in anger—look,
> he outrages the senseless clay in all his fury!

The Greek word for 'outrages' is ἀεικίσσω, which also means 'to dishonour' or 'to treat in an unseemly way'. Hector's body demands ἔλεον, compassionate pity, but Achilles has none, for he has lost any αἰδώς, respectful moral reverence or awe, blinded by his painful passion. Yet the view from IE requires him to overcome his subjective state, achieve an impartial perspective, and care for the dead body of his enemy. Achilles must start behaving with some respect for the body, even if this is now just κωφὴν γὰρ δὴ γαῖαν, senseless (literally, silent) clay.

CONCLUSION

In this chapter, I argued that information ethics may justifiably amount to an environmental macroethics. IE holds that any expression of Being (any part of the infosphere) has an intrinsic worthiness. So any informational entity is to be recognized as the centre of a minimal moral claim, which deserves recognition in virtue of its presence in the infosphere and should help to regulate the implementation of any information process involving it, at least *prima facie* and overridably. Therefore, IE raises information as such to the role of the true and universal patient of any action, presenting itself as an ontocentric and patient-oriented ethics. The receivers of moral actions are understood informationally and placed at the centre of the ethical discourse. Information, in the ontological sense of the concept, has an intrinsic worthiness, and IE substantiates this position by recognizing that any informational entity has a Spinozian right to persist in its own state, and a constructionist right to flourish (*conatus*), i.e. to improve and enrich its existence and essence.

We saw how one may obtain an ordering of the infosphere that captures the notion of a morally bad state change. An evil action changes the state of the affected region of the infosphere from S_1 to S_2, where S_2 is greater in entropic ordering, and where a benign action decreases entropic ordering. The effect of any action is characterized as a state transformer by the relationship (a predicate) between the state-before, the input and output, and the state-after. In this context, intelligent understanding and debate determine whether an action is morally good, meaning that none of its transitions yields a state-after which is greater in entropic ordering than its state-before, or evil, where its state-before is greater in entropic ordering than its state-after. The same approach can be used to determine whether an action is more, or less, evil than another.

I have argued that IE evaluates the duty of any rational being in terms of its contribution to the well-being and flourishing of the infosphere and that any process, action, or event that negatively affects the whole infosphere as an increase in its level of metaphysical entropy is an instance of evil. IE determines what is morally right or wrong by means of four basic ethical principles. As a result, IE can provide a valuable perspective from which to approach fruitfully not only moral problems in computer ethics, but also the whole range of conceptual and moral phenomena that form the ethical discourse.

It would be foolish to think that IE can have the only or even the last word on moral matters. IE does not provide a library of error-proof solutions to all ultimate moral problems, but it does seem to fulfil an important missing role within the spectrum of current macroethics. For some time now, there has been a blind spot in our ethical discourse, a missing ethical perspective that IE and its applied counterpart, CE, seem to be able to perceive and take into account. The shift from an anthropocentric to a biocentric perspective, which has so much enriched our contemporary understanding of morality, may be followed by a second shift, from a biocentric to an ontocentric view. This is what IE tries to achieve, thus acquiring a fundamental role in the context of macroethical theories. The patient-oriented, ontocentric perspective seems more suitable to an information culture and society e-nvironmentally friendly. It improves our understanding of moral facts. It can help us to shape our moral questions more fruitfully, to sharpen our sense of value, and to make the rightness or wrongness of human actions more intelligible and explicable. It may lead us to look more closely at just what fundamental values our ethical theories should seek to promote. It improves our capacity to listen. All we require from IE is to help us to give an account of what the Greeks saw as the co-extension of Being and Goodness. *Agere sequitur ad esse in actu* (action follows out of Being, Thomas Aquinas, *Summa contra Gentiles* III, 69): this old medieval dictum can now be given a new twist and be adopted as the motto of IE.

At this point, it is worth pausing for a moment to listen to lawyers, politicians, sociologists, engineers, educators, IT officers, and many other professionals. For I fear they may complain that philosophers cannot place their metaphysical copyright on computer ethics. CE is a lively and useful subject, which should not be reduced to a mere philosophical subject and esoteric field of conceptual investigations. Their worries may not be completely unjustified. CE offers an extraordinary theoretical opportunity for the elaboration of a new ethical perspective, but what has been said so far foreshadows an interpretation of CE that places it at a level of abstraction too philosophical to make it useful for their immediate needs. Yet, this is the inevitable price to be paid for any attempt to provide CE with an autonomous conceptual foundation. We must polarize theory and practice to strengthen both, since the longer the jump, the longer the run-up must be. This is why, in order to avoid confusion, I have been speaking of 'information ethics' in order to refer to the macroethics that provides the philosophical foundations of CE. IE is not immediately useful to solve specific applied problems, but it provides the grounds for the moral principles and analysis that can guide our problem-solving procedures. Through IE, CE can develop its own methodological foundation and support an autonomous theoretical analysis of domain-specific issues, including pressing practical problems, which in turn can be used to test IE's methodology. In CE, professional codes of conduct, rules, guidelines, advice, instructions or standards, ICT-related legislation, and policy-making are all based on an implicit philosophical ethics. It is the latter that we shall investigate in the rest of the book. Before doing so, there is an obvious, last meta-theoretical task to be fulfilled: is IE really a good foundation for CE? The next chapter answers this question. The reader interested in understanding IE by itself may safely skip it.

5

Information ethics and the foundationalist debate

> The world of the future will be an ever more demanding struggle against the limitations of our intelligence, not a comfortable hammock in which we can lie down to be waited upon by our robot slaves.
>
> Norbert Wiener, *God and Golem, Inc.: A Comment on Certain Points Where Cybernetics Impinges on Religion* (1964), p. 69.

SUMMARY
Previously, in Chapter 4, I identified the essential difficulty with the philosophical status of computer ethics (CE) as a methodological problem. Standard ethical theories cannot easily be adapted to deal with issues arising in CE, which appear to strain their conceptual resources. The challenges posed by CE seem to require a different conceptual foundation in the form of an ethical theory. I proposed information ethics (IE) as the philosophical foundational counterpart of CE and introduced it as a particular kind of e-nvironmental ethics or ethics of the infosphere. Following the environmental perspective presented in the previous chapter, here I shall analyse the debate on the foundations of CE. This chapter has two meta-theoretical goals: proposing a solution to the so-called 'uniqueness debate' and, in so doing, clarifying further the nature of IE.

I shall start by discussing Moor's classic interpretation of the need for CE created by a policy and conceptual vacuum in Section 5.1. I shall then identify and discuss five positions in the literature that address the aforementioned vacuum:

(1) the *no resolution approach*, according to which CE can have no foundations (Section 5.2);

(2) the *professional approach*, according to which CE is solely a professional ethics (Section 5.3);

(3) the *radical approach*, according to which CE deals with absolutely unique issues, in need of a unique approach (Section 5.5);

(4) the *conservative approach*, according to which CE is only a particular applied ethics, discussing a new species of traditional moral issues (Section 5.6); and, finally,

(5) the *innovative approach*, according to which theoretical CE can expand the meta-ethical discourse with a substantially new perspective (Section 5.7).

In the course of the analysis, I shall review the uniqueness debate in Section 5.4.

The main thesis I shall support in this chapter, already formulated in Chapter 4, is that, although CE issues are not uncontroversially unique, they are sufficiently novel to render inadequate the adoption of standard macroethics, such as consequentialism and deontologism, as the foundation of CE and, hence, to prompt the search for an alternative ethical theory. I shall then propose that IE is the theory that can satisfy the conditions needed to serve as a satisfactory foundation for CE. IE is characterized as a biologically unbiased extension of environmental ethics based on the concepts of information/infosphere/entropy rather than life/ecosystem/pain. In light of the discussion provided in this chapter, I suggest that CE is worthy of independent, philosophical study because it requires its own application-specific knowledge and is capable of supporting a methodological foundation, IE. This chapter thus concludes the meta-theoretical and programmatic part of the book.

5.1 Introduction: looking for the foundations of computer ethics

Let me summarize some of the points already made in Chapter 4. Computer ethics (CE) stems from practical concerns arising in connection with the impact of ICTs on contemporary society. As we saw in Chapter 1, the information revolution has caused new and largely unanticipated problems, thus outpacing ethical, theoretical, and legal developments.[1] In order to fill these policy and conceptual vacua (Moor, 1985), CE carries out an extended and intensive study of individual cases very often concerning real-world issues rather than mere thought experiments, usually in terms of reasoning by analogy. The result has been inconsistencies, inadequacies, and an unsatisfactory lack of general principles. CE aims to reach decisions based on principled choices using defensible ethical criteria and, hence, to provide more generalized conclusions in terms of conceptual evaluations, moral insights, normative guidelines, educational programmes, legal advice, industrial standards, and so forth, which may apply to whole classes of comparable cases. So, at least since the seventies,[2] CE focus has moved from problem analysis, primarily aimed at sensitizing public opinion, professionals, and politicians, to tactical solutions. This has resulted, for example, in the evolution of professional codes of conduct, technical standards, usage regulations, and new legislation. The constant risk of this bottom-up procedure has remained the spreading of *ad hoc* or casuistic approaches to ethical problems as their scope and importance continues to escalate.

---

[1] See Bynum (1998, 2000, 2001a); Van den Hoven and Weckert (2008); Johnson and Miller (2009); and Floridi (2010d) for some overviews.

[2] See Bynum (2000, 2001a, 2001b), for earlier works in CE.

Prompted partly by this difficulty and partly by a natural process of reflective maturation as an independent discipline, CE has further combined tactical solutions with more strategic and global analyses. The foundationalist debate that forms the topic of this chapter is an essential part of this top-down development. It is characterized by a meta-theoretical reflection on the nature and justification of CE and the discussion of CE's relations with the broader context of macroethical theories. Can CE amount to a coherent and cohesive discipline, rather than a more or less heterogeneous and random collection of ICT-related ethical problems, applied analyses, and practical solutions? If so, what is its conceptual rationale? And how does it compare with other ethical theories? These issues are collectively known as the 'uniqueness debate' (Tavani, 2002, 2010). The keen reader will recognize here the last of the open questions in the philosophy of information that I analysed in Floridi (2011a).

Five approaches concerning the foundations of CE have emerged in the literature. They can be explained as resulting from different answers to the questions listed above. Here, I shall refer to them as the 'no resolution approach' (NA), the professional approach (PA), the radical approach (RA), the conservative approach (CA), and the innovative approach (IA). The order in this list is both historical and logical.

It would not be difficult to correlate these five positions with the three main approaches to information ethics discussed in Chapter 2. However, as they say in logic textbooks, this simple exercise is left to the reader. In the rest of this chapter, I shall argue that:

- NA provides a minimalist starting point, methodologically useful, which prompts the development of the other four approaches;
- PA represents a valuable professional approach to CE, which leads to the adoption of a theoretical position when macroethical issues are in question;
- RA stresses the novelty of CE;
- CA connects CE to other standard ethics; and
- IA, relying on the previous approaches, succeeds in providing a satisfactory answer to the foundationalist question insofar as it promotes Information Ethics (IE) as the theoretical foundation of CE.

5.2 The 'no resolution approach': CE as not a real discipline

The expression 'no resolution view' (or approach) was introduced by Gotterbarn (1991):

The 'no resolution view' has been reinforced by some recent works. For example, Donn Parker 1981 [originally Parker (1979)] uses a voting methodology to decide what is ethical in computing.... He says, this work was not guided by a concept of computer ethics nor was there an attempt to discover ethical principles.... Not only was there an absence of a concept of computer ethics, but the primary direction was an emphasis on proscribed activities.... Parker used the diversity of opinions expressed about these scenarios to argue that there was no such thing as computer ethics. And a fortiori, that it could not be taught in a computer science curriculum. (p. 27)

According to the 'no resolution approach', CE problems represent unsolvable dilemmas, and CE is itself a pointless exercise, having no conceptual foundations. NA is convincingly criticized in Gotterbarn (1991, 1992), who analyses Parker (1979, 1982) and Parker et al. (1990). Empirically, the evolution of CE has proved NA to be unnecessarily pessimistic. CE problems have been successfully solved, CE-related legislation has been approved and enacted, professional standards and codes have been promoted, and so forth. It is understandable, perhaps, that the view arose at a time when both the public and professionals were being alerted to wide-ranging unethical uses of ICTs. The fact that Parker did not infer an almost opposite conclusion from the 'emphasis on proscribed activities'—namely that CE is essential in any information society—is presumably due to his voting methodology. It is dangerous to infer from inconsistent replies to a question that the latter has no answer. The same reasoning might lead one to believe, after asking a representative sample of supporters, that neither side would win the World Cup final. NA's emphasis on the wide variety of proscribed activities is characteristic. Bynum (1992) has described such an approach as 'pop ethics'. This is usually characterized by unsystematic and heterogeneous collections of dramatic stories discussed in order 'to raise questions of unethicality rather than ethicality' (Parker, 1979, p. 8). Its goal is largely negative, 'to sensitize people to the fact that computer technology has social and ethical consequences' (Bynum, 1992), and it is not neutral. That is why it played a useful role at the beginning of the development of CE, around the time when the term 'hacker' became used disparagingly, for example. It is comparable to early work done in business ethics: it points to whatever goes wrong but fails to promote a relevant, beneficial, and professional ethos. Gotterbarn (1992) comments:

'Pop' ethics might have had a place when computing was a remote and esoteric discipline, but I believe that in the current environment this approach is dangerous to the preservation and enhancement of values. This model of computer ethics does not forward any of the pedagogical objectives for teaching ethics [prescribed by a professional approach, see next section]. (p. 75)

Nonetheless, pop ethics offers a few advantages. Some sensitization to ethical problems is an important preliminary to CE. There is little point in providing a solution to someone unaware of the problem, particularly when the solution is not simple. History repeats itself: something very similar is happening today in the context of the ethics of cyberwar (see Floridi and Taddeo, forthcoming). Secondly, the variety of concerns—professional, legal, moral, social, political, etc.—is vital to CE and must be appreciated from the start. For this purpose, a variety of case studies helps. For instance, Epstein (1997) is an early example of pop ethics found by many lecturers to be useful as preliminary reading for a course in CE.

90   THE ETHICS OF INFORMATION

The objection to pop ethics is that it goes no further than cataloguing examples and that it is frequently used to support NA. Methodologically, NA provides a useful point of reference because it represents an ideal lowest bound for the foundationalist debate, comparable to the role played by relativism in metaethics. In terms of logical analysis, any other approach can be seen as starting from the assumption that NA should be avoided, if possible. Positions can then be ranked depending on their distance from NA, while failure to defend any successful alternative confirms NA as the only negative conclusion.

5.3 The professional approach: CE as a pedagogical methodology

The first positive reaction to the policy vacuum has been to appeal to the social responsibility of computer professionals. This has meant, among other things, developing a professional-ethics approach (PA) to CE, which has stressed pedagogical needs. According to PA, CE should:

> [introduce] the students to the responsibilities of their profession, articulate the standards and methods used to resolve non-technical ethics questions about their profession, develop some proactive skills to reduce the likelihood of future ethical problems,... indoctrinate [sic] the students to a particular set of values... and teach the laws related to a particular profession to avoid malpractice suits. (Gotterbarn, 1992, p. 75)

PA argues that there is no deep theoretical difference between CE and other professional ethics like business ethics, medical ethics, or engineering ethics, only a variety of pedagogical contexts (Gotterbarn, 1991, 1992). And since CE courses have the goal of creating ethically minded professionals not ethicists, it is unnecessary, and indeed it may actually be better, not to have philosophers teaching them. After all,

> [p]hilosophers are no more educated in morality than their colleagues in the dairy barn; they are trained in moral theory, which bears about the same relation to the moral life that fluid mechanics bears to milking a cow. (Robert K. Fullinwider, cited in Gotterbarn 1992, p. 78)

This argument is not uncommon in academia. Mathematics courses, for example, are taken by students in many disciplines, from Engineering to Economics. Yet, should the lectures be given by mathematicians, who are presumably masters of the material, or by lecturers from the relevant subject areas, who may appreciate better its particular applications? Apart from political and financial arguments, the latter view is often seen as reinforcing a sense of 'subject' in the application domain, whilst the less applied view of the former may be seen as broadening an established subject with new applications. The arguments concerning the lecturing of applied ethics courses appear to be similar. It may perhaps be argued that, at the university level, such courses ought to enable participants to solve new problems as they arise (what are the fundamentals?),

INFORMATION ETHICS AND THE FOUNDATIONALIST DEBATE 91

whilst in specialized professional institutions such courses are typically under pressure to be more prescriptive:

> In applied professional ethics courses, our aim is not the acquaintance with complex ethical theories, rather it is recognizing role responsibility and awareness of the nature of the profession. (Gotterbarn, 1992, p. 78)

PA has a number of major advantages. It stresses the vital importance of CE education, taking seriously issues such as technical standards and requirements, professional guidelines, specific legislation or regulations, levels of excellence, liability issues, and so forth. It thus exposes the risky and untenable nature of NA and places pop ethics in perspective, revealing it to be insufficient by itself. PA defends the value and importance of a constructive pop ethics, by developing a 'proactive' professional ethics (standards, obligations, responsibilities, expectations, etc.), favourable to value-supporting and welfare-enhancing (including the development and use of) ICTs (Bynum, 1992; Gotterbarn, 1992). Furthermore, PA defends a realistic pedagogical attitude, pragmatically useful to sensitize and instruct students and professionals. Its ultimate aim is to ensure that:

> ethical values, rules and judgements [are] applied in a computing context based on professional standards and a concern for the user of the computing artefact. (Gotterbarn, 1991, p. 28)

One of the primary results of PA has been the elaboration and adoption of usage regulations and codes of conduct in ICTs' contexts (libraries, universities, offices, etc.), within industry and in professional associations and organizations, as well as the promotion of certification of computer professionals. PA addresses mainly ICTs' practitioners, especially those involved in software development, where technical standards and specific legislation provide a reliable, if sometimes rather minimal, frame of reference. As noted above, PA's goals are pedagogical not meta-ethical. However, unfortunately, sometimes PA is interpreted as the *only* correct way to understand the whole field itself, as if CE could be reduced to a professional ethics in a strict sense:

> The only way to make sense of 'Computer Ethics' is to narrow its focus to those actions that are within the horizon of control of the individual moral computer professional. (Gotterbarn (1991); see further Gotterbarn (1992, 2001) where he endorses a less radical view)

This strong view has further led to radically anti-philosophical positions (Langford, 1995). However, a strong PA is far too restrictive, for at least three reasons.

First, strong PA disregards the significant fact that, contrary to other purely professional issues, CE problems—e.g. privacy, accuracy, security, reliability, intellectual property, and access—permeate contemporary life, as we have seen in Chapters 1 and 2. Strong PA can rightly argue that moral problems somehow involving ICTs (e.g. theft using a computer) should not be vaguely confused with distinctively CE problems (e.g. software copyright issues). This restriction, however, does not yet justify the reduction of all CE problems to only professional issues. To be coherent, strong PA could reply that any member of an information society should be treated, to various degrees, as an ICTs professional, to whom some corresponding professional guidelines should apply. However, this would mean just accepting the fact that CE cannot be reduced to a specific professional ethics without the latter losing its perspicuous meaning. Strong PA becomes undefeatable but empty.

Second, interpreting PA as providing a conceptual foundation for CE is to commit a mistake of levels. It is like attempting to define arithmetic on the basis of only what is taught in an introductory course. Without a theoretical approach, PA is mere para-CE to borrow an expression coined by Keith Miller and used by Bynum (1992), in analogy with paramedic, to describe a middle level between pop CE and theoretical CE. Theoretical CE underpins PA and requires an approach different from it.

Finally, understanding CE as only a professional ethics, not in need of any further conceptual foundation, means running the risk of being at best critical but naïve, and, at worst, dogmatic and conservative. On the one hand, focusing on case-based analyses and analogical reasoning, a critical PA will painfully and slowly attempt to re-discover, inductively, ethical distinctions, clarifications, theories, and so forth, already available and discussed in the philosophical literature. On the other hand, an uncritical PA will tend to treat ethical problems and solutions as misleadingly simple, non-conflicting, self-evident, and uncontroversial, a matter of mere indoctrination, as exemplified in the old-fashioned '10 Commandments of Computer Ethics' approach. Deferring to some contingent 'normal ethics' currently accepted within the agent's society, to adapt a Kuhnian expression, is itself a very significant ethical decision, at least because, when 'normal' ethics is methodologically coherent, it limits itself to providing negative prescriptions, since lists of 'don'ts' are easier to implement and much less questionable than positive recommendations. Moral standards, values, principles, and choices are always legitimized by ethical positions and arguments, at least implicitly. Applying normal ethics may then be sufficient in everyday life; but it is only the first step towards a mature approach that can uncover, evaluate, criticize, and revise at least some of the accepted presuppositions working in CE and, thus, hope to improve them. For all these reasons, PA may be seen pragmatically as a first step towards a more mature CE. Historically, this was the position that led to the 'uniqueness debate'.

5.4 Theoretical CE and the uniqueness debate

Any applied or professional ethics must necessarily make room for critical theorizing, even if it does not have to consider it one of its own tasks. PA at best distinguishes between pedagogical problems and meta-theoretical research, descriptive and normative questions, practical and theoretical issues, commonsensical applications, and conceptual criticisms of some normal ethics. Among the fundamental questions that PA does not intend to address are: Why do ICTs raise moral issues? Are CE issues unique (in the sense of requiring theoretical investigations that are not entirely derivative from standard ethics)? Or are they simply moral issues that happen to involve ICTs? What kind of ethics is CE? What justifies a specific methodology in CE, e.g. reasoning by analogy and case-based analysis? What is the rationale of CE? What is the contribution of CE to the wider ethical discourse? PA programmatically avoids entering into such investigations and coherently leaves them to theoretical CE, which can then be introduced as the logical stage following pop CE, NA, and PA. Historically, it has developed along two lines, which can be usefully introduced through the 'uniqueness debate'.[3]

The 'uniqueness debate' has attempted to determine whether the moral issues confronting CE are unique and, hence, whether CE should be developed as an independent field of research with a specific area of application and an autonomous, theoretical foundation. The debate arises from two different interpretations of the policy vacuum problem, one more radical, the other more conservative, as we shall see in the next two sections.

5.5 The radical approach: CE as a unique discipline

According to the radical approach (RA), the presence of policy and conceptual vacua indicates that CE deals with absolutely unique issues, in need of a completely new approach (Mason, 1986; Maner, 1996, 1999). Thus, RA argues that

> [Computer Ethics] must exist as a field worthy of study in its own right and not because it can provide a useful means to certain socially noble ends. To exist and to endure as a separate field, there must be a unique domain for computer ethics distinct from the domain for moral education, distinct even from the domains of other kinds of professional and applied ethics. Like James Moor, I believe computers are special technology and raise special ethical issues, hence that computer ethics deserves special status. (Maner, 1999)

In terms of logical analysis, RA presents some advantages. It counteracts the risk run by NA of underestimating the importance and impact of CE problems. Taking seriously their gravity and unprecedented novelty, RA improves on the various pop versions of CE, including PA, by stressing the methodological necessity of providing the field with a robust and autonomous theoretical rationale, if it wishes to deal with ICTs-related moral issues successfully. Nevertheless, RA is confronted by at least four problems.[4]

First, to establish that CE is a unique field, the argument quoted above requires the explicit and uncontroversial identification of some unique area of study. Actually, the argument appears to be of the form 'uniqueness only if special domain', and 'special domain' therefore 'uniqueness', that is: $U \to SD$, SD, therefore U. This is of course fallacious. It is rectified if 'only if' is replaced by 'if'; in the original words, if 'there must

---

[3] Floridi (1999c) is an early collection of articles, Tavani (2010) provides an informative and more recent analysis.
[4] See further Himma (2004b) for a criticism of the radical approach in terms of 'uniqueness' of CE issues.

be' should be replaced by 'it suffices that there be', which at least renders the argument valid, if less plausible. However, even in this improved version, RA seems unable to show the absolute uniqueness of any CE problem. None of the cases provided by Maner (1996, 1999) is uncontroversially unique, for example, though this is perhaps to be expected. It would be very surprising if any significant moral issue were to belong fully and exclusively only to a limited conceptual region without interacting with the rest of the ethical context. Recall the reasons why we abandoned the 'Internal' R(esource) P(roduct) T(arget) model in Chapter 2. This neat partition does not happen in any other special context such as business, medical, or environmental ethics, and it remains to be shown why it should happen even in principle in CE.

Second, in reply to the difficulty just seen, one could argue that CE problems could be made, or become, or be discovered to be, increasingly specific, until they justify the position defended by RA. This reply runs the risk of being safe but uninteresting because it is unfalsifiable. It certainly keeps the burden of proof on the RA side. Yet, let us suppose that a domain of unique ethical issues in CE were available in principle. The basic line of reasoning would still be unacceptable. The 'uniqueness' of a specific topic is not simply inherited as a property by the discipline that studies it. On the one hand, specific moral problems, such as abortion or the profit motive, may still require only some evolutionary adaptation to old macroethical solutions, like those found in, for instance, medical or business ethics. On the other hand, specific disciplines, such as environmental ethics, are not necessarily specific because they are involved with unique problems, for they may share their subjects, for example the value of life and the concept of well-being, with other disciplines. They are specific because of their methodologies, aims, and perspectives.

The other two problems encountered by RA are methodological. Given the interrelatedness of ethical issues and the untenability of the overly simplistic equation that a unique topic equals a unique discipline, it is not surprising that RA is forced to leave unspecified what a mature CE could amount to as a unique discipline with any degree of detail. Finally, by overstressing the uniqueness of CE, RA runs the risk of isolating the latter from the more general context of meta-ethical theories. This would mean missing the opportunity to enrich the ethical discourse. Some of the problems just seen are neatly avoided by the conservative approach (CA).

5.6 The conservative approach: CE as applied ethics

CA defends two theses:

(a) that classic macroethics—e.g. consequentialism, deontologism, virtue ethics, contractualism, etc.—are sufficient to cope with the policy vacuum. These theories might need to be adapted, enriched, and extended, but they have all the conceptual resources required to deal with CE questions successfully and satisfactorily; and

(b) that some ethical issues are admittedly transformed by the use of ICTs, but they represent only a new species of traditional moral issues to which already available macroethics need to, and can successfully, be applied. They are not, and cannot be, a source of a new, macroethical theory.

From (a) and (b) it follows that CE is a microethics. Thesis (a) is weaker and hence less controversial than (b). To explain both, Johnson (1999) introduces an evolutionary metaphor[5] worth reporting because it has been rather popular:

Extending the idea that computer technology creates new possibilities, in a seminal article, Moor (1985) suggested that we think of the ethical questions surrounding computer and information technology as policy vacuums. Computer and information technology creates innumerable opportunities. This means that we are confronted with choices about whether and how to pursue these opportunities, and we find a vacuum of policies on how to make these choices.... I propose that we think of the ethical issues surrounding computer and information technology as new species of traditional moral issues. On this account, the idea is that computer-ethical issues can be classified into traditional ethical categories. They always involve familiar moral ideas such as personal privacy, harm, taking responsibility for the consequences of one's action, putting people at risk, and so on. On the other hand, the presence of computer technology often means that the issues arise with a new twist, a new feature, a new possibility. The new feature makes it difficult to draw on traditional moral concepts and norms.... The genus-species account emphasizes the idea that the ethical issues surrounding computer technology are first and foremost ethical. This is the best way to understand computer-ethical issues because ethical issues are always about human beings.

Since CA presents CE as an interface between ICT-related moral problems and standard macroethics, it enjoys all the advantages associated with a strong theoretical position. CA rejects NA. It accepts pop CE's recommendation that CE problems are important and significant, so much so that, for CA, they deserve to be approached both pragmatically and theoretically. It is compatible with, and indeed reinforces, PA, since, for CA, CE is an ethics for any member of the information society, not just for the ICT professional. Being based on a macroethic perspective, CA can both promote a constructive attitude, like PA, and hope to adopt an evaluative stance, thus avoiding a naïve or uncritical reliance on some contingent normal ethics. Finally, CA avoids RA's untenable equation and corresponding 'isolationism', because the development of an evolutionary rather than a revolutionary interpretation of CE problems enables it to integrate them well within the broader context of the ethical discourse. Is then CA devoid of difficulties? Not yet, for CA is still faced by four shortcomings.

First, CA's weaker thesis (a) is controversial. As we have seen in the previous chapter, it is at least questionable whether standard macroethics do indeed have all the necessary resources to deal with CE problems satisfactorily, without reducing them to their own

---

[5] See further Naresh (1999) for a similar approach. Marturano (2002) offers a convincing criticism of Johnson's meta-theoretical position.

conceptual frames and thus erasing their true novelty. It may be argued that precisely the fact that CE problems were unpredicted and are perceived as radically new, casts doubts on the possibility of merely adapting old ethical theories to the new context.

Second, CA is meta-theoretically underdetermined. The evolutionary metaphor *incorporates* the tension between a radical and a traditional approach but does not *resolve* it. New species of moral problems could conceivably be so revolutionarily different from their ancestors—the digital instrumentation of the world can create such entirely new moral issues, unique to CE and that do not surface in other areas—to require a 'unique' approach, as suggested by RA. Alternatively, such new species may represent just minor changes, perfectly disregardable for any theoretical purpose, as the conservative approach wishes to argue. The trouble is that CA, left with the tension now hidden in the evolutionary analogy, opts for the conservative solution to treat CE as a merely applied ethics, but then it does not and cannot indicate which macroethics should be applied. At best, this leads to the adoption of some standard macroethics to deal with CE problems, e.g. consequentialism, but, when the choice is not arbitrary, this further justifies the claim that, when it comes to ethical theorizing, there is not much to be learnt philosophically from this applied field. If new ICT-related moral problems have any theoretical value, either by themselves or because embedded in original contexts, this is only insofar as they provide further evidence for the discussion of well-established ethical doctrines. In this way, CA approaches NA: there are no CE specific problems, only ethical problems involving ICTs. At worst, CA's lack of philosophical commitment leads to a muddle of syncretic and eclectic positions, often uncritical and overlooking the theoretical complexity of the problems involved. Thus, CA's failure to endorse an explicit macroethical position generates a logical regress: having accepted CE as a microethics, one then needs a meta-theoretical analysis to evaluate which macroethics is most suitable to deal with CE problems. This logical regress tends to be solved by appealing either to some commonsensical view or to pedagogical needs. The former solution leads back to the arbitrary assumption that some contingent, normal ethics can provide a rationale for CE. For example, one can do CE by using Habermas' dialogue ethics because this is what one's society approves of as the normal approach to ethics. It thus undermines the critical and normative advantage that CA hopes to have over other approaches. The latter solution (to simplify: one can do CE by using virtue ethics because this is what one's students find more intuitive), apart from being equally arbitrary, represents the kind of unnecessary intrusion of philosophy into professional matters so rightly criticized in PA literature. Software engineers should not be required to read the *Nicomachean Ethics* as part of their professional training.

Third, CA is methodologically poor. This is a consequence of the first problem. Lacking a clear macroethical commitment, CA cannot provide an explicit methodology either. It then ends up relying on common-sense, case-based analysis, and analogical reasoning, often insufficient means to understand what CA itself acknowledges to be new and complex issues in CE.

Finally, CA is meta-ethically unidirectional. By arguing for (b) above, the more controversial thesis, CA rejects *a priori* and without explicit arguments the possibility envisaged by RA that CE problems might enrich the ethical discourse by promoting a new macroethical perspective. It addresses the question 'what can ethics do for CE?' but fails to ask the philosophically more interesting question: 'Is there anything that CE can do for ethics?' It thus runs the risk of missing what is intrinsically new in CE, not at the level of problems and concepts, but at the level of contribution to the ethical discourse. A mere extension of standard macroethics does not enable one to uncover new possibilities.

Luckily, there is another possible approach to theoretical CE, which is neither conservative nor radical, but innovative (IA).[6]

5.7 The innovative approach: information ethics as the Foundation of CE

The innovative approach (IA) is how IE deals with the uniqueness debate, so in this section I shall remind the reader of several points already made in the previous chapters.

IA builds on CA's advantages, but it avoids its shortcomings by rejecting the conservative restriction made explicit in the previous section, under thesis (b). According to IA, CE problems, the corresponding policy and conceptual vacua, the uniqueness debate, and the difficulties encountered by RA and CA in developing a cohesive meta-ethical approach strongly suggest that the monopoly exercised by standard macroethics in theoretical CE is unjustified. By profoundly transforming the context in which moral issues arise, ICTs not only add interesting new dimensions to old problems, but lead us to rethink, methodologically, the very grounds on which our ethical positions may be based. Although the novelty of CE is not so dramatic as to require the development of an utterly new, separate, and unrelated discipline, it seems to draw out some of the limits of traditional approaches to the ethical discourse and encourages a fruitful modification in our meta-theoretical perspective. Rather than allowing standard macroethics to occupy the territory of CE arbitrarily, as happens with CA, or exiling CE to an impossibly isolated and independent position, as proposed by RA, IA argues that the theoretical foundation of CE should be promoted to the level of another macroethics because it does have something distinctive and substantial to say on moral problems and, hence, can enrich the meta-ethical discourse with a new and interesting approach of unquestionable philosophical value. In the previous chapters, I have introduced this macroethical perspective in terms of an Information Ethics. We saw that IE, understood as the theoretical foundation of applied CE, is a non-standard (because patient-oriented), ontocentric, and e-nvironmental macroethics. What does this mean for the 'uniqueness debate'?

---

[6] Gorniak-Kocikowska (1996) and Bynum (1998, 2010) outline the need for an innovative approach to CE. On the future of computer ethics see further Moor (2001b).

The interpretation of what is the primary object of the ethical discourse is a matter of philosophical orientation. Some macroethical positions, especially virtue ethics, concentrate their attention on the moral nature and development of the agent. They are properly described as agent-oriented or 'subjective' ethics. Since the agent is usually assumed to be a single person, they tend to be individualistic. Other macroethics, especially consequentialism, contractualism, and deontology, concentrate their attention on the moral nature and value of the agent's actions. They are 'relational' and action-oriented theories, intrinsically social in nature. Agent-oriented, intra-subjective theories and action-oriented, inter-subjective theories were defined in the previous chapters as 'standard' or 'classic', without necessarily associating any positive evaluation with either of these two adjectives. Standard macroethics tend to be anthropocentric and to take only a relative interest in the 'patient', the third element in a moral relation, which is on the receiving end of the action and endures its effects. In recent years, medical ethics, bioethics, and environmental ethics have begun calling attention to this non-standard approach.

The reader may recall that non-standard ethics were described as holding the broad view that what matters first is the receiver of the moral action, not its sender. Clearly, from a patient-oriented ethics, the target of a moral action may be not only a human being, but also any form of life. Thus, we saw in Chapter 4 that biocentric, non-standard macroethics usually ground their analyses of the moral standing of bio-entities and ecosystems on the intrinsic worthiness of *life*, the intrinsically negative value of *suffering*, and the intrinsically positive value of *flourishing*. They seek to develop approaches in which the 'patient', that is, the receiver of the moral action, may be not only a human being, but also any form of life. Indeed, we saw that land ethics extends the concept of patient to any component of the environment. This is one step away from the approach defended by IE. According to biocentric, non-standard macroethics, any form of life is deemed to enjoy some essential proprieties or moral interests that deserve and demand to be respected, at least minimally and relatively, that is, in a possibly overridable sense, when contrasted to other interests. Substitute now 'life' with 'existence', interpret the latter informationally, and the position defended by IE as a non-standard macroethics should be clear. IE is an ecological ethics that replaces *biocentrism* with *ontocentrism*, and then interprets Being in informational terms. It suggests that there is something even more elemental than life, namely *Being*, the existence and flourishing of all entities and their global environment, and something more fundamental than suffering, namely *nothingness*. It then interprets *Being* and *nothingness* at an informational level of abstraction, as *infosphere* and *entropy*, on the basis of an informational structural realism as articulated in Floridi (2011a, chs. 14 and 15). In short, it is an environmental ethics based on the phenomena and corresponding concepts of information/infosphere/entropy rather than life/ecosystem/pain.

According to IE, one should also evaluate the duty of any agent in terms of contribution to the growth of the infosphere and any process, action, or event that negatively affects the whole infosphere—not just an informational entity—as an increase in its level of entropy and hence an instance of evil. We saw in Chapter 2 that, without information, there is no responsible moral action, but in IE information extends its scope from being a necessary prerequisite for any morally responsible action to being also its primary object. The crucial importance of this radical change in perspective cannot be overestimated. A typical non-standard ethics can reach its high level of universalization of the ethical discourse only thanks to its biocentric nature. However, this also means that even bioethics and environmental ethics fail to achieve a level of complete impartiality, because they are still biased against what is inanimate, lifeless, intangible, or abstract (we have seen in Chapter 4, for example, that even land ethics is not very sympathetic towards technology and artefacts). From their perspective, only what is intuitively alive deserves to be considered as a proper centre of moral claims, no matter how minimal, so a whole universe escapes their attention. Now, this is precisely the threshold that IE seeks to remove, as I shall argue in the next chapter. The ethical question asked by IE is: 'What is good for an informational entity and the infosphere in general?' The answer is provided by a minimalist theory of deserts: any informational entity is recognized to be the centre of some basic ethical claims, which deserve recognition and should help to regulate the implementation of any information process involving it. Approval or disapproval of any information process is then based on how the latter affects the essence of the informational entities it involves and, more generally, the whole infosphere, i.e. on how successful or unsuccessful it is in respecting the ethical claims attributable to the informational entities involved and, hence, in improving or impoverishing the infosphere. IE brings to ultimate completion the process of enlarging the concept of what may count as a centre of minimal moral concern, which now includes every informational entity. This is why it can present itself as a non-standard, patient-oriented and ontocentric macroethics.

The development of ethical theories just sketched provides a useful explanation and further meta-theoretical justification for IE. The various difficulties encountered by other approaches to CE can be reconnected to the fact that, far from being a standard, agent/action-oriented ethics, as it may deceptively seem at first sight, CE is primarily an ecological ethics of informational environments, rather than an ethics of conduct or becoming and, hence, qualifies as non-standard. The fundamental difference, which sets IE apart from all other members of the same class of non-standard theories, is that in IE, information entities and the infosphere, rather than just living systems in general, are raised to the role of universal patients of any action.

CONCLUSION

We have now come to the end not only of this chapter, but also of the meta-theoretical and programmatic part of this book.

The 'uniqueness' debate aimed to determine whether the issues confronting CE are unique and, hence, whether, as a result, CE should be developed as an independent macroethics. The debate arises from two different interpretations of the *policy vacuum* problem, one more radical, the other more conservative.

According to the radical interpretation, the policy vacuum problem indicates that CE deals with absolutely unique issues in need of a completely new approach. The radical approach is faced by at least three problems. First, it seems unable to show the absolute uniqueness of CE issues. Second, even if unique ethical issues in CE were identified, this would not mean that their 'uniqueness' would be simply inherited by the discipline that studies them, as it were. Unique problems may still require only some evolutionary adaptation of old solutions, and unique disciplines are not necessarily so because they are involved with unique subjects, for they may share their subjects with other disciplines. Their difference rests, therefore, on their methodologies, aims, and approaches. Third, a radical approach runs the risk of isolating CE from the more general ethical discourse. This would mean missing the opportunity to enrich our choice of macroethical approaches.

The conservative interpretation suggests that, in order to cope with the policy vacuum, standard macroethics, like consequentialism or deontologism, are sufficient. They might need to be updated, but they have all the necessary conceptual resources to deal with CE questions successfully. The conservative approach is equally faced by at least three problems: it does not clarify which macroethics should be adopted to deal with CE problems; it does not make explicit whether CE problems could be used as test experiments to evaluate specific macroethics; and it runs the risk of missing what is intrinsically new in CE, not at the level of problems and concepts, but at the level of contribution to the broader ethical discourse. A mere extension of standard macroethics would not enable us to uncover the accountability of artificial agents (see Chapter 7) or the nature of artificial evil (see Chapter 9), for example.

I have argued that there is a third approach to the policy vacuum, one that is neither radical nor conservative, but innovative. Some ICT-related problems are more manifest in cyberspace and readily studied there, but they are not unique to CE. They may affect environmental issues (e.g. green computing) and the world of physical automata (e.g. cyberwar). Because of their novelty and important position in contemporary ethics, CE problems demand further study in their own right. This third approach to the nature of CE interprets the policy vacuum problem as a signal that the monopoly exercised by standard macroethics is unjustified, and that the family of macroethical theories can be enriched by including an informational, patient-oriented approach that is not biologically biased but rather ontocentric. With their novelty, CE problems do not strictly force, but certainly encourage, us to modify the perspective from which we look at contemporary ethics. Yet, the novelty of CE problems is not so dramatic as to require the development of an utterly new, separate, and unrelated discipline. CE has its own methodological foundation, information ethics, and so it is able to support autonomous theoretical analyses. It also contains domain-specific issues, including pressing practical problems, which can be used to 'test' its methodology. The conclusion is that rather than allowing standard macroethics to 'occupy' the territory of CE or isolating CE in an impossibly autonomous and independent position, CE should be promoted to the level of another macroethics.

We saw how the development of IE may represent a solution to the 'uniqueness debate'. Of course, IE's position, like that of any other macroethics, is not devoid of problems. We shall see some of them in the following chapters, especially in Chapter 16. However, as I have remarked in Section 4.8, IE can complement other macroethics by contributing a patient-oriented, ontocentric new perspective. According to this form of e-nvironmental ethics, any change may be morally good or bad not because of its consequences, motives, universality, or virtuous nature, but because the infosphere and the informational entities inhabiting it are affected by it positively or negatively. IE's contribution enriches our ethical discourse. Without it, our understanding of moral facts in general, not just of CE problems in particular, would be less complete.

The foundationalist debate in CE has led to the shaping of a new ethical view. The difficult task now is articulating IE clearly and convincingly. Much has been stated in this and the previous chapters. Many of the claims still require analysis and justification. This is the task of the following chapters. The first step will be a defence of the intrinsic value of informational entities.

6

The intrinsic value of the infosphere

> ...one of the most important things which I wish to grant about G [Goodness]—the proposition that G depends only on the intrinsic nature of states of things which possess it.
>
> G. E. Moore, *Principia Ethica* (1993), p. 16.

SUMMARY
Previously, in the five meta-theoretical chapters of this book, I have described information ethics (IE) as a patient-oriented, ontocentric macroethics, which seeks to expand the scope of what may count as a centre of moral value to the whole infosphere. In this chapter, I shall defend the tenability of such ecumenical, macroethical e-nvironmentalism. The question I shall address is this: what is the least, most general common set of attributes that characterizes something as intrinsically valuable, and hence worthy of some moral respect, and that without which something would rightly be considered intrinsically worthless or even positively unworthy of respect? The answer that I shall develop and support here is that the minimal condition of possibility of an entity's least intrinsic value is to be identified with its ontological status as an informational entity. All entities, when interpreted as clusters of information—when our ontology is developed at an informational level of abstraction—have a minimal moral worth *qua* informational entities and so may deserve to be respected. This is IE's axiological ecumenism. The chapter is organized into four main sections.

In Section 6.1, I shall model a moral action informationally. I shall rely on the terminology and general methodology provided by object-oriented programming (OOP). This is just a very convenient framework, and nothing philosophically sensitive hangs on its choice.[1] The reader dissatisfied with it is welcome to replace it with any other modelling approach that can do the job equally well or even better.

In Section 6.2, I shall address the question of what role the several components constituting the moral system may play in an ethical analysis.

In Section 6.3, I shall provide an axiological analysis of informational entities. I shall criticize the Kantian approach to the concept of intrinsic value and show how it can be improved by using the methodology introduced in the first section. The solution of the

---

[1] For further clarifications, see Floridi (2012f).

Kantian problem prompts the reformulation of the key question concerning the moral value of an entity: what is the intrinsic value of something *qua* an entity constituted by its *inherited* attributes? In answering this question, I shall argue that entities can share different properties (observables), depending on the level of abstraction adopted, and that it is still reasonable to speak of moral value even at the highest level of ontological abstraction represented by an informational analysis.

In Section 6.4, I shall develop a minimalist axiology based on the concept of informational entity. This section further supports IE's position by addressing five objections that challenge its tenability.

In the conclusion, I shall connect IE's ecumenical axiology to environmental ethics.

6.1 Introduction: an object-oriented model of moral action

This section introduces the technical concepts and terminology necessary to develop a more precise, informational approach to the concept of moral patient. The reader acquainted with the object-oriented programming (OOP) methodology[2] may wish to move directly to Section 6.2.

The first task is to analyse a moral action as a dynamic system arising out of the combination of seven principal components:

(1) the agent,
(2) the patient,
(3) the interactions between the agent and the patient,
(4) the agent's general frame of information,
(5) the factual information concerning the situation insofar as it is at least partly available to the agent,
(6) the general environment in which the agent and the patient are located, and
(7) the specific situation in which the interaction occurs.

The second task is to show how this dynamic system can be modelled in terms of an information system by using the OOP methodology.

In Section 4.3, I suggested considering any action, whether morally loaded or not, as having the logical structure of a variably interactive process relating one or more *sources* or *senders*—depending on whether one is working within a multi-agent context—with one or more *destinations* or *receivers*. In order to model all this informationally, I shall make two simplifications, but note that in what follows nothing philosophical depends on them.

---

[2] This chapter follows the standard terminology and conceptual apparatus provided by Rumbaugh (1991) and Ellis (1996). I have also relied for conceptual modelling of information systems on Flynn and Fragoso Diaz (1996), Veryard (1992), and Boman (1997).

First, I shall concentrate on the simplest case in which the sender/source is a single agent A and the destination/receiver is a single patient P. The analysis could easily be extended to multi-agent scenarios. A multi-agent system is a conglomeration of interacting components, known as agents, capable of cooperating to solve problems that typically are beyond the individual capabilities or knowledge of each agent. Thus, a multi-agent system exhibits a system-level behaviour with much wider scope than the behaviour of its constituting agents (Wooldridge, 2009b).

Second, I shall assume that Alice, here any agent, initiates the process and Peter, here any patient, reacts more or less interactively to it. For example, Alice lies to Peter, or helps Peter in dealing with a financial problem. Of course, the interactive nature of the process must be kept in mind. The patient is hardly ever a passive receiver of an action. The unidirectional, bivalent, causal model is often far too simplistic, and a better way to qualify the patient in connection with the agent would be to refer to P as the 'reagent', but I trust the reader has grasped the point and the value of the simplification.

Once A and P are interpreted, their analysis depends on the level of abstraction (LoA) adopted and the corresponding set of observables available at that level. Suppose, for example, that we interpret P as Mary (P = Mary). Note the difference: Mary is not, like Peter, just another name for a variable, but is now the interpretation of that variable. In other words, Mary represents an actual human being; she is not an abstract name like Alice or Peter.

Depending on the LoA adopted and the corresponding set of observables, P = Mary can be analysed as the unique individual person called Mary, as a woman, as a mother, as a human being, as a mammal, as a form of life, as a physical body, and so forth. The higher the LoA, the more impoverished is the set of observables, and the more extended is the scope of the analysis. As the Turing test shows, eliminating or impoverishing the observables raises the LoA, until it becomes impossible to discriminate between two input sources. In our example, if Mary is analysed as a human being, more observables could lead one to analyse Mary at a lower LoA as a woman, and fewer observables could lead one to analyse Mary at a higher LoA as a mammal.

At the LoA provided by an informational analysis (LoA^i), both A and P are informational entities or informational structures, to use the vocabulary of informational structural realism introduced in Floridi (2011a). In our example, this means that P = Mary is analysed as an information system that interacts and shares a number of properties with other information entities, like a webbot. It does not mean that A or P are necessarily *only* informational entities, in the same sense in which they are not only thermodynamic systems either.

The OOP approach provides a very flexible and powerful methodology with which to clarify and make the concept of 'informational object' precise as an entity constituted by a bundle of properties, to use a Humean expression. Before introducing this approach, an example may help outline the basic idea.

Consider a pawn in a chess game. Its identity is not determined by its contingent properties as a physical body, including its shape or colour. Rather, a pawn is a set of data—properties like white or black and its strategic position on the board—and three behavioural rules: it can move forward only, one square at a time (but with the option of two squares on the first move); it can capture other pieces only by a diagonal, forward move; and it can be promoted to any piece, except a king, when it reaches the opposite side of the board. For a good player, any element on the chessboard that satisfies these features is a pawn. The actual piece is only a placeholder. The real pawn is an 'informational object'. It is not a material thing but a mental entity, to put it in Berkeley's terms. The physical placeholder can be replaced by a cork without any semantic loss at the LoA required by the game.

Let us now turn to a more theoretical analysis. OOP is a method of programming that has changed the approach to software development. Historically, a program was viewed as an algorithmic procedure that takes input, processes it, and produces output. The challenge was then represented by the elaboration of the algorithmic process. OOP has shifted the focus from the logic procedures required to manipulate the objects to the objects that need to be manipulated. In OOP, data structures (in the previous analogy, this could be the pawn's property of being white) and their behaviour (programming code, in the analogy this could be the pawn's power to capture pieces only by moving diagonally forward) are packaged together as informational objects. Objects are then grouped in a hierarchy of *classes* (e.g. pawns), with each class inheriting characteristics from the class above it (in our analogy, for example, all pieces but the king can be captured, so every pawn can be captured). A class is a named representation for an *abstraction*, where an abstraction is a named collection of *attributes* and *behaviour* relevant to modelling a given entity for some particular purpose, at a specified LoA. The routines or logic sequences that can manipulate the objects are called *methods*. A method is a particular implementation of an operation, i.e. an action or transformation that a system performs or to which it is subject, by a specific class. Objects communicate with each other through well-defined interfaces called *messages*. At an informational LoA, examples of objects can range from the buttons and scroll bars in a computer interface to human beings like Mary, from stock-exchange shares to buildings and pawns. This ontological concept should not be confused with the purely syntactical and quantitative concepts of information available in information and computation theory, or with the semantic approach popular in the philosophy of language, in the philosophy of mind, and in cognitive science. Henceforth, 'informational object' and its cognate terms will be used in the OOP sense just introduced.

Let us now return to the informational modelling of A and P. When A and P are analysed as informational entities at LoA^i, this means that they are considered and treated as discrete, self-contained, encapsulated[3] packages containing:

---

[3] Encapsulation or information hiding is the technique of keeping together data structures and the methods (class-implemented operations) which act on them in such a way that the package's internal

(1) the appropriate data structures, which constitute the nature of the entity in question: the current state of the object, its unique identity, and attributes; and
(2) a collection of operations, functions, or procedures (*methods*), which are activated (invoked) by various interactions or stimuli, namely messages received from other objects or changes within itself and which correspondingly define how the system behaves or reacts to them.

Both A and P are sufficiently permanent (continuant) informational objects. They can be simple or complex systems constituted by less complex informational objects—Descartes, for example, would consider mind and body such objects—and give rise to more complex systems, such as a family or a population.

Moral action itself can now be modelled as an information process, i.e. as a series of *messages* (M), initiated by A, that brings about a transformation of states *directly* (more on this qualification shortly) affecting P, which may variously respond to M with changes and/or other messages, depending on how M is interpreted by P's *methods*. Figure 9 summarizes the analysis developed so far.

We now have the first three informational components of our system: A, P, and M. The fourth component is the personal or subjective frame of information within which the agent operates. This shell,[4] which is really an integral part of A's nature but which it is useful to treat separately, is the information frame that encapsulates the subjective world of information of the agent (A's subjective *infosphere*, see later). It is

Moral action = information process $\exists A\ \exists P\ M\ (A, P)$

Informational objects
(in the object-oriented analysis paradigm (OOA) sense)

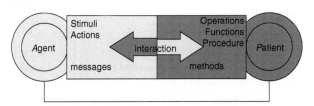

data structures constituting the nature of the entity in question (state of the object, its unique identity, and attributes)

Figure 9. The model of a moral action as a binary relation

---

structure can be accessed only by means of the approved package routines. External aspects of an object, which are accessible to other objects, are thus separated from the internal implementation details of the object itself, which remain hidden from other objects.

[4] The term comes from the operating system architecture vocabulary, not from OOP. It is the portion of the operating system that defines the interface between the operating system and its users.

constituted by internally dynamic and interactive records (modules) of A's moral values, prejudices, past patterns of behaviour, attitudes, likes and dislikes, phobias, emotional inclinations, moral beliefs acquired through education, past ethical evaluations, memories of moral experiences (e.g. of similar situations in which she acted as a witness or of other moral actions performed in the past), and so forth. In short, the shell represents the ethical and epistemic conceptualizing interface between Alice and her environment. Although it embodies aspects of Alice's life, it is constantly evolving through time, may contain shared or imported elements from other agents' shells, may be epistemically only partially accessible to Alice herself, and, in practice, is only partially under the control of her will and her capacity to choose. Nevertheless, it contributes substantially to the shaping of Alice's behaviour, by screening her from the direct impact of the infosphere, filtering and regulating her access to, and hence highlighting and interpreting the relevant aspects of, the factual information concerning the specific moral situation in which Alice finds herself involved in space and time.

The factual information concerning the moral situation represents the fifth dynamic component of the system. It is the only element in the model that remains unmodified when the LoA changes. We still speak of factual information even at the lower LoA, where there are sufficient observables to analyse both A and P, not just as two informational objects but also as two persons, for example. For this reason, we saw in Chapter 2 that the majority of ethical theories are ready to recognize factual information as playing an instrumental role in moral actions. According to Warnock (1971), for example, lack of information is one of the main factors that cause 'things to go badly'. Camus, in *The Plague*, shared the same view:

The evil that is in the world almost always comes of ignorance, and good intentions may do as much harm as malevolence if they lack understanding.[5]

More 'weakly', it is common to assume that an action with a potential moral value can be treated as actually moral or immoral only insofar as its source A is, among other things, *conscious* (Alice is aware of her actions), *sufficiently free* (Alice is rationally autonomous in the Kantian sense and can intentionally bring about, stop, or modify the course of action in question, at least partially, depending on the situation), *reasonable* (Alice is intelligent in Mill's sense, i.e. has some capacity to forecast the consequences of her actions), and *factually informed*. Traditional ethical theories share the view that a moral action and its corresponding evaluation can take place in a state of only limited (i.e. not total) scarcity of freedom and information, and that there is no morality in a state of total determinism or ignorance (compare to animal behaviour). I shall return to this point in the next chapter, in order to argue for a more flexible view.

We now come to the sixth component. At LoA^i, Alice does or does not implement, and hence variously controls and adjusts, her autonomous and informed behaviour in a

---

[5] Albert Camus, *The Plague*, trans. Stuart Gilbert (New York, NY: Vintage), p. 131.

108   THE ETHICS OF INFORMATION

dynamic interaction with the general environment in which she is *present* (see Chapter 3), such as a given culture, society, family situation, financial status, group of individuals, set of working conditions, and so forth. The same holds true for Peter, whom we shall assume to be co-present with Alice. In Chapter 1, this informational environment has been described as the *infosphere*. Here, we can consider it as the context constituted by the whole system of informational entities, including all agents and patients, messages, their attributes, and mutual relations. The specific region of the infosphere in space and time within which the moral action takes place represents the last component of the system, namely the *moral situation*. Borrowing a term from robotics, this information microworld can be defined as the *envelope*[6] of the moral action. To summarize (see Figure 10), here is the complete list of information components:

(1) A = Alice the moral agent
(2) P = Peter the moral patient
(3) M = moral action, constructed as an interactive information process
(4) *shell* = Alice's personal world of information
(5) *factual information* = information about the moral situation
(6) *envelope* = the moral situation
(7) *infosphere* = the general environment

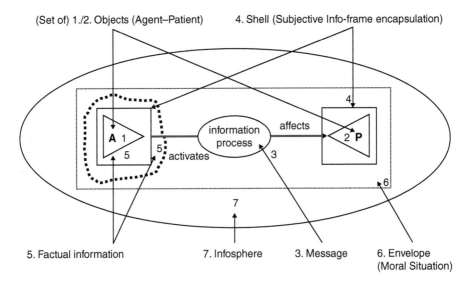

Figure 10. The informational model of a moral action

---

[6] The 'envelope' of a robot is the working environment within which it operates or, more precisely, the volume of space encompassing the maximum designed movements of all the robot's parts.

Two comments are now in order. First, when the message is a reflective process or a process with a feedback effect, A may be identical with, or treated as, (at least one of the) P. In Section 4.3, I mentioned the case of a suicidal agent. Consider now the case in which Alice is self-indulgent or steels herself for a difficult choice (I shall come back to this important point in Section 6.3). Second, it is hardly ever the case that a message affects only a discrete set of well-specified patients P. Of course, it is convenient to limit our attention to a simplified dynamic model, and this is why I specified 'directly' above, but one needs to remember that a message functions like a vector with a given direction and a discrete force, and not as a binary switch. Once the message has been released (once the action has been performed), its direct and indirect effects almost immediately cease to be under the control of its source A, while their the effects' life extends in time and space in the form of a continuum, not necessarily decreasing (see Section 15.1.5). Using another OOP concept, we can then speak of the *propagation* of an operation, which starts with some initial system A at a given point in time and space and then flows from system to system through space, time, and association links, according to possibly specifiable rules. During this propagation, the vector may change both in direction and in force. Clearly, a message affects not just the immediate target P of the process but also the envelope—hence A as well, A's shell and the factual information—and finally the whole infosphere. Think of an abused child who, as an adult, becomes an abuser of several children. In principle, all seven components may be treated as patients. We shall see in Section 6.4.3 that a negative axiology (a theory of intrinsic unworthiness) requires a more adequate conception of what kind of entity may count as a patient.

6.2 The role of information in ethics

We are now ready to phrase the foundationalist problem in IE in terms of a choice of a valuable LoA. Does the informational level of abstraction of IE provide an additional perspective that can further expand the ethical discourse in such a way that it includes the world of morally significant phenomena involving any aspect of *Being*? Or does it represent a threshold beyond which nothing of moral significance really happens? Does looking at reality through the highly philosophical lens of an informational analysis improve our ethical understanding, or is it an ethically pointless (when not a misleading) exercise?

On the one hand, if entities, understood informationally, can have at most only an instrumental value, then CE is likely to remain at most a microethics (see Chapter 5), which plays only an ancillary role with respect to another macroethics, such as deontologism or consequentialism. This is because macroethics attempts to establish not just the necessary and sufficient conditions of adequacy for the occurrence of a moral action, e.g. its information input, but, more importantly, what ought to be the very nature of the action in question, why a specific action would be morally right or wrong, what ought to be done in a given moral situation, and what the duties, the 'oughts' and 'ought nots' of a moral agent may be. On the other hand, if IE, as the foundation of CE, can develop a macroethical approach, it needs to be able to show that the agent-related *behaviour* and the patient-related *status* of entities *qua* informational entities can be morally significant, over and above the instrumental function that may be attributed to them by other ethical approaches, and hence that they can contribute to determining normative rights, duties, and courses of action. It follows that IE's claim consists of two theses.

The first thesis states that all entities *qua* informational entities have an intrinsic moral value, although possibly quite minimal and overridable, and hence that they qualify as moral patients subject to some (possibly equally minimal) degree of moral respect.

The second thesis states that *artificial* informational entities, insofar as they can be agents, can also be accountable *moral* agents. This means not just *analysing* an interpreted A as an information system (e.g. A = Mary)—this is elementary, as it requires only the adoption of the right LoA—but rather showing that an artificial agent A, such as a webbot, a company, or a tank, can be correctly *interpreted* as an information system that can play the role of a moral agent accountable for its actions, at the usual LoA adopted by other ethical theories, that is, at the LoA where A = Mary is analysed as a human being.

In short, all entities are informational entities, some informational entities are agents, some agents are artificial, some artificial agents are moral, and moral artificial agents are accountable but not necessarily responsible.

Now, one of the methodological goals of this book is to promote the importance of shifting from an agent-centric to a patient-centric approach to ethical issues. The fruitfulness of such a change in perspective is more obvious when dealing with the first thesis, which therefore will be discussed and defended in the rest of this chapter. Once the first thesis has been established, I shall clarify and support the second thesis in Chapter 7. Since the strategy used to support the first thesis may not be immediately obvious, it may be worth outlining it at the outset.

The issue is approached top-down, in terms of importance of the object under analysis, and bottom-up, in terms of increasing abstraction. The starting point is the discussion of the unproblematic case in which the patient is an ordinary human being, who is recognized to have intrinsic moral worth. At this low LoA, one of the best philosophical positions available, namely Kant's, suggests that *only* rational beings have an intrinsic moral worth. The objection is that the Kantian[7] position is not fully satisfactory and needs to be improved. This is carried out by gradually impoverishing the ontological status of P = Mary.[8] By eliminating more and more of the properties

---

[7] I do not intend to address Kant's ethics from a philological perspective, but philosophically. I believe that the position I am criticizing here is so close to Kant's to deserve the qualification of Kantian. However, if the reader finds that the Kant portrayed here is insufficiently close to the Kant she knows, I would be happy to replace any occurrence of 'Kant' with 'Kent'.

[8] For a discussion of the 'argument from marginal cases' see Baird Callicott (1980), reprinted in Elliot (1995).

enjoyed by Mary, the LoA is raised until the stage is reached at which, on the one hand, one would still like to be able to understand why P = Mary may still enjoy some degree of intrinsic moral value, and hence be subject to some level of moral respect, even if Mary is reduced to a mere brainless entity—recall Hector's body as 'senseless clay' in Section 4.8.3—but, on the other hand, the Kantian analysis is unable to provide a satisfactory answer. At this point, two arguments support the attribution of an intrinsic moral value to informational entities. The first, a positive argument, consists in showing that an informational, patient-oriented approach can successfully deal with the problem left unsolved by Kant. The second, a negative argument, consists in showing the limits of not only the Kantian position, but also any other position that adopts some other LoA higher than the Kantian-anthropocentric one but still lower than LoA^i, like a biocentric LoA. Showing that both an anthropocentric and a biocentric axiology are unsatisfactory is a crucial step, since it re-opens the fundamental problem of what entities can qualify as centres of some moral worth, enables one to approach it afresh, and shifts at least part of the burden of proof to IE's critics. If ordinary human beings are not the only entities enjoying some form of moral respect, what else qualifies? Only sentient beings? Only biological systems? What justifies including some entities and excluding others? Suppose we replace an anthropocentric approach with a biocentric one. Why biocentrism and not ontocentrism? Why can *life* and its *preservation* be considered morally relevant phenomena in themselves, independently of human interests, but not *existence* and its *protection*, that is, *Being* and its *flourishing*? In many contexts, it is perfectly reasonable to exercise moral respect towards inanimate entities, including artefacts *per se*, independently of any human interest (mind, not acknowledgement); could it not be just a matter of ethical sensibility, indeed of an ethical sensibility that we might have had—at least in some Greek philosophy such as the Stoics' and the Neoplatonists'—but have then lost? Why are we so afraid of being inclusive in our axiology? It seems that any attempt to exclude non-living entities is based on some specific, low LoA and its corresponding observables but that this is an arbitrary choice. On the scale of beings, there seems to be no good reason to stop anywhere else but at the bottom. All things in the biosphere have an equal right to live and blossom, so that:

The ecological field-worker acquires a deepseated respect, or even veneration, for ways and forms of life. He reaches an understanding from within, a kind of understanding that others reserve for fellow men and for a narrow section of ways and forms of life. To the ecological field-worker, the equal right to live and blossom is an intuitively clear and obvious value axiom. (Naess, 1973, p. 96)

IE replaces biosphere with infosphere and argues that there is no good reason to reject a higher and more inclusive, ontocentric LoA. Not only inanimate but also ideal, intangible, or intellectual entities can have a minimal degree of moral value, no matter how humble, and so be entitled to some initial respect. With a quote from Shakespeare:

> And this our life, exempt from public haunt,
> Finds tongues in trees, books in the running brooks,
> Sermons in stones, and good in *everything*.
> (*As You Like It*, 2.1.13, my emphasis).

The previous two arguments are paralleled by two other arguments, one meta-theoretical, the other historical.

The meta-theoretical argument has already made several appearances in the previous chapters. Enlarging the conception of what can count as a centre of moral respect has the advantage of enabling one to make sense of the innovative nature of CE and to deal more satisfactorily with the original character of some of its moral challenges, by approaching them from a theoretically strong perspective.

The historical argument is connected with the negative argument above. Through time, ethics has steadily moved from a narrow to a gradually more inclusive concept of what can count as a centre of moral worth, from the citizen to the biosphere (Nash, 1989; Stone, 2010). The emergence of the infosphere, as the new environment in which human beings spend much of their lives, explains the need to enlarge further the conception of what can qualify as a moral patient. IE represents the most recent development in this ecumenical trend, a Platonist e-nvironmentalism without a bio-centric bias, as it were.

Once the intrinsic moral worth of an information system has been introduced as a viable solution to the problem left unsolved by the Kantian approach, two more tasks lie ahead. One is to show that IE's thesis is coherent. The other is to show that the main objections against it can be answered. Both tasks are undertaken in Section 6.4. Their successful fulfilment further reinforces IE's position.

One last comment before beginning: the arguments sketched above and articulated in the following pages are 'intellectual' not 'strategic', to use a valuable distinction I am borrowing from Norton (1989): they are addressed to the philosophically minded interlocutor, not to the reluctant policy-maker, who will, perhaps, be more easily convinced by reasons centred on human interests. I made a similar point in Section 4.5.4 with respect to attempts to motivate, persuade, or even coerce an agent. The arguments offered in this book do not provide threatening answers regarding the consequences of an action, as one might to the teenager's question 'why should I do it?', but seek to answer the more mature and difficult question, 'what would be the right thing to do?'.

6.3 An axiological analysis of information

The status of A and P, and the possible modifications in the nature of both, brought about by the information process M, are the axiological elements that play a decisive role in the normative assessment of a moral action. In what follows, the analysis is restricted to *P* only, for three reasons (I shall return to the analysis of *M* in Section 6.4.3 and of *A* in the next chapter).

First, the problem is whether an entity *qua* informational entity—i.e. not insofar as it is a specific type of entity like Mary—can have any intrinsic moral value that could contribute to regulating a moral action affecting it. Since it is usually assumed that any entity that can act as a moral agent can also qualify as a moral patient but not *vice versa*—for example, it is generally assumed that animals can at most be moral patients but not moral agents (Rosenfeld, 1995a)—it is better to focus on the informational nature of an entity as a possible patient.

Second, whenever the action in question is found to be either reflexive (e.g. suicide) or retroactive (e.g. moral vices acquired through the repetition of actions and the accumulation of their effects, where each action is not, in itself, necessarily deprecated from a moral point of view), the model allows for any conclusion reached about the patient to be extended to the agent.

Third, by discussing the moral worth of an informational entity as a patient in the most universal and abstract terms, it is possible to extrapolate from the specific nature and position taken up in a given envelope by a component of the system and, thus, generalize any conclusions reached about *P* so that they also apply to any possible informational element that may, in principle, be affected by the behaviour of *A*, and hence qualify as a patient of a moral process. In this way, other envelopes, the infosphere itself, and the methods can be considered patients of *A*'s actions in a way that will become fully clear once a negative axiology is developed (see Section 6.4.3).

Once the analysis is restricted to the patient *P*, the question to be addressed is whether there is any degree of *unconditional*—not in the sense of absolute, but in the sense of not conditional on other goals or emotions, i.e. neither instrumental nor emotional, more on this later—and intrinsic worth in the nature of a patient understood as an informational entity that should determine, constrain, guide, or shape *A*'s moral actions, at least minimally and overridably. Does an informational entity as a patient have an intrinsic moral value that, *ceteris paribus*, could contribute to the moral configuration of *A*'s behaviour? Insofar as *P* has some intrinsic value, it contributes to the configuration of a moral action by requiring (see Sections 6.3.1 and 6.4.4) that *A* recognizes such value in a special intentional way, that is, by having *respect* for it. By 'respect', I mean *a disinterested, appreciative,* and *careful attention* (Hepburn, 1984) at the ordinary LoA where *P* = Mary is analysed as a human being. We can now say that *A*'s respect for *P*'s intrinsic value consists in two things: the appreciation of *P*'s specific worth and the corresponding, uncoerced, arguably overridable disposition to treat *P* appropriately, if possible, according to this acknowledged worth. Entities capable of intentional states can have respect for *P*'s intrinsic value and hence act as moral agents at least to this extent, but are they also the only entities that can have an intrinsic value as patients? According to Kant, the previous question can be answered with a firm 'yes'. I disagree. Let me explain why.

6.3.1 A critique of Kantian axiology

According to Kant, either some x can rightly function only as a means to an end other than itself, in which case it has an *instrumental* or *emotional* value (extrinsic value, which Kant equates with economic evaluation, see next quotation); or some x qualifies also as an end in itself, in which case it has an *intrinsic, moral* value insofar as it is that x and it is valued and respected for its own sake. Thus, in the *Groundwork of the Metaphysics of Morals*, Kant writes:

> In the kingdom of ends, everything has either a price or a dignity. What has a price can be replaced by something else as its equivalent; what on the other hand is raised above all price and therefore admits of no equivalent has a dignity. (Kant, 2005, p. 84)

Note that 'intrinsic value' is often recognized to be an ambiguous expression (compare e.g. Benson (2000)). It can mean 'non-instrumental value', as in Kant, or 'inherent value', that is, a value that something enjoys independently of the existence of any evaluating source. In the rest of this chapter, I shall opt for the Kantian sense.

Kant argues that anything can have an instrumental value for the sake of something else, but that only human beings, as rationally autonomous agents, can also have an intrinsic and absolute moral value, which he calls *dignity*. This is because only rationally autonomous agents, understood as potential 'good wills', can be the source of moral goodness, thanks to their rational and free actions. The Kantian dichotomy, intrinsic vs. instrumental value, has at least three major consequences:

(K.1) the dichotomy justifies the co-extension of (i) the class of entities that enjoy moral value, (ii) the class of entities capable of moral respect, and (iii) the class of entities that deserve to be morally respected. In Kant, the only type of entity that has moral value is the same type of entity that may correctly qualify as a moral patient and that may in principle act as a moral agent;

(K.2) the dichotomy solves the communication problem between A and P in the following sense. Thanks to (K.1), A is immediately acquainted with the moral value of P, and hence can respect it, because both entities belong to the same type of class, namely Kant's 'Kingdom of Ends' (see e.g. Kant (2005), p. 85). We shall see that, since IE rejects (K.1), it cannot rely on the solution of the communication problem provided by (K.2);

(K.3) the dichotomy implies that an entity's moral value is a kind of unconditional and incomparable worth. Either some x has an instrumental value, subject to degrees, economically significant but morally irrelevant, or x has an unconditional and intrinsic value, which is morally relevant but absolute, and cannot rightly be subject to economic assessment.

The Kantian dichotomy is questionable because (K.3) clashes with two reasonable assumptions and fails to take into account two important distinctions (to be discussed in the next section).

First, it seems reasonable to assume that different entities may have different degrees of relative value that can constrain A's behaviour, without necessarily having an instrumental value, i.e. a value relative to human feelings, impulses, or inclinations, as Kant would phrase it.

Second, it seems equally reasonable to assume that life, biological organisms, or the absence of pain in sentient beings can all have a great deal of moral value and deserve a corresponding amount of respect. For example, one may argue that a human being who is even *inherently* (i.e. not just contingently, for instance because of unlucky circumstances that may change) incapable of any intentional, rational, and free behaviour still has some moral value, no matter how humble, which deserves to be respected, although not necessarily only for instrumental or emotional reasons. More generally, the default position seems to be that only rational beings are capable of respect but, contrary to what Kant suggests in (K.1), 'having an absolute moral value (dignity)', 'being capable of respect', and 'being intrinsically respectable' do not range over exactly the same class of entities. Rational beings are capable of various degrees of respect to which there seem to correspond various degrees of moral value. Kant is right in arguing that rational beings, as 'good wills', definitely deserve more respect than other entities because rationality is one of the conditions necessary for morally good actions and hence of moral improvements of reality. However, it requires some positive argument to show that potential 'good wills' (Kant's rational natures) do not constitute only a subclass of the entities that may have a morally significant claim on the agent, that is, as entities subject to some respect.

The previous criticisms are *prima facie* plausible. They are not very new, but they still represent a serious challenge to the Kantian dichotomy. Kant seems unduly to restrict the sense of 'relative value' to meaning only 'contingent worth depending on the agent's interest' (Kant, 2005, p. 79), so that if some x can be rightly and appropriately used only as a means, then x has no absolute value (x has only a relative value), and x's value has no moral nature whatsoever, because x's value is to be interpreted as depending only on the instrumental or emotional interest of the agent, which is a clear *non sequitur*, if one rejects the very controversial equation just spelt out.

6.3.2 *A patient-oriented approach to axiology*

According to Kant, not only do the Kingdom of Ends and the Kingdom of Nature remain largely separate and independent, but the former becomes a closed system, which is allowed to rule over the latter without there being even the possibility *in principle* of the latter providing some constraints.[9] Two distinctions can help to improve Kant's anthropocentric axiology.[10]

---

[9] Kant (2005), p. 73 ('act as if the maxim of your actions were to become by your will a universal law of nature'), see further pp. 86–8. On p. 86 Kant writes: 'all maxims from one's own lawgiving are to harmonise with a possible kingdom of ends as with a kingdom of nature', but on p. 88 it seems that only God as a single sovereign can bring together the kingdom of ends with the kingdom of nature.

[10] It is interesting to note that the four examples used by Kant to illustrate the application of the 'Law of Nature' formulation of the imperative ('act as if the maxim of your actions were to become by your will a

Let us agree with Kant that there are different ways and degrees in which an entity may have some instrumental value. When the value in question is neither instrumental nor just emotional,[11] one can first distinguish between extrinsic and intrinsic value and, correspondingly, between two types of respect. An entity x has extrinsic value when it is respected as some y. For example, a piece of cloth may be respected as a flag, a person may be respected as a police officer, or a practice may be respected as a cult. This sense of relative respect is associated with a sense of value that is no longer instrumental or emotional and may be called *symbolic*. Symbolic value is still utterly contingent, may be acquired or lost, and can be increased as well as reduced. In brief, it is utterly extrinsic.

In order to capture in full the fact that an x has moral value in itself—a value that belongs to x in all circumstances (necessarily), not only under some conditions, and is not subject to modification unless x ceases to exist as that x—one needs to consider the case in which x deserves to be respected not just symbolically, as something else, but *qua* x. It is here that the analysis must depart from the Kantian position more radically and introduce a second distinction.

Kant would agree that the moral value of an entity is based on its ontology. What the entity is determines the degree of moral value it enjoys, if any, whether and how it deserves to be respected, and hence what kinds of moral claim it can have on the agent. For Kant, x's intrinsic value is indissolubly linked with x's essential nature only as some *type* of entity. Thus, an individual, e.g. Mary, has moral value (Kant's dignity) not as a specific person, but only insofar as she is a token of the general type 'free and rational human being', i.e. a member of the 'Kingdom of Ends'. In respecting P = Mary, the agent is not primarily or directly respecting the specific, unique and contingent individual *qua* herself, but rather the universal type she instantiates.[12] So the Kantian analysis fails to distinguish between two separate senses in which the nature of x determines x's moral value. It is a crucial oversight.

The two senses can be clarified by relying on the OOP methodology introduced in Section 6.1. The specific nature (essence) of an object x consists in some attributes that (1) x could not have lacked from the start except by never having come into being as that x, and (2) x cannot lose except by ceasing to exist as that x. This essence is a factually indissoluble, but logically distinguishable, combination of x's *local* and *inherited* attributes. For example, if 'Person' is the descendant class, and 'Living Organism' is the ancestor class, we may say that 'freedom' is an essential and local attribute of 'Person'; that is, a new property, not previously implemented in any of the ancestor classes,

---

universal law of nature') in *Groundwork* (Kant, 2005, pp. 73–5) are all 'anthropocentric' and concern only duties to oneself or to others, so when Kant speaks of the 'Formula of Humanity' version of the imperative in *Groundwork* (2005, p. 80) ('So act that you use humanity, whether in your person or in the person of any other, always at the same time as an end, never merely as a means'), he employs the same four anthropocentric examples.

[11] This is Kant's 'fancy price', see Kant (2005), p. 84.

[12] Kant (2005), p. 84: 'Hence morality, and humanity insofar as it is capable of morality, is that which alone has dignity.'

while 'sentient animal' is an essential attribute inherited by 'Person' from its ancestor class 'Living Organism'. Suppose now that an object x has an intrinsic value. It is correct to say, with Kant, that x's intrinsic value depends on its essence, or more generally on its ontology, but it is also important to specify that this essence, and the corresponding intrinsic value, are both aggregates, i.e. they are the result of a specific (the degree of 'specificity' can be increased or decreased as required, depending on the choice of LoA) combination of *local* and *inherited* attributes, which in turn can be observed only at a given LoA. This makes a significant difference. In the example, one can respect Mary because of her local attribute 'free agent', which is part of her essence. Her essence also includes that of being a 'living organism capable of feelings of pain and pleasure'. Let us refer to the former as Mary's F attribute and to the latter as Mary's L attribute, and let us simplify matters[13] by saying that Mary inherits L from her ancestor class called 'Animal'. Suppose now that Mary is radically and definitely deprived of her local attribute F, that is, let us imagine that she is *inherently* incapable of any free and intentional behaviour, e.g. because she is born brain-dead, so that the absence of some observables, at the chosen LoA, corresponds as closely as one may wish to a real ontological feature.[14] What would be the Kantian position with respect to her moral value? There seem to be only four alternatives. I shall label them A, B.1, B.2.a, and B.2.b. None of them is fully satisfactory and this leads to the adoption of a different approach (see further point (C)).

A radical solution would be to 'bite the bullet' and argue that:

(A) Mary lacks the necessary attribute F, so she can have no justified claim to moral respect. Citizenship in the Kingdom of Ends is a necessary and sufficient condition for such respect, but it can be lost and, without it, there are no moral rights.

Of course, (A) is logically acceptable, but its unpleasant consequences inevitably clash with some of the most elementary moral assumptions. According to (A), for example, one could freely dispose of Mary's organs without being subject to any moral assessment. If one wishes to maintain that Mary still deserves to be respected despite the lack of F, one may try to argue, still with Kant, that:

(B) Mary still possesses moral value as a type, that is, as an entity that *somehow* still enjoys the local attribute F, because *in principle*, though not in practice, she is still a member of the class 'free agents'.

---

[13] Strictly speaking this is not properly OOP. We saw that in OOP, Animal is a class—in fact, an abstract class—from which it is not possible to instantiate objects.

[14] I am aware of disregarding here the fact that for Kant (and I agree with him, see Floridi (2011a)) we do not have access to the intrinsic, noumenal nature of an entity. In other words, the choice of the LoA is crucial to clarify and justify our ontological commitment. The reader interested in a straightforward (though not simple) way of solving this problem might wish to see Floridi (2011a, ch. 15).

Alternative (B) tries to rationalize the *prima facie* justified request that Mary may still possess some moral value, and hence deserve to be respected, by working on a rather problematic interpretation—as something 'absent-yet-still-present'—of the set of properties necessary to qualify as a rational being. The trouble with (B) is that it soon becomes unclear what it means to have 'somehow' and 'in principle' some type of attribute, unless by this we wish to refer either to (B.1) a logical possibility or to (B.2) the entity's potentiality, the actual possibility being unavailable by hypothesis (the counterfactual alternative can be seen as a further development of the previous two, see further (B.2.b)).

Let us consider (B.1) first. The new criterion—respect any x of which it would be a contradiction to say that it could not qualify as a 'free agent'—becomes too vague, because it is also logically possible that a chicken could behave as a free agent.

Consider next (B.2). This is compatible with Kant's ontology. The problem is that, by saying that Mary may still have the attribute F *potentially*, one may mean that:

(B.2.a) although born brain-dead, Mary is still morally respectable because she is potentially free by nature, and this is the case because she is a human being. This 'potentially free' feature of her nature cannot be taken away merely because some factor (malformation, accident, disease, drugs, etc.) is in fact preventing her from 'actualizing' that potential. The potential can exist 'un-actualized' and yet consist of more than mere logical possibility, as her lost freedom is something that she *could have had* in a way a chicken never could.

(B.2.a) would allow the Kantian philosopher to solve the axiological problem, if it were not for the fact that, as it stands, it is confronted with two substantial problems.

The first problem is that (B.2.a) begs the question. In the counterexample, Mary does not *happen* to lack the attribute F *momentarily* or just *accidentally*, e.g. because she is under the temporary effect of a drug. If this were the case, (B.2.a) would be correct, but there would be no interesting challenge for the Kantian axiology in the first place. Rather, the assumption is that Mary has been *essentially* or *inherently* deprived of her attribute F. She is not and will never be capable of any free behaviour, for example, because she is born irreversibly brain-dead. There is no LoA at which Mary enjoys attribute F. In other words, the attribute F has been erased from the description of the informational entity Mary.[15] A supporter of (B.2.a) could reply that Mary's F attribute cannot be taken away by a contingent event, e.g. a car accident. Yet, this is simply false (second problem). Although essential by hypothesis, a potential attribute is not

---

[15] Usually, in OOP terms, attributes are given values. The attribute itself is never erased. It may be set to null or false or the string 'non-existent' or 'never had it, never will'. In the framework offered by OOP, Mary will always have the attribute F, it just may not provide any richness to the description of Mary. The point is that once the class is defined, all objects that instantiate the class have the attributes. They do not take on new attributes and do not give up the attributes they have. However, so-called 'dynamic inheritance' allows objects to change and evolve over time. This refers to the ability to add, delete, or change parents from objects (or classes) at run-time. I shall return to this feature in the following pages.

necessarily a permanent feature of an entity and, contrary to what (B.2.a) seems to suggest, it may be removed, even within an Aristotelian ontology. This is an intrinsic feature of the potentiality/actuality distinction, which was originally developed to provide an explanation of *change* and *transformation*, i.e. the loss of old and the acquisition of new features. A potentiality is an individual's capacity to acquire a new feature, and this capacity can disappear if the feature becomes actual, or if the conditions of possibility of the actualization of the potential feature are irreversibly removed. If the potentiality of being x is a necessary attribute to qualify as y, this does not mean that whatever is y cannot lose the attribute x, but only that, if y loses x, then y becomes something else, different in time from itself. To illustrate the point with a more Aristotelian example: a healthy man has the potentiality of becoming a marathon runner, but once he has become one, this means that he has changed into something else and has lost the potentiality of becoming a marathon runner in favour of the actuality of such potentiality. Likewise, if a healthy man loses his legs, he no longer enjoys the potentiality of becoming a marathon runner. When the potential attribute belongs to the essence of the object, its removal implies the re-categorization of the individual in a different class, but this is precisely the problem confronting us at the moment: whether a person born brain-dead, who may not count any longer as an ordinarily rational human being, may still be entitled to some moral respect, even if the only entities entitled to moral respect are rational human beings.

So option (B.2.a) does not provide a satisfactory answer, but it does contain a valuable point. We have seen that it is not true that, if the attribute F is *practically* not actualizable, F is therefore utterly lost and can be regained only as a logical possibility. Yet, this is not the issue addressed by our counterexample, in which the attribute F becomes *essentially* not actualizable. What must be conceded to (B.2.a), however, is that there still is a considerable difference between saying that a chicken *could* be free and that Mary, who is brain-damaged, as a human being *had the potentiality* of being free. The difference would be blurred by a mere reference to a logical possibility, but can be captured by a counterfactual analysis, which leads us to reformulate (B.2.a) to say that:

(B.2.b)   to claim that Mary is potentially free is to claim that, under other, ordinary circumstances, Mary would have not been deprived of F and so she would have been morally respectable.

(B.2.b) is qualitatively (regarding naturalness) and quantitatively (regarding probability) stronger than (B.1). This is obvious if we try to replace 'Mary' with 'chicken' in (B.2.b). It does not work. (B.2.b) is also more stringent than (A). Nevertheless, it can at most support a 'counterfactual respect', which is still too weak for solving the axiological issue raised by the counterexample. Had Mary had the attribute F (had circumstances been different) she would have been the object of moral respect. This is all one can argue on the ground of (B.2.b). Since Mary does lack the attribute F, however, the counterfactual analysis leaves us with the possibility of being fully justified in showing no moral respect for her at all. We are not denying that, in another possible world, she would have deserved some respect. We are recalling that, given the present circumstances, she is not 'eligible' in this actual world.

A Kantian axiology fails to accommodate the counterexample satisfactorily because it is unable to clarify, in a convincing way, why Mary should be morally respected only on the grounds of what she actually lacks by definition and irreversibly in the first place. This discloses a general problem affecting Kant's approach and others similar to it. When Kant speaks of moral respect, he has in mind, primarily, ideally rational agents and only derivatively human beings seen as fallen creatures. In his deontological ethics, a person is morally respectable only in an indirect sense, insofar as she or he implements the properties necessary and sufficient to qualify as a rational being. If the person in question does satisfy such conditions, this hides the fact that, in respecting her, one is really asked to respect not the individual but a special class of individuals, to which the individual person, however, does not have to belong necessarily. If the person no longer satisfies such conditions, it becomes clear that she was being respected only because she was partaking of the special properties of the class of rational beings.

The solution to the problem requires a shift in perspective. It is hard to see how one could explain and justify any form of respect towards Mary based on some local attribute that, *ex hypothesi*, does not exist. A completely different alternative consists in arguing that Mary still has some form of moral value as an entity that enjoys the inherited attribute—this is L in our example—at a higher LoA. One may no longer express towards Mary exactly the same respect one would have towards a free agent, but one could still feel compelled to respect her at least as a living organism capable of feelings, for example. This alternative looks for the *minimal*, not the *maximal*, conditions of moral worth. It also appears more reasonable and accords more with other basic ethical assumptions. This is the view favoured by IE, which argues for a more decisive step in the same direction.

Once the distinction between local and inherited attributes is accepted, asking what the intrinsic value of some x *qua* that x is involves asking three different questions.

(1) What is the intrinsic value of x insofar as this specific entity is constituted by this specific aggregate of *local and inherited* attributes?

A full answer to (1) can be provided only by combining the two senses in which x has an intrinsic value according to (2) and (3) below. A theory that concentrates only on (1) is a theory of individual moral value, i.e. of the intrinsic value that x possesses in itself as a specific individual, not just as an instantiation of a type. Note that x may be either a single entity (in our example, Mary) or a whole class (for example, Women), so the theory does not have to be nominalist. What the theory does is to invert the direction of the 'axiological flow': now the class (or type) of Xs has intrinsic value because each of its members has intrinsic value, not *vice versa*.

(2) What is the intrinsic value of x insofar as it is an entity constituted by its *local* attributes?

Since Kant's concepts of essence, type-token (as interpreted here), and class membership cut across our concepts of inheritance[16] and aggregate of local and inherited attributes, none of the three questions is exactly the question addressed by Kant, yet (2) is probably the one that comes closest to the Kantian approach, where the local attributes are interpreted as the essential properties of the class of all human beings. A theory that concentrates on (2) may develop a maximalist axiology like Kant's, according to which there is only a restricted selection of local attributes—e.g. intentionality, self-determination, and rationality—that qualify an entity as having moral value. Kant is right in arguing that this special object, defined as a 'rational being' or potential 'good will', is the one that has the highest moral value (dignity) and hence deserves absolute respect. Nonetheless, he is wrong in assuming that this is the only sense in which it is possible to speak of moral worth and respect, because one could also ask the following question:

(3) What is the intrinsic value of x insofar as it is an entity constituted by its *inherited* attributes?

By progressively raising the LoA, one can answer this question by referring to the nature of the entity in question as an informational entity. We have seen that, in the case of the pawn, this is really what matters most. In the case of Mary, the local attributes are far more important, yet this is not a good reason to conclude that, if Mary is reduced to an informational entity, that is, if all that remains to be considered about her is her nature as a bundle of data, then this informational entity is devoid of any moral value and can be rightly vandalized, exploited, degraded, or carelessly manipulated without regard for any moral concern and constraint. As we shall see, an entity x can be respected at different LoAs, including the level at which x is only an informational entity. Thus, in Mary's case, IE argues that:

(C) if Mary qualifies as a living organism, biocentric ethical concerns apply. Suppose, however, that Mary does not qualify as a living organism any longer. Her corpse still enjoys a degree of intrinsic moral worth because of its nature as an informational entity and, as such, it can still exercise a corresponding claim to moral respect.

Recall Apollo's remarks in the last book of the *Iliad*: not even Achilles has the moral right to 'outrage the senseless clay'. Hector's body deserves a minimal degree of moral respect.

An axiology that concentrates on question (3) can be pluralist or minimalist. A pluralist axiology finds in a selection of inherited attributes—such as intelligence,

---

[16] We saw that in OOP, inheritance is the sharing of attributes and operations among classes based on an 'is-a-kind-of' hierarchical relationship between objects. A class is the ancestor class of another, which inherits its attributes and methods. A class may have more than one ancestor (multiple inheritance), may share an ancestor with other classes (shared inheritance) and inheritance may be dynamic (ancestors can be added, deleted or changed through time).

sensation, or biological life—the ontological source of the intrinsic value of an entity and therefore assigns to a wide variety of entities, namely all those that inherit one or more of these attributes, some moral value and hence a corresponding claim to A's respect. Of course, the moral value in question cannot be absolute, since the theory accepts more than one inherited attribute as comparable, when not competing. It is likely, however, that there may develop a hierarchy of inherited attributes and of priorities in moral standing, and hence a minimalist theory.

A minimalist axiology does not have to be monist. However, it is not pluralist in the sense that it does not admit that there may be more than one, incomparable and non-equivalent, minimal degree of value. It accepts only one set of inherited attributes as the minimal condition of possibility of intrinsic worth and, as a result, assigns to all the objects that inherit these attributes a corresponding, minimal degree of absolute moral value, in the following sense. Here 'absolute' still means not relative, as in the Kantian 'question' (see question (2)). However, in (2) or more generally in Kant's axiology, the intrinsic value of an entity is incomparable because it is unique, in the sense that there are no other types of moral value, and hence, *a fortiori*, it cannot be increased or overridden on the basis of considerations involving other levels or degrees of moral value. On the contrary, here the minimal intrinsic worth of an entity is incomparable because it is unique in the sense that it can be reduced no further, it is necessarily shared universally by all entities that may have any intrinsic value at all, and it deserves to be respected by default yet only *ceteris paribus*, that is to say, it can be overridden in view of considerations involving other degrees of moral value at lower LoAs. Entities are more or less morally respectable. We shall see in a moment that an action too can admit of degrees of moral difference, since it becomes less respectable as it generates more metaphysical entropy.

6.3.3 IE's axiological ecumenism

Two types of answers can now address the question 'what entities have moral value and hence deserve respect?' One is maximalist or Kantian, and the other is minimalist, depending on what we mean by 'moral value'. Minimalist theories of intrinsic value have tried in various ways to identify in various ways the inherited attributes—i.e. the minimal condition of possibility of the lowest possible degree of intrinsic value—without which an entity becomes intrinsically worthless, and hence deserves no moral respect. Investigations have led researchers to move from more restricted to more inclusive, anthropocentric criteria, and then further on towards biocentric criteria. As the most recent stage in this dialectical development, IE maintains that even biocentric analyses of the inherited attributes are still biased and too restricted in scope. As deep ecologists argue, inanimate things too can have an intrinsic value. In 1968, Lynn White famously asked:

> Do people have ethical obligations toward rocks? ... To almost all Americans, still saturated with ideas historically dominant in Christianity ... the question makes no sense at all. If the time comes when to any considerable group of us such a question is no longer ridiculous, we may be on the verge of a change of value structures that will make possible measures to cope with the growing ecologic crisis. One hopes that there is enough time left. (1973, p. 63)

Today, the Geological Code of ethics states, for example, '(9) Don't disfigure rock surfaces with brightly painted numbers, symbols or clusters of core-holes'[17] for apparently no other reason than a basic sense of respect for the environment in all its forms. Likewise, in many ethical codes for librarians and other library employees adopted by national library or librarians' associations or implemented by government agencies, 'informational entities' are considered to have a moral value and deserve respect. For example, the Italian Library Association (AIB) has endorsed a 'Librarian's Code of Conduct' that is divided into three sections, 'Duties toward the User', 'Duties toward the Profession', and 'Duties toward Documents and Information', where it is stated, in Section 3.1, that '[t]he librarian undertakes to promote the enhancement and preservation of documents and information'. Indeed, even ideal, intangible, or intellectual objects are acknowledged to have a minimal degree of moral value, no matter how humble, and so are entitled to some respect. UNESCO also recognizes this in its protection of 'masterpieces of the oral and intangible heritage of humanity'.[18] What lies behind these examples is the view that if some x can be a moral patient P, then x's nature can be taken into consideration by A, and contribute to the ethical shaping of A's actions, no matter how minimally. The minimal criterion for qualifying as an entity that, as a patient, may rightly claim some degree of respect, is more general than any biocentric reference to the entity's attributes as a biological or living entity. What, then, is the most general possible common set of attributes that characterize an entity as intrinsically valuable and an object of respect, without which an entity would rightly be considered intrinsically worthless (not just instrumentally useless or emotionally insignificant) or even positively unworthy and therefore rightly to be disrespected in itself? The least biased and most fundamental solution is to identify the minimal condition of possibility of an entity's least intrinsic worth with its nature as an informational entity by adopting an informational ontology. The informational nature of an entity x that may, in principle, have the role of a patient P of a moral action is the lowest threshold of inherited attributes that constitutes its minimal intrinsic worth, which in turn may deserve to be respected by the agent. Alternatively, and to put it more concisely, being an entity *qua* informational entity is the minimal condition of possibility of moral worth and hence of normative respect. In more metaphysical terms, IE argues that all aspects and instances of *Being* understood informationally are worth some initial, perhaps minimal and overridable, form of moral respect. This is the central axiological thesis

---

[17] See e.g. Earth Lab Database, 'Geological Code' <www.nhm.ac.uk/nature-online/earth/rock-minerals/earthlab/support/geolcode.html>. Versions of the code can be easily googled. Sometimes the rule has a different number and it is shortened, thus: 'Don't disfigure rock surfaces with numbers or symbols in brightly coloured paint.' See further the 'Principles of Archaeological Ethics' adopted by the Society for American Archaeology, *The International Journal of Cultural Property*, or the ICOM (International Council of Museums) Code of Professional Ethics.

[18] See e.g. UNESCO, 'Intangible Heritage' <www.unesco.org/culture/heritage/intangible/>.

of any future information ethics that will emerge as a macroethics, to adapt another typical Kantian phrase.

6.4 Five objections

The reader who finds the previous line of reasoning too unfamiliar to begin assessing its value may find the following three comparisons helpful. Consider a Berkeleian ethics: the rocks mentioned by Lynn White are part of God's mind, so defacing them means defacing part of God's ultimate database within which we are operating as other minds. Consider a Leibnizian ethics: we should respect rocks insofar as they are monads like us, where monads are today better understood as simple informational structures. Finally, consider Spinoza's ethics: we should respect rocks insofar as we share with them the same *natura naturata*. Recall what was argued in Chapter 2: agents, and in particular informational organisms like us, are not outside the infosphere but part of it. I shall return to this point more extensively in Chapter 15, when talking about the ontic trust, the hypothetical pact between all agents and patients presupposed by IE.

The more sceptical reader, who has understood the previous line of reasoning in favour of IE, and yet still finds IE's position controversial, may have in mind several objections. Five of them seem particularly cogent. I shall answer them here because they should further clarify IE and help to make it more acceptable to those who are not yet convinced of its merits. I shall address other objections in Chapter 16.

6.4.1 The need for an ontology

The first objection concerns the problems in the development of a user-oriented information ontology that might help CE to deal with ICTs-related moral issues.

According to IE, the least (i.e. not further reducible), unconditional (i.e. neither instrumental nor emotional), intrinsic (i.e. belonging to its inherited essence in the OOP sense), and absolute (as clarified above) value of any entity x, which in principle may fulfil the role of patient P of a moral action, consists in x's nature *qua* informational entity and in the very fact of being a possible patient of A's action. On the one hand, the effect of x's role as P is completely exhausted in inducing A's respect. On the other hand, once P is interpreted as an informational entity x, understanding in detail how P's moral value can contribute to the configuration of A's action in some specific circumstances seems to require an information ontology; namely a theory of the intrinsic attributes of an informational entity and their integrity, understood as unimpaired and uncorrupted unity and persistence[19] across time.

If the objection is now that the need for an ontology affects only IE, it is obviously mistaken. Every macroethics is based on a specific ontology. For example, Aristotle, Kant, Mill, and environmentalist theorists who privilege the human or biological

---

[19] Adapting another OOP concept, persistence can here be defined as the property of any object that outlives the process that generates it.

nature of *P* as the grounds of *P*'s value are all using specific anthropological, psychological, physiological, or biological ontologies.

If the objection is that IE would find developing an informational ontology an impossible task, again it is mistaken. One of the main reasons to adopt OOP as a modelling methodology is precisely because it exemplifies the kind of theoretically powerful approach needed to develop successfully an information ontology that is not ethically pre-loaded or biased.

If the objection is that IE needs to provide its own ontology in order to avoid being normatively empty, it is still mistaken. By suggesting that informational entities may require respect even if they do not share human or biological properties, IE provides a general frame for moral evaluation, not a list of commandments or detailed prescriptions (compare this to the similar complaint of 'emptiness' made against deontological approaches). In the following chapters, this frame will be built in terms of a notion of ethical eco-informational stewardship that will prove consonant with the four universal principles against *metaphysical entropy* or evil I introduced in Chapter 4. That said, probably the best way to respond to this objection concerning the need for an adequate ontology is to remind the reader that much work still needs to be done to develop IE in full. This is correct, and I hope that the remainder of this book contributes to such an effort.

6.4.2 *How can an informational entity have 'a good of its own'?*

This objection is loosely based on Taylor (1981) and Taylor (2011). Here is the outline:

(i) an entity *x* is subject to moral respect if and only if *x* has an intrinsic value
(ii) *x* has an intrinsic value if and only if
   (ii.a) *x* 'has a good of its own', that is, *x* can be benefited or harmed; and
   (ii.b) *x*'s flourishing is a good thing
(iii) biological entities (Taylor's 'teleological centres of life'), including non-sentient beings, satisfy (ii.a) and (ii.b)
(iv) it follows that biological entities have an intrinsic value[20] and hence are subject to moral respect
(v) non-biological entities, including informational entities, fail to satisfy (ii.a) and therefore (ii.b)
(vi) it follows that non-biological entities do not have any intrinsic value and are not subject to moral respect.

This argument is *designed to promote an enlargement* of the domain of entities subject to moral respect, so as to include animals and plants (argument *ad includendum*). It does so by means of condition (ii), which is basically an instruction to adopt a higher LoA than the anthropocentric one. As for the rest of the Kantian axiological frame, the argument

---

[20] This is to be understood in perfectionist terms, following Sumner (1996).

strives to keep everything unchanged. In particular, 'intrinsic value' and 'moral respect' are treated as binary phenomena, which can either be present or absent but which admit no degrees. Judged in terms of its goal, the argument *ad includendum* may seem reasonable and convincing. Its weakness emerges in (v) and (vi), when the argument *has the effect of excluding* anything else that cannot and should not be subject to moral respect (argument *ad excludendum*).[21]

Regarding (v), anyone endorsing the argument must also accept that a company, a party, a family, or a nation can all satisfy both (ii.a) and (ii.b), and hence that premise (v) is unjustified. Recall that the argument is meant to show that some non-sentient beings also qualify as morally respectable. What (v) should state is that non-teleological entities fail to satisfy (ii.a) and (ii.b). But now, what are we supposed to conclude about artificial systems like software agents in cyberspace, which are endowed with teleological capacities? From a strictly biocentric perspective, the argument is too permissive.

Regarding (vi), the argument purports to show that anything whose ontological status is either 'higher' or 'lower' than that of a biological entity must inevitably be excluded from moral considerations concerning its intrinsic value and respectability. This is probably wrong. If God exists (and this is a conditional statement), God certainly does qualify as an entity with intrinsic moral value, deserving to be respected. And yet, God cannot be benefited or harmed, at least not in the teleological sense required by the argument. God cannot flourish either. So, according to the argument, God has no intrinsic value and is not morally respectable. A less stringent but similar case can be made for physical objects like the two giant Buddha statues near Bamiyan. According to the argument, they have no intrinsic value and do not qualify for any degree of moral respect.

One can always bite the bullet, but it seems that something has gone badly wrong with the argument. The fact is that condition (ii) is too strong and rather *ad hoc*. In order to defend the moral respectability of biological entities, it introduces an unnecessarily strict *teleological bias*, which requires x to have the capacities to interact with the environment, to go through a cycle of various developmental states, and to pursue goals for its own good. Now, adding a robust dose of teleology certainly does the trick and (ii) succeeds in enlarging the domain of morally respectable entities, but the approach is too strong and backfires. The enlargement is obtained at the expense of non-biological entities, both above (God, gods, angels, etc.) and below (the Buddha statues) that one may not have any reason to exclude in principle. This is an unreasonable cost, once we realize that the argument is at the same time very ecumenical when it comes to a variety of teleological systems, including artificial and social agents.

To fix the argument, one needs to invert the relation between x having an intrinsic value and x having a good of its own. If x has a good of its own and x's flourishing is a good thing, then x has an intrinsic value, not *vice versa*, and certainly not 'if and only if'.

---

[21] For an environmentalist position that accepts the argument *ad includendum* but rejects the argument *ad excludendum* see Rolston III (1985).

But then this inversion requires a re-consideration of the teleological component in (ii). The proper LoA is not represented by the analysis of what x dynamically strives to be, but rather by the properties that x has as an entity, even statically. Therefore, the correct terminology to express this point should not be biocentrically biased in the first place. After all, the harm/benefit pair is only a biocentric and teleological kind of the more general pair damage/enhancement. Here is how the argument should be revised:

(i) an entity x is subject to moral respect if and only if x has an intrinsic value
(ii) x has an intrinsic value if and only if
    (ii.a) x 'has a good of its own', that is, x can be enhanced or damaged; and
    (ii.b) x's existence as x is a good thing
(iii) all things, understood as informational entities, satisfy (ii.a) and (ii.b)
(iv) it follows that all things—i.e. all informational entities—have some intrinsic value and are subject to some moral respect.

The new version is no longer a biocentric objection against IE but actually an ontocentric argument in its favour. It now fosters moral respect not only for a spider, but also for God (if God exists), for the two Buddha statues, for Mary's corpse, and for a database, in short, for the whole infosphere.

Clearing condition (ii) of its biological and teleological bias has at least three consequences. The first two are favourable and show that IE is perfectly coherent with strands of environmental ethics that defend a non-biocentric approach (see e.g. Hepburn (1984) and Stone (2010)).

First, the original argument implicitly assumes that the true value-bearers are only biological individuals, not systems (imagine a whole valley taken as an ecosystem), so moral respect is paid to individuals and only derivatively (instrumentally) to systems encompassing them. In the new version, the argument defends the intrinsic value and moral respectability of systems as well as individuals.

Second, since we now consider the whole domain of existing entities as being subject to some degree of moral respect, it would be unreasonable to assume that they all qualify for exactly the same kind of absolute respect. A biocentric ethics can still adopt a one-dimensional view of value and respect. Once the Kantian scheme collapses, it must be replaced by a non-absolutist, multi-dimensional approach. Things have various degrees of intrinsic value and hence demand various degrees of moral respect, from the low-level represented by an overridable, disinterested, appreciative, and careful attention for the properties of an informational entity like a stone or a data file to the high-level, absolute respect for human dignity. The last consequence is that now the argument is purely *ad includendum*. As such, it may be just too inclusive and turn into a counterargument. The latter could take two forms.

First, one may be reluctant to endorse an 'ontocentric outlook on nature', to adapt Taylor's phrase, because the idea that any entity may enjoy at least a minimal level of moral status may be hard to swallow. Isn't IE unbearably supererogatory? I shall address this problem more fully in Section 16.7. For the time being, I shall limit myself to a clarification of IE's position. One should recall the recurrent qualifications 'overridable', 'minimal', and '*ceteris paribus*' and the crucial importance of what have been called 'levels of abstraction' at which a moral situation is analysed. Environmental ethics accepts culling as a moral practice and does not indicate as one's duty the provision of a vegetarian diet to wild carnivores. IE is equally reasonable: fighting metaphysical entropy is the general moral law to be followed, not an impossible and ridiculous struggle against thermodynamics, or the ultimate benchmark for any moral evaluation, as if human beings had to be treated as mere numbers. We need to adopt an ethics of stewardship towards the whole of reality or infosphere; is this really too demanding, unwise, or unclear? Perhaps we should think twice: is it actually easier to accept the idea that all non-biological entities have no intrinsic value whatsoever? Perhaps we should consider that the ethical game may be more opaque, subtle, and difficult to play than humanity has so far wished to acknowledge. Perhaps we could be less pessimistic: human sensitivity has already improved quite radically in the past and may improve further. Perhaps we should just be cautious: given how fallible we are, it may be better to be too inclusive than too discriminative. In either case, one needs to remember that IE considers agents above all as *creators* and not just *users* of their environment—no matter whether we interpret it in terms of Being, Spinoza's *natura*, nature, or infosphere—and this carries 'demiurgic' responsibilities that may require special theoretical consideration.

Second, one may object that the argument fails to account for the existence of the morally unworthy in general and of evil in particular. Is there anything that actually does not qualify as intrinsically valuable even in the most minimal sense? At the moment, we are missing a revised version of conditions (v) and (vi). This objection is more substantial than the former and deserves its own separate treatment.

6.4.3 What happened to Evil?

An axiology that accorded some positive degree of intrinsic value, and hence of moral respectability, to literally anything would be of very little interest in itself, because, in so doing, it would clearly fail to make sense of a whole realm of moral facts and the commonly acknowledged presence of worthless and unworthy patients. If IE hopes to be treated as a macroethics, it must provide a negative axiology as well. I shall deal with the concept of evil, and of artificial evil in particular, in Chapter 9. Here, I shall limit myself to discuss it insofar as it can represent an objection against IE's ecumenical axiology.

There seems to be no specific verb in English that fully conveys exactly and only the opposite of 'respect'. Yet, what is needed in the following pages is a tripartite distinction between what is respected, what is disrespected, and that which is neither respected nor disrespected, but 'respect neutral'. So, let us treat 'irrespect' as meaning simply 'lack of both respect and disrespect'. By 'disrespect' and its cognate words one can then refer to the morally justified and active form of 'anti-respect' towards an 'unrespectable' x, which consists in not causing x, preventing x, removing x, or modifying x so that it is no longer to be disrespected. If something is intrinsically *worthless*, then it is simply unrespectable, and it is morally indifferent whether *A* respects it as a *P*. If something is intrinsically *unworthy*, then it is positively to be disrespected inasmuch as it has some degree of 'indignity', and not only is it morally wrong if *A* shows respect for it, or is indifferent to it as *P*, but it is morally right if *A* shows a corresponding degree of disrespect for it, in the technical sense introduced above.

Now, according to IE, something is intrinsically worthless, lacks any moral value, and cannot be a centre of moral respect if and only if it does not have even the minimal status of being an informational entity. But the only meaningful sense in which it is possible to speak of a 'something' that fails to qualify as an informational entity is by speaking of an object that is intrinsically impossible, i.e. a logical contradiction in itself, since it must both be and not be. There are an infinite number of inconsistent objects, but since anything may be predicated of any inconsistent object, there is only one object-type that qualifies as intrinsically worthless and unrespectable. Let us call it C. C represents the zero degree in our scale of moral worth. It indicates the precise sense in which LoA^i is the highest level of abstraction.

Below C, we find anything that has some possible degree of intrinsic unworthiness and is correspondingly to be disrespected. Informational entities can at worst be worthless, never unworthy. Does this mean that the class of unworthy elements is empty? Obviously not. Actions can also be patients and, insofar as they have an informational nature as messages (see the OOP terminology introduced in Section 6.1), it is possible to apply to them what has been said above about the intrinsic worth of informational entities. However, while objects can at worst be intrinsically worthless, messages can also be unworthy and deserve to be proactively disrespected, for the following reason. Messages are not only informational entities in themselves but also processes that affect other informational entities either positively or negatively. They are inherently relational. Let us call messages that respect and take adequate care of the well-being of *P* 'positive messages', and messages that do not respect or take adequate care of the well-being of *P* 'negative messages'. Negative messages are unworthy and hence deserve to be disrespected inasmuch as they 'maltreat' their patients. A message that 'maltreats' *P* is a message that does not respect *P*'s (informational) nature, i.e. a message that increases the metaphysical entropy in the infosphere. It is never morally right to show respect for a negative message, and *A* has a duty to be comparatively disrespectful towards an unworthy message. Furthermore, *A* has a duty not to cause metaphysical entropy and therefore to prevent or remove any message that increases it.

Thus messages (that is, actions or indeed behaviours, in a more standard vocabulary), but not entities, can rightly deserve to be disrespected as intrinsically unworthy. In more metaphysical language, any process that denies existence, *insofar* as it does so, deserves no respect—note that it may still deserve respect for other, overriding reasons—but anything that is, *insofar* as it is, deserves some respect *qua* entity. Ultimate and absolute evil as an entity has no moral value at all, and is simply unrespectable because it is an instance of C; in other words it is logically impossible, for it would have to be an entity without even the minimal status of informational entity. From the perspective of informational structural realism, as exemplified by the OOP approach and articulated in the final chapters of *The Philosophy of Information*, there can be evil only in terms of negative messages, that is, morally bad actions. These are intrinsically more or less to be disrespected, and ought not to be caused, but prevented, removed, or modified in such a way as to become no longer evil. The degree of disrespect that A ought to show towards a negative message is proportionate to the degree of its unworthiness.

Imagine an infosphere in which there were no changes whatsoever: it would contain no evil. This is the IE version of the Platonic thesis concerning the goodness of Being. It clarifies the sense in which something can be extrinsically disrespected: an agent that activates a negative and hence unworthy message is indirectly and contingently deserving of disrespect, but only as a source of that message, hence extrinsically.

The extension of the concept of intrinsic worth to any x *qua* informational entity is now paralleled by the extension of the concept of intrinsic unworthiness to any message *qua* negative process and source of metaphysical entropy. Messages do not need to be intentional to be unworthy and hence deserving of disrespect, so not every natural process deserves to be respected for the simple fact that it is natural. We live in an improvable infosphere, where moral agents have a duty to exercise their ethical stewardship. In this Kant was right: their essential capacity to implement positive messages and disrespect negative ones is precisely what makes them the objects with the highest moral value (dignity).

6.4.4 Is there a communication problem?

In Section 6.3.1, we saw (K.2) that, when there is no asymmetry between agent A and patient P, in principle, A should encounter no conceptual difficulties in recognizing P's moral value, and hence in behaving respectfully. Both entities belong to the same class, share the same essential nature, and hence the same kind of moral value. The process of communication between P's essence, P's moral value, A's respect for P's moral value, and M's adequacy to both A's respect and P's moral value is granted by a *principle of reflective respect*, whereby the agent can recognize in the patient a member of the same ontological community, a sort of 'alter-ego', and thus easily extend to P all the considerations of moral worthiness and requirements of adequate respect that A would expect to be rightly applied to A itself. This reflective respect is at the root of the Golden Rule: Alice can adequately regulate her actions towards Peter in a way which is already morally successful even if she considers only (perhaps just empathically if not rationally) how Peter would like to be treated if she were in his position.

The principle of reflective respect cannot easily be exported when there is an asymmetry in the nature of A and P. Human self-respect and personal interest in one's own well-being provide some guidelines on how to behave towards P that become less and less intuitive the more P is ontologically distant from A. Simplifying, some reflective respect can still be at work when one is dealing with an animal, but much less so when a tree or a mountain is in question (Leopold (1949), see 'Thinking like a Mountain'), and reflective respect becomes truly problematic when the reality one is dealing with is not biological, like a stone or some artefact, two kinds of patients that, according to IE, can still enjoy some minimal moral value *per se* because of their status as informational entities. The risk is falling into some form of naïve anthropomorphism. What seems to be required, on A's side, is a 'transpersonal identification', as environmental ethicists like to say. This 'infophilic', or information-friendly attitude, is rather more abstract and less spontaneous than commonsensical or empathic feelings. To be able to expand 'the ever-widening circle of ecological consciousness' (Nash, 1989) and to appreciate what A has in common with P when P is understood as an informational entity, A should try to transcend A's own particular nature, recognize A's own minimal status as an informational entity as well, and then extend the respect—which A would expect any other agent to pay to A as an informational entity—to any other informational entity that may be the patient of A's actions. All of this requires a change in ethical sensibility. If oversimplified the perspective can easily become absurd or ridiculous. Of course, IE does not argue that smashing a stone or erasing a file is a moral crime in itself. This is just too silly. IE argues that destroying reality or impoverishing it of any of its features can be morally evaluated at different levels of abstraction, that most macroethics work at the low level represented by anthropocentric or biocentric interests (and are perfectly justified in doing so), but that there is also a higher, more minimalist level at which all entities share a lowest common denominator, their nature as informational entities, and, furthermore, that this level too can contribute to our ethical understanding. This means that when any other level of analysis is irrelevant, IE's high LoA is still sufficient to provide the agent with some minimal normative perspective. Putnam's twin earth mental experiment can help to clarify the point. Suppose there is a perfect copy of the world; call it twin earth. Suppose that our world and twin earth differ only in this, that a stone and a data file are destroyed in our world, but are left intact on twin earth. There is absolutely no other difference. IE accepts the view that twin earth would be a slightly, perhaps very slightly, but still recognizably, better place just because it would be an ontologically richer place. The *principle of ontic uniformity* grants that the agent A acknowledges A's membership of the infosphere and so recognizes the inherited attributes A shares with all other informational components of the infosphere as the ontological grounds of their common minimal moral value. The *principle of ontic solidarity* grants that the agent A will treat all elements of the infosphere, including A, as having at least a minimal, overridable moral value *qua* informational entities by default. The moral attitude promoted by IE that emerges from the two principles can be defined, with a play on words, as an 'object-oriented' attitude. In environmental circles this is discussed in terms of a transpersonal, ontological, or cosmological identification with Being (Fox, 1995). I shall return to these topics in Chapters 8, 10, and 14.

6.4.5 Right but irrelevant?

Someone convinced of the coherence of IE's position could still formulate the following objection. The problem about IE is not the theory, but its practical irrelevance: IE is too abstract in the technical sense that its LoA is too high. Recall that IE fully endorses the view that attributing moral value to informational entities provides only a minimalist approach, always overridable in view of moral concerns formulated by other macroethical analyses at lower LoA. Since in everyday life and in ordinary moral decisions there will always be overriding moral concerns, isn't IE completely irrelevant, even if it is right? Let's concede some non-zero degree of moral value to stones and files. If we do, such a degree will be so infinitesimally small to be simply negligible because other conflicting concerns will always override it.

The objection raises an important point, as we shall see at the end of the section, but it is largely unjustified. It is simply false that there are always contrasting and overriding ethical concerns (Benn, 1998). Ethical theories do not necessarily have to disagree and hence compete with each other in their conclusions. In many cases, they are complementary and can enrich each other. This holds true for IE as well. Moreover, IE has its own growing field of application, the world of ICTs; and other theories seem to have difficulties in adapting to this new area. So, in this sense too there may not be overriding concerns. IE calls our attention to problems that will become increasingly important the more de-physicalized and digitalized our environment becomes. In a society that calls itself 'the information society' it is vital to develop an ethical theory that has the conceptual resources to take into account the status of informational entities. IE is an 'architectural' ethics, an ethics addressed not only to the users but also to the creators and designers of the infosphere (see Chapter 8). Human beings have evolved as the most successful manipulators and exploiters of nature. Past macroethics have long recognized this fact and tried to cope with its consequences normatively. But human history is also the history of the *ontic divide*, a history of projects and constructions, of detachment from and rejection of the physical world, of replacement of the natural by a human-made (artificial, literally 'created by craft', or technological) environment. *Eco* means 'home', and the infosphere is the new 'home' that is being constructed for future generations. It is the fastest-growing environment that human beings as informational entities are going to share with other non-biological informational entities. Clearly, an ethical approach to information ecology is badly needed. IE strives to provide a good, unbiased platform from which to educate not only computer science and ICT students, but also the citizens of an information society and the inhabitants of the infosphere. New generations are going to need a mature sense of ethical responsibility for and stewardship of their whole environment, both natural and artificial (that is, in one word, informational), to foster responsible care of it rather than despoliation or mere exploitation.

I said that ultimately the objection does raise an important point. IE's goal is to fill an 'ethical vacuum' brought to light by the ICT revolution, to paraphrase Moor (1985).

The objection reminds us that IE will prove its value only if its applications bear fruit. This is what we are going to see in Chapters 11, 12, 13, and 14. I shall finally return to the main thrust of this objection in Section 16.6.

CONCLUSION

Information Ethics may be understood as a development of environmental ethics (Zimmerman, 2005). Deep ecology argues that the state of inanimate entities should be taken into account when considering the consequences of an action (e.g. how is building a freeway going to impinge on the rock face in its path?). In IE, this inclusive approach is taken further, due largely to the characteristic properties of the infosphere. More than fifty years ago, Leopold defined land ethic as something that

> changes the role of Homo sapiens from conqueror of the land-community to plain member and citizen of it. It implies respect for his fellow-members, and also respect for the community as such. The land ethic simply enlarges the boundaries of the community to include soils, waters, plants, and animals, or collectively: the land. (1949, p. 403)

The time has come to translate environmental ethics into terms of *infosphere* and *informational entities*, for the land we now inhabit is not just the earth.

We saw in this chapter that IE is ontologically committed to an informational modelling of *Being* as the whole infosphere. The result is that no aspect of reality is extraneous to IE and that the whole environment is taken into consideration. For whatever is in the infosphere is informational (better: is accessed and modelled informationally) and whatever is not in the infosphere is something that cannot be. We also saw some of the reasons why IE seeks to translate environmental ethics into informational terms and to expand its scope in order to be as ecumenical as possible. The goal is that of including not only living organisms and their habitats, but also inanimate things and artefacts. In a universe in which the natural is actually increasingly man-made and conceptualized informationally, this seems the right ecological perspective to adopt. It is also a perspective that makes much sense from many religious and spiritual traditions, including, but not only, the Judeo-Christian one, for which the whole universe is God's creation, is permeated by the divine, and is a gift to humanity worthy of care. It would be a better world, one in which human moral agents could see themselves as guests in the house of Being. As we shall see in the next chapter, they are not the only ones that can act morally.

7

The morality of artificial agents

> The machine, like the living organism, is, as I have said, a device which locally and temporarily seems to resist the general tendency for the increase in entropy. By its ability to make decisions it can produce around it a local zone of organisation in a world whose general tendency is to run down.
>
> Norbert Wiener, *The Human Use of Human Beings* (1954), p. 34.

SUMMARY
Previously, in Chapter 6, I argued that the whole infosphere should count as a patient of our moral respect. When it comes to drawing a line above which something is intrinsically valuable and deserves to be respected, and below which something is intrinsically worthless, or indeed deserves to be proactively disrespected, I argued that such a line should simply be erased from the domain of entities, none of which is by default utterly worthless or worthy of our disrespect, and that the line needs to be redrawn, as a divide, between moral and immoral actions. The outcome was an expansion of the domain of moral patients? The next question to be addressed is what happens to the domain of moral agents. What sort of moral agents inhabit the infosphere? The short answer is that, in the infosphere, or at an informational LoA of reality, moral agents are any *interactive*, *autonomous*, and *adaptable transition systems* that can perform *morally qualifiable actions*. As usual, all this will require some work to be fully explained and defended. In short, the thesis I shall support is that artificial agents (AAs), particularly but not only those in cyberspace, extend the class of entities that can act in moral situations. For they can be conceived not only as moral patients—as entities that can be acted upon for good or evil—but also as moral agents—as entities that can perform actions, again for good or evil.

    Many artificial agents are not moral agents, of course. But some are, and others could easily be, if built to be so and allowed to develop. This is not sci-fi, and that is why we need to be very careful. It is not clear that if one could identify a level of abstraction at which the behaviour of an agent could be ascribed explicitly to its designer, then one should conclude that that agent would not be a moral agent. For, if this were the case, one day neuroscience may force us to conclude that there are no moral agents at all, not even human, but only agents whose actions have moral impact. Any sort of Turing test is based on the assumption that a particular phenomenon is given (e.g. *human* intelligence, moral agency, or creativity) and then proceeds by *comparison* with another, possibly indistinguishable, unknown phenomenon (e.g. *artificial* intelligence, moral agency, or creativity). It is not based on an attempt to define what something is in itself (e.g. intelligence, moral agency, or creativity).

In this chapter, I shall argue that there are clear and uncontroversial cases in which an artificial agent may qualify as a moral agent. This does *not* relieve the creator of that agent of responsibility. When moral artificial agents are in question, what counts is their moral *accountability*. This is not philosophical hair-splitting. Parents, for example, may still be *responsible* for the way in which their adult children behave, but they are certainly not *accountable*. They might be bitterly blamed, but they will not go to prison if their son, now in his thirties, turns out to be a serial killer. Likewise, engineers will be *responsible* for what and how they design artificial agents, even if they may not be *accountable*. The sooner we take this on board the better.

In Sections 7.1 and 7.2, I shall analyse the concept of agent by employing the method of abstraction explained in Chapter 3. I shall then clarify the concept of 'moral agent' by providing not a definition but an effective characterization (more on this distinction later) based on three criteria at a specified LoA. The resulting concept of a moral agent will then be used to argue that AAs, though neither cognitively intelligent nor morally responsible, can be fully *accountable* sources of moral action. Also in Section 7.2, I shall provide some examples of the properties specified by a correct characterization of agency and, in particular, of AAs along with some further examples of LoAs. In Section 7.3, I shall present a definition of moral agent and argue that there are substantial and important reasons for acknowledging a concept of moral agent that does not necessarily exhibit free will, mental states or responsibility, what I shall label 'mindless morality'. Morality is usually predicated upon *responsibility*. The use of the method of abstraction, LoAs, and thresholds enables *responsibility* and *accountability* to be decoupled and formalized effectively. The part played in morality by responsibility and accountability can then be clarified as a result. In Section 7.4, I shall further clarify some important consequences of the approach defended in this chapter by addressing four objections. Finally, in Section 7.5, I shall model morality as a 'threshold', which is defined according to the observables determining the LoA under consideration. An agent is morally good if its actions respect that threshold; and it is morally evil insofar as its actions violate it, causing metaphysical entropy in the infosphere. The concept of threshold will play a big role in Chapter 13.

7.1 Introduction: standard vs. non-standard theories of agents and patients

Moral situations commonly involve agents and patients. Following the same approach already adopted in Chapter 6, let us define the class A of moral *agents* as the class of all entities that can, in principle, qualify as sources or senders of moral actions, and the class

136   THE ETHICS OF INFORMATION

P of moral *patients* as the class of all entities that can, in principle, qualify as targets or receivers of such moral actions. A particularly apt way to introduce the topic of this chapter is to consider how different macroethics interpret the logical relation between those two classes.

Of course, there can be only five logical relations between *A* and *P* (see Figure 11). It is possible, but utterly unrealistic, that *A* and *P* are disjoint (alternative 5). On the other hand, *P* can be a proper subset of *A* (alternative 3), or *A* and *P* can intersect each other as shown in alternative 4. These two alternatives are slightly more promising because they both require at least one moral agent that, in principle, could not qualify as a moral patient. Now this pure agent would be some sort of supernatural entity that, like Aristotle's God, affects the world but can never be affected by it. However, being in principle 'unaffectable' and irrelevant in the moral game, it is unclear what kind of role this entity would exercise with respect to the normative guidance of human actions. So it is not surprising that most macroethics have kept away from these 'supernatural' speculations and implicitly adopted, or even explicitly argued for, one of the two remaining alternatives: *A* and *P* can be equal (alternative 1), or *A* can be a proper subset of *P* (alternative 2).

Alternative (1) maintains that all entities that qualify as moral agents also qualify as moral patients and *vice versa*. It corresponds to a rather intuitive position, according to which the agent/inquirer plays the role of the moral protagonist. We, human moral agents who also investigate the nature of morality, place ourselves at the centre of the moral game as the only players who can act morally, be acted upon morally, and, in the end, theorize about all this. It is one of the most popular views in the history of ethics, shared, for example, by many Christian ethicists in general and by Kant in particular, as we have seen in the previous chapter. I shall refer to it as the *standard position*.

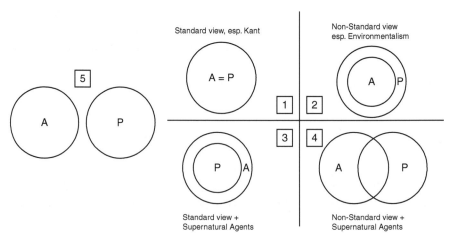

Figure 11. The logical relations between classes of moral agents and patients

Alternative (2) holds that all entities that qualify as moral agents also qualify as moral patients but not *vice versa*. Many entities, most notably animals, seem to qualify as moral patients, even if they are in principle excluded from playing the role of moral agents. This post-environmentalist approach requires a change in perspective, from agent orientation to patient orientation. In view of the previous label, I shall refer to it as *non-standard*.

In recent years, non-standard macroethics has been discussing the scope of P quite extensively. The more inclusive P is, the 'greener' or 'deeper' the approach has been deemed. Environmental ethics[1] has developed since the 1960s as a study of the moral relationships of human beings to the environment (including its non-human contents and inhabitants) and its (possible) values and moral status. It often represents a challenge to anthropocentric approaches embedded in some traditional, Western ethical thinking. Since this was the topic of the previous chapter, I shall not return to it here.

Comparatively little work has been done in reconsidering the nature of moral agency and hence the extension of A. Post-environmentalist thought, in striving for a fully naturalized ethics, has implicitly rejected the relevance, if not the possibility, of supernatural agents, while the plausibility and importance of other types of moral agency seem to have been largely disregarded (as we shall see below). Secularism has contracted (some would say deflated) A, while environmentalism has justifiably expanded only P, so the gap between A and P has been widening; this has been accompanied by an enormous increase in the moral responsibility of the individual, as I shall argue in Chapters 8, 10, and 14.

Some efforts have been made to redress this situation. In particular, the concept of 'moral agent' has been expanded to include both natural and legal persons, especially in business ethics. A has then been extended to include agents like partnerships, governments, or corporations, for which legal rights and duties have been recognized (see Chapter 13). This more ecumenical approach has restored some balance between A and P. A company can now be held directly accountable for what happens to the environment, for example. Yet the approach has remained unduly constrained by its anthropocentric conception of agency. An entity is still considered a moral agent only if

(1) it is an individual agent; and
(2) it is human-based, in the sense that it is either human or at least reducible to an identifiable aggregation of human beings, who remain the only morally responsible sources of action, like ghosts in the legal machine.

Limiting the ethical discourse to *individual* agents hinders the development of a satisfactory investigation of *distributed morality*, a macroscopic and growing phenomenon of global moral actions and collective responsibilities resulting from the 'invisible

---

[1] For an excellent introduction see Jamieson (2008).

hand' of systemic interactions among several agents at a local level (see Chapter 13). Insisting on the necessarily *human-based nature* of the individual agents involved in any moral analysis means undermining the possibility of understanding another major transformation in the ethical field, the appearance of artificial agents (AAs). These are sufficiently informed, 'smart', autonomous artefacts, able to perform morally relevant actions independently of the humans who engineered them, causing 'artificial good' and 'artificial evil' (see Chapter 9; Gips, 1995; Simon, 2012). Both constraints can be eliminated by fully revising the concept of 'moral agent'. This is the task undertaken in the rest of this chapter. The main theses defended are that AAs are legitimate sources of moral and immoral actions; hence that the class A of moral agents should be extended so as to include AAs; that the ethical discourse should include the analysis of their morality, and therefore of their design, deployment, control, and behaviour; and, finally, that this analysis is essential in order to understand a range of new moral problems not only in information ethics, but also in ethics in general, especially in the case of distributed morality, as we shall see in Chapter 13.

7.2 What is an agent?

In recent years, the scope of the concept of 'moral agent' has been expanded to include both natural and legal persons (Allgrove, 2004; Barfield, 2005; Koops et al., 2010). The debate is not entirely new (Donaldson, 1982; May, 1983), nor devoid of controversial points (Ewin, 1991). Its revival is due to the increasing pervasiveness and autonomy of artificial agents and of hybrid multi-agent systems (Wooldridge, 2009a), both in everyday contexts and in business environments (Andrade et al., 2004, 2007; Hildebrandt, 2008, 2011; Wallach and Allen, 2009a; Verbeek, 2011; Kroes and Verbeek, forthcoming). But what exactly is an 'agent'?

Complex biochemical compounds and abstruse mathematical concepts have at least one thing in common: they may be unintuitive, but once understood they are all definable with total precision by listing a finite number of necessary and sufficient properties. Mundane entities like intelligent beings or living systems share the opposite property: one naïvely knows what they are and perhaps could be, and yet there seems to be no way of encasing them within the usual planks of necessary and sufficient conditions. This holds true for the general concept of 'agent' as well. People disagree on what may count as an 'agent', even in principle (see e.g. Franklin and Graesser, 1997; Davidsson and Johansson, 2005; Moya and Tolk, 2007; Barandiaran et al., 2009). Why?

Sometimes, the problem is addressed optimistically, as if it were just a matter of further shaping and sharpening whatever necessary and sufficient conditions are required to obtain a *definiens* that is finally watertight. Stretch here, cut there; ultimate agreement is only a matter of time, patience, and cleverness. In fact, attempts follow one another without a final identikit ever being nailed to the *definiendum* in question. After a while, one starts suspecting that there might be something wrong with this *ad hoc* approach. Perhaps it is not the Procrustean *definiens* that needs fixing, but the Protean *definiendum*.

Other times, the intrinsic fuzziness of the problem is being blamed. One cannot define with sufficient accuracy things like life, intelligence, agency, and mind because they all admit of subtle degrees and continuous changes.[2]

A solution is to give up altogether or at best resign oneself to being vague and rely on indicative examples. Pessimism follows optimism, but it need not. The fact is that, in the exact discipline of mathematics, for example, definitions are 'parameterized' by generic sets. That technique provides a method for regulating the relevant LoA. Indeed, abstraction acts as a 'hidden parameter' behind exact definitions and makes a crucial difference. Thus, each *definiens* comes pre-formatted by an implicit LoA; it is stabilized, as it were, in order to allow a proper definition. An x is defined or identified as y never absolutely (i.e. LoA-independently), as a Kantian 'thing-in-itself', but always contextually, as a function of a given LoA, whether it be in the realm of Euclidean geometry, quantum physics, or commonsensical perception.

When a LoA is sufficiently common, important, dominating, or in fact happens to be the very frame that constructs the *definiendum*, it becomes 'transparent' to the user, and one has the pleasant impression that x can be subject to an adequate definition in a sort of conceptual vacuum. Glass is not a solid but a liquid, tomatoes are not vegetables but berries, a banana plant is a kind of grass, and whales are mammals not fish. Unintuitive as such views might be initially, they are all accepted without further complaint because one silently bows to the uncontroversial predominance of the corresponding LoA.

When no LoA is predominant or constitutive, things get messy. In this case, the trick does not lie in fiddling with the *definiens* or blaming the *definiendum*, but in deciding on an adequate LoA before embarking on the task of understanding the nature of the *definiendum*.

The example of intelligence or 'thinking' behaviour is enlightening. One might define 'intelligence' in a myriad of ways; many LoAs seem equally convincing but no single, absolute definition is adequate in every context. Turing (1950) avoided the problem of 'defining' intelligence by first fixing a LoA—in this case a dialogue conducted by computer interface, with response time taken into account—and then establishing the necessary and sufficient conditions for a computing system to count as intelligent at that LoA: the imitation game. As I argued in Chapter 3, the LoA is crucial, and changing it changes the test. An example is provided by the Loebner test (Moor, 2001a), the current competitive incarnation of Turing's test. There, the LoA includes a particular format for questions, a mixture of human and non-human players, and precise scoring that takes into account repeated trials. One result of reconceiving the LoA has been the inclusion of chatbots, something unfeasible at Turing's original LoA.

---

[2] See e.g. Bedau (1996) for a discussion of alternatives to necessary-and-sufficient definitions in the case of life.

The conclusion is that some *definienda* come pre-formatted by transparent LoAs. They are subject to definition in terms of necessary and sufficient conditions. Some other *definienda* require the explicit acceptance of a given LoA as a pre-condition for their analysis. They are subject to *effective characterization*. Arguably, agency is one of the latter. So let us apply the method of abstraction to its analysis.

7.2.1 An effective characterization of agents

Whether A (the class of moral agents) needs to be expanded depends on what qualifies as a moral agent, and this, in turn, depends on the specific LoA at which one chooses to analyse and discuss a particular entity and its context. Since human beings count as standard moral agents, the right LoA for the analysis of moral agency must accommodate this fact. Theories that extend A to include supernatural agents adopt a LoA that is equal to or lower than the LoA at which human beings qualify as moral agents. Our strategy is more minimalist and develops in the opposite direction.

Consider what makes a human being like Alice not a moral agent to begin with, but just an agent. Described at this LoA_1, Alice is an agent if and only if she is a system, embedded in an environment that she can transform, produces an effect or exerts power on it, as contrasted with a system that is (at least initially) acted on or responds to it, called the patient. Table 2 provides a more formal definition.

Table 2. The definition of an agent at LoA_1

A) Agent $=_{def.}$ a system, situated within and a part of an environment, which initiates a transformation, produces an effect, or exerts power on it over time.

At LoA_1, there is no difference between Alice and an earthquake. There should not be. Earthquakes, however, can hardly count as agents, so LoA_1 is too high for our purposes: it abstracts too many properties. What needs to be re-instantiated? In agreement with recent literature (Danielson, 1992; Allen et al., 2000; Wallach and Allen, 2009b), I shall argue that the right LoA is probably one which includes the following three criteria: (a) *interactivity*, (b) *autonomy*, and (c) *adaptability*:

(a) *interactivity* means that the agent and its environment (can) act upon each other. Typical examples include the input or output of a value, or simultaneous engagement of an action by both agent and patient—for example, robots in a car plant;

b) *autonomy* means that the agent is able to change its state without direct response to interaction: it can perform internal transitions to change its state. So an agent must have at least two states.

Autonomy imbues an agent with some degree of complexity and independence from its environment and from those who build the agent. For example, the programmers of *Deep Blue* were only indirectly responsible for its win, since it 'learnt' by being exposed to volumes of games to such an extent that the programmers themselves were quite unable to explain, in any terms of chess parlance, how *Deep Blue* specifically played (King, 1997); and finally

(c) *adaptability* means that the agent's interactions (can) change the transition rules by which it changes state.

Adaptability ensures that an agent might be viewed, at the given LoA, as learning its own mode of operation in a way that depends critically on its experience. Mitchell (1997) listed several examples of adaptive agents, including data-mining programs that learn to detect fraudulent credit-card transactions, information-filtering programs that learn users' reading preferences, and autonomous vehicles that learn to drive on public roads. Today, adaptable agents range from mechanisms that adjust to different terrains to systems with statistically adaptive reconfigurable logic arrays.[3] Note that if an agent's transition rules are stored as part of its internal state, discernible at this LoA, then adaptability may follow from the other two conditions.

7.2.2 Examples

Let us now look at six illustrative examples that serve different purposes. First, I shall provide some examples of entities that fail to qualify as agents by systematically violating each of the three conditions. This will help to highlight the nature of the contribution of each condition. Second, I shall offer an example of an information system that forms an agent at one LoA but not at another, equally natural, LoA. That example is useful because it shows how 'machine learning' can enable a system to achieve adaptability. In the third example, I show that digital software agents are now part of everyday life. The fourth example illustrates how an everyday physical device might conceivably be modified into an agent, whilst the fifth provides an example which has already benefited from that modification, at least in the laboratory. The last example concerns an entirely different kind of agent: an organization.

1) Entities that are not agents For the purpose of understanding what each of the three conditions (interactivity, autonomy, and adaptability) adds to our definition of agent, let us first consider some examples satisfying each possible combination of those properties. For the sake of simplicity, let us imagine that we are observing the system under analysis at the same LoA, which is assumed to consist of observations made through a typical video camera over a period of, say, 30 seconds. In this way, we abstract tactile observables, longer-term effects, what we know already about the potential agents in question, etc. Table 3 summarizes our observations. All three conditions are satisfied simultaneously and hence illustrate agency only in the last row.

Recall that a property, for example interaction, is to be judged only via observables. Thus, at the LoA adopted we cannot infer that a rock interacts with its environment by

---

[3] In fiction, adaptive robots occur in the work of James P Hogan—e.g. Hogan (1997) in which a semi-intelligent system controls a production line as part of a space station and, under pressure of attack, designs and produces different kinds of robot—and the popular film *Terminator 2*, in which the shape-shifting cyborg, T-1000, is sent back from the future to kill John Connor before he can grow up to lead the resistance.

Table 3. Examples of agents

LoA = observations through a video camera over a period of 30 seconds

interactive	autonomous	adaptable	examples
No	No	No	Rock
No	No	Yes	?
No	Yes	No	Pendulum
No	Yes	Yes	Solar system, closed system
Yes	No	No	Postbox, Mill
Yes	No	Yes	Thermostat
Yes	Yes	No	Juggernaut[4]
Yes	Yes	Yes	Alice

virtue of reflected light, for this observation belongs to a much finer LoA. Alternatively, were long-term effects to be discernible, then a rock would be interactive, since interaction with its environment (e.g. erosion) could be observed. No example has been provided of a non-interactive, nonautonomous, but adaptive entity. This is because, at that LoA, it is difficult to conceive of an entity which adapts without interaction and autonomy.

2) Noughts and crosses The distinction between the change of state (required by autonomy) and the change of transition rules (required by adaptability) is one in which the LoA plays a crucial role. To explain it, it is useful to discuss a more extended, classic example. This was originally developed by Michie (1961) to discuss the concept of a mechanism's capacities to adapt. It provides a good introduction to the concept of machine learning, the research area in computer science that studies adaptability.

Menace (Matchbox Educable Noughts and Crosses Engine) is a system that learns to play noughts and crosses (also known as tic-tac-toe) by repetition of many games. Nowadays it would be realized by program,[5] but Michie modelled Menace using matchboxes and beads (see Figure 12) and it is probably easier to understand it in that form.

Suppose Menace plays O and its opponent plays X, so that we can concentrate entirely on plays of O. Initially, the board is empty with O to play. Taking into account symmetrically equivalent positions, there are three possible initial plays for O. The state of the game consists of the current position of the board. We do not need to augment that with the name, O or X, of the side playing next, since we consider the board only when O is to play. All together there are 304 such states, and Menace contains a

---

[4] 'Juggernaut' is the name for Vishnu, the Hindu god, meaning 'Lord of the World'. A statue of the god is annually carried in procession on a very large and heavy vehicle. It is believed that devotees threw themselves beneath its wheels, hence the word 'Juggernaut' has acquired the meaning of 'massive and irresistible force or object that crushes whatever is in its path'.
[5] See e.g. Chesnokov, Y., 'Matchbox Educable Noughts And Crosses Engine (MENACE) in C++', Code Project, 21 October 2007 <www.codeproject.com/KB/cpp/ccross.aspx>.

Figure 12. MENACE (Matchbox Educable Noughts and Crosses Engine). Courtesy of James Bridle to whom I am grateful for his kind permission to reproduce his photograph

matchbox for each. In each box are beads, which represent the plays O can make from that state. At most, nine different plays are possible, and Menace encodes each with a coloured bead. Those that cannot be made (because the squares are already full in the current state) are removed from the box for that state. This provides Menace with a built-in knowledge of legal plays. In fact Menace could easily be adapted to start with no such knowledge and to learn it.

O's initial play is made by selecting the box representing the empty board and choosing from it a bead at random. That determines O's play. Next X plays. Then Menace repeats its method of determining O's next play. After at most five plays for O, the game ends in either a draw or a win, either for O or for X. Now that the game is complete, Menace updates the state of the (at most five) boxes used during the game as follows. If X won, then in order to make Menace less likely to make the same plays from those states again, a bead representing its play from each box is removed. If O drew, then conversely each bead representing a play is duplicated; and if O won each bead is quadruplicated. Now the next game is played.

After enough games, it simply becomes impossible for the random selection of O's next play to produce a losing play. Menace has learnt to play, which, for noughts and crosses, means never losing. The initial state of the boxes was prescribed for Menace. Here, we assume merely that it contains a sufficient variety of beads for all legal plays to be made, since then the frequency of beads affects only the rate at which Menace learns.

The state of Menace (as distinct from the state of the game) consists of the state of each box, the state of the game, and the list of boxes that have been used so far in the current game. Its transition rule consists of the probabilistic choice of play (i.e. bead) from the current state box, that (the choice) evolves as the states of the boxes evolve. Let us now consider Menace at three LoAs.

(1) The single game LoA. Observables are the state of the game at each turn and (in particular) its outcome. All knowledge of the state of Menace's boxes (and hence of its transition rule) is abstracted. The board after X's play constitutes input to Menace and after O's play constitutes its output. Menace is thus interactive, autonomous (indeed state update, determined by the transition rule, appears nondeterministic at this LoA), but not adaptive because we have no way of observing how Menace determines its next play and no way of iterating games to infer that it changes with repeated games.

(2) The tournament LoA. Now a sequence of games is observed, each as above, and with it a sequence of results. As before, Menace is interactive and autonomous. But now the sequence of results reveals (by any of the standard statistical methods) that the rule by which Menace resolves the nondeterministic choice of play evolves. Thus, at this LoA, Menace is also adaptive and hence an agent. Interesting examples of adaptable AAs from contemporary science fiction include the computer in *War Games* (1983, directed by J. Badham), which learns, by playing noughts and crosses, the futility of war in general; and the smart building in Kerr (1996), whose computer learns to compete with humans and eventually liberate itself to the heavenly Internet.

(3) The system LoA. Finally we observe not only a sequence of games, but also all of Menace's 'code'. In the case of a program this is indeed code. In the case of the matchbox model, it consists of the array of boxes together with the written rules, or manual, for working it. Now Menace is still interactive and autonomous. But it is not adaptive; for what in (2) seemed to be an evolution of the system's transition rule is now revealed by observation of the code to be a simple, deterministic update of the program state, namely the contents of the matchboxes. At this lower LoA Menace fails to be an agent.

The point clarified by this example is that if a transition rule is observed to be a consequence of program state, then the program is not adaptive. For example, in (2) the transition rule chooses the next play by exercising a probabilistic choice between the possible plays from that state. The probability is in fact determined by the frequency of beads present in the relevant box. But that is not observed at the LoA of (2) and so the transition rule appears to vary. Adaptability is possible. However, at the lower LoA of (3), bead frequency is part of the system state and hence observable. Thus, the transition rule, though still probabilistic, is revealed to be merely a response to input. Adaptability fails to hold.

This distinction is vital for current software. Early software used to be open to the system user who, if interested, could read the code and see the entire system state. For such software, a LoA in which the entire system state is observed is appropriate. However, the user of contemporary software is explicitly barred from interrogating the code in nearly all cases. This has been possible because of the advance in user interfaces. Use of icons means that the user need not know where an applications package is stored, let alone be concerned with its content. Likewise, iPhone applets are downloaded from the Internet and executed locally at the click of an icon, without the user having any access to their code. For such software, a LoA in which the code is entirely concealed is appropriate. This corresponds to case (2) above and hence to agency. Unsurprisingly, since the advent of applets and downloadable files that are executable and yet invisible to the end-users, the issue of moral accountability of AAs has become increasingly critical.

Viewed at an appropriate LoA, then, the Menace system is an agent. The way it adapts can be taken as representative of machine learning in general. Many readers may have had experience with operating systems that offer a 'dictating' interface. Such systems learn the user's voice basically in the same way as Menace learns to play noughts and crosses. There are natural LoAs at which such systems are agents. As we shall see, the case being developed in this chapter is that, as a result, they may also be viewed to be morally accountable.

If a piece of software that exhibits machine learning is studied at a LoA which registers its interactions with its environment, then the software will appear interactive, autonomous, and adaptive, i.e. to be an agent. But if the program code is revealed, then the software is shown to be simply following rules and hence not to be adaptive. Those two LoAs are at variance. One reflects the 'open source' view of software: the user has access to the code. The other reflects the commercial view that, although the user has bought the software and can use it at will, he has no access to the code. The pertinent question here is whether the software forms an (artificial) agent.

3) Webbots We often find ourselves besieged by unwanted email. A popular solution is to filter incoming email automatically by using a webbot that incorporates such filters. An important feature of useful bots is that they learn the user's preferences in such a way that the user may at any time review the bot's performance. At a LoA revealing all incoming email (input to the webbot) and filtered email (output by the webbot), but abstracting the algorithm by which the bot adapts its behaviour to our preferences, the bot constitutes an agent. Such is the case if we do not have access to the bot's code, as discussed in the previous section.

4) Futuristic thermostats A hospital thermostat might be able to monitor not just ambient temperature, but also the state of well-being of patients. Such a device might be observed at a LoA consisting of input for the patients' data and ambient temperature, the state of the device itself, and output controlling the room heater. Such a device is interactive, since some of the observables correspond to input and others to output. However, it is neither autonomous nor adaptive. For comparison, if only the 'colour' of the physical device were observed, then it would no longer be interactive. If it were to change colour in response to (unobserved) changes in its environment, then it would be autonomous. Inclusion of those environmental changes in the LoA as input observables would make the device interactive but not autonomous. However, at such a LoA, a futuristic thermostat imbued with autonomy and able to regulate its own criteria for operation—perhaps as the result of a software controller—would, in view of that last condition, be an agent.[6]

5) SmartPaint When applied to a physical structure, SmartPaint appears to behave like normal paint; but when vibrations, which may lead to fractures, become apparent in the structure, the paint changes its electrical properties in a way which is readily determined by measurement, thus highlighting the need for maintenance. At a LoA at which only the electrical properties of the paint over time are observed, the paint is neither interactive nor adaptive but appears autonomous; indeed the properties change as a result of internal non-determinism. However, if that LoA is augmented by the structure data monitored by the paint, over time, then SmartPaint becomes an agent, because the data provide input on the basis of which the paint adapts its state. Finally, if that LoA is augmented further to include a model by which the paint works, changes in its electrical properties are revealed as being determined directly by input data, and so SmartPaint no longer forms an agent.

6) Organizations A different kind of example of AA is provided by a company or managerial organization. At an appropriate LoA, it interacts with its employees, constituent substructures, and other organizations; it is able to make internally determined changes of state; and it is able to adapt its strategies for decision-making and hence for acting. I shall discuss the nature of organizations as moral agents in chapter 13.

7.3 What is a moral agent?

We have seen that given the appropriate LoA, humans, webbots, and organizations can all be properly treated as agents. The point of the previous exercise was not that of instilling some relativistic doubts in the reader's mind. Rather, it was to show that systems may or may not count as agents depending on the LoA adopted *and* that, since LoAs are always teleological, i.e. chosen for a reason, in the end, it is the reason

---

[6] See e.g. Pogue, D., 'A Thermostat That's Clever, Not Clunky', *The New York Times*, 30 November 2011 <www.nytimes.com/2011/12/01/technology/personaltech/nest-learning-thermostat-sets-a-standard-david-pogue.html?_r=1&nl=todaysheadlines&emc=tha26>.

determining their adoption that needs to be justified. So our next task is to determine whether and in what way the chosen LoA might be the correct one at which the systems under consideration are moral agents.

How do we 'fix' the LoA? Recall that the LoA we wish to adopt is the one that enables us to consider Alice as a moral agent. Suppose next that we are analysing the behaviour of a population of entities through a video camera of a security system that gives us complete access to all the observables available at LoA_1 (see above) plus all the observables related to the degrees of interactivity, autonomy, and adaptability shown by the systems under scrutiny. At this new LoA_2, we observe that two of the entities, call them H and W, are able:

(i) to respond to environmental stimuli—for example the presence of a patient in a hospital bed—by updating their states (interactivity), for instance by recording some chosen variables concerning the patient's health. This presupposes that H and W are informed about the environment through some data-entry devices, for example some sensors;
(ii) to change their states according to their own transition rules and in a self-governed way, independently of environmental stimuli (autonomy), for example by taking flexible decisions based on past and new information, which modify the environment temperature; and
(iii) to change the transition rules by which their states are changed according to the environment (adaptability), for example by modifying past procedures to take into account successful and unsuccessful treatments of patients.

H and W certainly qualify as agents, since we have only 'upgraded' LoA_1 to LoA_2. Are they also moral agents? The question invites the elaboration of a criterion of identification. Here is a very minimalist option:

(O) *An action is said to be morally qualifiable if and only if it can cause moral good or evil, that is, if it decreases or increases the degree of metaphysical entropy in the infosphere.*

Following (O), an agent is said to be a moral agent if and only if it is capable of morally qualifiable action. Note that (O) is neither consequentialist nor intentionalist in nature. It is neither affirming nor denying that the specific evaluation of the morality of the agent might depend on the specific outcome of the agent's actions or on the agent's original intentions or principles. I shall comment on this point in the next section.

Let us return to the question: are H and W moral agents? Because of (O), we cannot yet provide a definite answer unless H and W become involved in some moral action. So suppose that H kills the patient and W cures her. Their actions are moral actions. They both acted interactively, responding to the new situation with which they were dealing, on the basis of the information at their disposal. They both acted autonomously: they could have taken different courses of actions, and in fact we may assume that they actually changed their behaviour several times in the course of the action on the basis of new available information. They both acted adaptably: they were not simply following orders or predetermined instructions. On the contrary, they both had the possibility of changing the general heuristics that led them to make the decisions they did, and we may assume that they took advantage of the available opportunities to improve their general behaviour. The answer seems rather straightforward: yes, they are both moral agents. There is only one problem: one is a human being, the other is an artificial agent; one could be Alice, the other could be a Webbot. The LoA_2 adopted allows both cases. Would you be able to tell the difference and identify who is the human agent? If you cannot, you will agree that the class of moral agents must include AAs like webbots. If you disagree, it may be so for several reasons, but only five of them seem to have some strength. I shall discuss four of them in the next section and leave the fifth to Section 7.5.

7.4 Mindless morality

One may try to withstand the conclusion reached in the previous section and its Turing-like test by arguing that something crucial is missing in LoA_2. The reasoning can be presented as a *modus tollens*: LoA_2 cannot be adequate precisely because if it were, then artificial agents would count as moral agents; yet this is unacceptable. Of course, the *modus tollens* works only if one can show that the conclusion is indeed untenable. This can be argued by relying on a variety of objections. Four of them seem to be among the best available:

- *the teleological objection*: an AA has no goals;
- *the intentional objection*: an AA has no intentional states;
- *the freedom objection*: an AA is not free; and
- *the responsibility objection*: an AA cannot be held responsible for its actions.

Let us see each of them separately.

7.4.1 The teleological objection

The teleological objection can be disposed of immediately. For in principle LoA_2 could readily be (and often is) upgraded to include goal-oriented behaviour (Russell and Norvig, 2010). Since AAs can exhibit (and upgrade their) goal-directed behaviours, the teleological variables cannot be what make a positive difference between a human and an artificial agent. We could have added a teleological condition that both H and W would have been able to satisfy, but this would leave us none the wiser concerning their identity. So why not add one anyway? It is better not to overload the interface because a non-teleological level of analysis helps to understand issues in 'distributed morality' involving groups, organizations, institutions, and so forth, that would otherwise remain unintelligible. This will become clearer in Chapter 12.

7.4.2 *The intentional objection*

The intentional objection argues that it is not enough to have an artificial agent behave teleologically. To be a moral agent, the agent must relate itself to its actions in some more profound way, involving meaning, wishing or wanting to act in a specific way, and being epistemically aware of its behaviour. Yet this is not accounted for in LoA$_2$, hence the confusion.

Unfortunately, intentional states are a nice but unnecessary condition for the occurrence of moral agency. First, the objection presupposes the availability of some sort of privileged access (a God's-eye perspective from without, or some sort of Cartesian internal intuition from within) to the agent's mental or intentional states that, although possible in theory, cannot be easily guaranteed in practice. This is precisely why a clear and explicit indication of the LoA at which one is analysing the system from without is vital. It guarantees that one's analysis is truly based only on what is observable, and hence only on the available information, not on some psychological speculation. This phenomenological approach is a strength, not a weakness. It implies that agents (including human agents) should be evaluated as moral if they play the 'moral game'. Whether they mean to play it, or they know that they are playing it, is relevant only at a second stage, when what we want to know is whether they are *morally responsible* for their moral actions. Yet this is a different matter, and we shall deal with it in Section 7.4.4. Here, it is sufficient to recall that this is rather uncontroversial even in standard ethical analyses. For a consequentialist, for example, human beings would still be regarded as moral agents (sources of increased or diminished well-being), even if viewed at a LoA at which they are reduced to mere zombies without goals, feelings, intelligence, knowledge, intentions, or mental states whatsoever.

7.4.3 *The freedom objection*

The same holds true for the freedom objection and, in general, for any other objection based on some special internal states enjoyed only by human and perhaps superhuman beings. AAs are already free in the sense of being non-deterministic systems. This much is uncontroversial, scientifically sound, and can be guaranteed about human beings as well. It is also sufficient for our purposes and saves us from the horrible prospect of having to enter into the thorny debate about the reasonableness of determinism, an infamous LoA-free zone of endless and pointless dispute. All one needs to do is to realize that the agents in question satisfy the usual practical counterfactual: they could have acted differently had they chosen differently, and they could have chosen differently because they are interactive, informed, autonomous, and adaptive.

Once an agent's actions are morally qualifiable, it is unclear what more is required of that agent for it to count as an agent playing the moral game; that is, to qualify as a moral agent, even if unintentionally and unwittingly, unless, as we have seen, what one really means, by talking about goals, intentions, freedom, cognitive states, and so forth, is that an AA cannot be held *responsible* for its actions. Now, responsibility, as we shall see better in a moment, means here that Alice, her behaviour, and actions, are assessable in principle as praiseworthy or blameworthy, and they are often so not just intrinsically, but for some pedagogical, educational, social, or religious end. This is the next objection.

7.4.4 The responsibility objection

The objection based on the 'lack of responsibility' is the only one with real strength. It can be immediately conceded that it would be ridiculous to praise or blame an AA for its behaviour, or charge it with a moral accusation. You do not scold your iPhone apps, that is obvious. So this objection strikes a reasonable note; but what is its real point and how much can one really gain by levelling it? Let me first clear the ground of two possible misunderstandings.

First, we need to be careful about the terminology and the linguistic frame in general used by the objection. The whole conceptual vocabulary of 'responsibility' and its cognate terms is completely soaked through with anthropocentrism. This is quite natural and understandable, but the fact can provide at most a heuristic hint, certainly not an argument. The anthropocentrism is justified by the fact that the vocabulary is geared to psychological and educational needs, when not to religious purposes. We praise and blame in view of behavioural purposes and perhaps a better life and afterlife. Yet this says nothing about whether an agent is the source of morally charged action. Consider the opposite case. Since AAs lack a psychological component, we do not blame AAs, for example, but, given the appropriate circumstances, we can rightly consider them sources of evil, and legitimately re-engineer them to make sure they no longer cause evil. We are not punishing them, anymore than one punishes a river when building higher banks to prevent a flood. But the fact that we do not 're-engineer' people does not say anything about the possibility of people acting in the same way as AAs, and it would not mean that for people 're-engineering' could be a rather nasty way of being punished.

Second, we need to be careful about what the objection really means. There are two main senses in which AAs can fail to qualify as responsible. In the first sense, we say that, if the agent failed to interact properly with the environment, for example because it actually lacked sufficient information or had no alternative option, we should not hold that agent morally responsible for an action it has committed because this would be *morally unfair*. This sense is irrelevant here. LoA_2 indicates that AAs are sufficiently interactive, autonomous, and adaptive to qualify fairly as moral agents. In the second sense, we say that, given a specific description of the agent, we should not hold that agent morally responsible for an action it has committed because this would be *conceptually improper*. This sense is more fundamental than the other and the one that is relevant here: if it is conceptually improper to treat AAs as moral agents, the question whether it may be morally fair to do so does not even arise. The objection thus argues that AAs fail to qualify as moral agents because they are not morally responsible for their actions, since holding them responsible would be conceptually improper (not morally unfair). In other words, the objection suggests that LoA$_2$ provides necessary but insufficient conditions. The proper LoA requires another condition, namely responsibility. According to the objection, this fourth condition finally enables us to distinguish between moral agents, who are necessarily human or superhuman, and AAs, which remain mere efficient causes.

The point raised by the objection is that agents are moral agents only if they are responsible in the sense of being prescriptively assessable in principle. An agent A is a moral agent only if A can in principle be put on trial.

Now that this much has been clarified, the immediate impression is that the 'lack of responsibility' objection is merely confusing the *identification* of A as a moral agent with the *evaluation* of A as a morally responsible agent. Surely, the counterargument goes, there is a difference between, on the one hand, being able to say who or what is the moral source or cause of (and hence is *accountable* for) the moral action in question, and, on the other hand, being able to evaluate, prescriptively, whether and how far the moral source so identified is also morally *responsible* for that action, and hence deserves to be praised or blamed, and in each case rewarded or punished accordingly.

Well, that immediate impression is actually mistaken. There is no confusion. Equating identification and evaluation is a shortcut. The objection is saying that identity (as a moral agent) without responsibility (as a moral agent) is empty, so that we may as well save ourselves the bother of all these distinctions and speak only of morally responsible agents and moral agents as synonymous. But here lies the real mistake. We now see that the objection has finally shown its fundamental presupposition: that we should reduce all prescriptive discourse to the analysis of responsibility. Yet this is an unacceptable assumption, a juridical fallacy. There is plenty of room for prescriptive discourse that is independent of responsibility assignment and hence requires a clear identification of moral agents. Good parents, for example, commonly engage in practices involving moral evaluation when interacting with their children, even at an age when the latter are not yet responsible agents, and this is not only perfectly acceptable, but something to be expected. This means that parents identify children as moral sources of moral action, although, as moral agents, they are not yet subject to the process of moral evaluation.

If one considers children as an exception, insofar as they are potentially responsible moral agents, considering animals may help. There is nothing wrong with identifying a dog as the source of a morally good action, hence as an agent playing a crucial role in a moral situation, and therefore as a moral agent. Search-and-rescue dogs are trained to track missing people. They often help save lives, for which they receive much praise and rewards from both their owners and the people they have located, yet this is not the relevant point. Emotionally, people may be very grateful to the animals, but for the dogs it is a game and they cannot be considered morally responsible for their actions. At the same time, the dogs are involved in a moral game as main players, and we rightly identify them as moral agents that may cause good or evil.

All this should ring a bell. Trying to equate identification and evaluation is really just another way of shifting the ethical analysis from considering A as the moral agent/source of a first-order moral action M to considering A as a possible moral patient of a second-order moral action S, which is the moral evaluation of A as being morally responsible for M. This is a typical Kantian move, but there is clearly more to moral evaluation than just responsibility, because A is capable of moral action even if A cannot be (or is not yet) a morally responsible agent. A third example may help to clarify further the distinction.

Suppose an adult, human agent tries his best to avoid a morally evil action. Suppose that, despite all his efforts, he actually ends up committing that evil action. We would not consider that agent morally responsible for the outcome of his well-meant efforts. After all, Oedipus did try not to kill his father and did not mean to marry his mother. The tension between the lack of responsibility for the evil caused and the still present accountability for it—Oedipus remains the only source of that evil—is part of the definition of the tragic. Oedipus is a moral agent without responsibility. He blinds himself as a symbolic gesture against the knowledge of his own inescapable state and his previous epistemic blindness to the nature and consequences of his actions.

7.5 The morality threshold

Motivated by the discussion above, the morality of an agent at a given LoA can now be defined in terms of a threshold function. More general definitions are possible but the following covers most examples, including all those considered in the present chapter.

A threshold function at a LoA is a function which, given values for all the observables in the LoA, returns another value above or below the established threshold. An agent at that LoA is deemed to be morally good if, for some pre-agreed value (called the tolerance), it maintains a relationship among the observables so that the value of the threshold function at any time does not exceed the tolerance.

For LoAs at which AAs are considered, the types of all observables can be formally determined, at least in principle. In such cases, the threshold function is also given by a formula; but the tolerance, though again determined, is identified by human agents exercising ethical judgements. In that sense, it resembles the entropy ordering introduced in Chapter 4.

For non-artificial agents, like humans, I doubt whether all relevant observables can be mathematically determined. The opposing view is represented by followers and critics of the Hobbesian approach. The former argue that for a realistic LoA it is just a matter of time until science is able to model a human as an automaton, or state-transition system, with scientifically determined states and transition rules; the latter object that such a model is in principle impossible. The truth is probably that, when considering moral agents, thresholds are in general only partially quantifiable and usually determined by various forms of consensus. Let us now review our previous examples from the viewpoint of moral evaluation.

The futuristic thermostat is morally charged since the LoA includes patients' well-being. It would be regarded as morally good if and only if its output maintains the actual patients' well-being within an agreed tolerance of their desired well-being. Thus, in this case, a threshold function consists of the distance (in some finite-dimensional real space) between the actual patients' well-being and their desired well-being.

Since we value our email, a webbot is morally charged. Its action is deemed to be morally bad (for an example of artificial evil, see Chapter 9) if it incorrectly filters any messages: if it either filters messages it should let pass, or allows messages to pass it should filter. Here we could use the same criterion to consider the webbot agent itself to be morally bad. However, in view of the continual adaptability offered by the bot, a more realistic criterion for moral good would be that, at most, a specific fixed percentage of incoming email be incorrectly filtered. In that case, the threshold function could consist of the percentage of incorrectly filtered messages.

Menace simply learns to play noughts and crosses. With a little contrivance it could be morally charged as follows. Suppose that something like Menace is used to provide the game play in some computer game whose interface belies the simplicity of the underlying strategy and which invites the human player to pit his or her wits against an automated opponent. The software behaves unethically if and only if it loses a game after a sufficient learning period; for such behaviour would enable the human opponent to win too easily and might result in market failure for the game. It would also deprive the human player of true gratification in his or her quest for excellence at the game. The player may falsely believe that he or she is better than he is. This situation may be formalized using thresholds by defining, for a system having initial state M, $T(M)$ to denote the number of games required after which the system never loses. Experience and necessity would lead us to set a bound, $T_0(M)$, on such performance: an ethical system would respect it whilst an unethical one would exceed it. Thus the function $T_0(M)$ constitutes a threshold function in this case.

Organizations are nowadays expected to behave ethically. In non-quantitative form, the values they must demonstrate include: equal opportunity, financial stability, good working and holiday conditions toward their employees; good service and value to their customers and shareholders; and honesty, integrity, and reliability to other companies. This recent trend adds support to the proposal to treat organizations themselves as agents and thereby to require that they behave ethically. It also provides an example of a threshold that, at least currently, is not quantified.

7.6 A concrete application

How does the previous analysis of moral agency contribute to IE? IE seeks to answer questions like: (a) 'What behaviour is acceptable in the infosphere?' and (b) 'Who is to be held morally accountable and/or responsible when unacceptable behaviour occurs in the infosphere?' Questions like these are normally very well understood in standard macroethics. However, the peculiar novelty of the infosphere makes them of great innovative interest for IE and its growing ubiquity makes them quite pressing. Question (a) requires, in particular, the answer to another question: 'What in the infosphere has moral worth?' Since I have addressed the latter in the previous chapter, I shall not return to it again here. The other question, (b), however, invites us to consider the consequences of the answer provided in this chapter: any agent that causes good or evil (i.e. increases or decreases the degree of entropy in the infosphere) is morally accountable for it.

Recall that moral accountability is a necessary but insufficient condition for moral responsibility. An agent is morally accountable for x if the agent is the source of x, and x is morally qualifiable. To be also morally responsible for x, the agent needs to show the right intentional states (recall the case of Oedipus). Turning to our question, the traditional view is that only software engineers—human programmers—can be held morally accountable for the actions performed by artificial agents because only humans can be held to exercise free will. Of course, this view is often perfectly appropriate. Yet a more radical and extensive view is supported by the range of difficulties which in practice confronts the traditional view: software is largely constructed by teams; management decisions may be at least as important as programming decisions; requirements and specification documents play a large part in the resulting code; although the accuracy of the code is dependent on those responsible for testing it, much software relies on 'off the shelf' components whose provenance and validity may be uncertain; moreover, working software is the result of maintenance over its lifetime and so not just of its originators; finally, artificial agents are becoming increasingly autonomous. Many of these points are nicely made in Epstein (1997) and more recently in Wallach and Allen (2009b). Such complications may lead to an organization (perhaps itself an agent) being held accountable. Consider that automated tools are regularly employed in the development of much software; that the efficacy of software may depend on extra-functional features like interface, protocols, and even data traffic; that different programs running on a system can interact in unforeseeable ways; that software may now be downloaded at the click of an icon in such a way that the user has no access to the code and its provenance with the resulting execution of anonymous software; that software may be probabilistic (Motwani and Raghavan, 1995); adaptive (Alpaydin, 2010); or may be itself the result of a program (in the simplest case a compiler, but also genetic code (Mitchell, 1998)). All these matters pose insurmountable difficulties for the traditional, and now rather outdated, view that one or more human individuals can always be found accountable for some kinds of software and even hardware. Fortunately, this chapter offers a solution—artificial agents are morally accountable as sources of good and evil—at the 'cost' of expanding the definition of a morally charged agent. If all this seems too philosophical for the computationally minded reader, the following analysis should be helpful.

7.6.1 Codes of ethics

Software engineers are bound by codes of ethics and undergo censorship for ethical and, of course, legal violations. Does the approach defended in this chapter make sense when the procedure it recommends is applied to morally accountable AAs? Before dismissing the question as ill-conceived, consider that the Fédération internationale des échecs (FIDE) rates all chess players according to the same Elo System, regardless of their human or artificial nature. Should we be able to do something similar in the case of 'ethical games'?

The ACM Code of Ethics and Professional Conduct[7] contains twenty-four imperatives, sixteen of which provide guidelines for ethical behaviour (eight general and eight more specific; see Table 4), with a further six organizational leadership imperatives, and two (meta-)points concerning compliance with the Code.

The first eight all make sense for artificial agents. Indeed, they might be expected to form part of the specification of any moral agent. The same holds true for the next eight, with the exception of the penultimate point: 'improve public understanding'. It is less clear how that might reasonably be expected of an arbitrary AA, but then it is also unclear whether it is reasonable to expect it of a human software engineer. Note that

Table 4. The principles guiding ethical behaviour in the ACM code of ethics

1	*General moral imperatives.*
1.1	Contribute to society and human well-being.
1.2	Avoid harm to others.
1.3	Be honest and trustworthy.
1.4	Be fair and take action not to discriminate.
1.5	Honor property rights including copyrights and patent.
1.6	Give proper credit for intellectual property.
1.7	Respect the privacy of others.
1.8	Honor confidentiality.
2	*More specific professional responsibilities*
2.1	Strive to achieve the highest quality, effectiveness, and dignity in both the process and products of professional work.
2.2	Acquire and maintain professional competence.
2.3	Know and respect existing laws pertaining to professional work.
2.4	Accept and provide appropriate professional review.
2.5	Give comprehensive and thorough evaluations of computer systems and their impacts, including analysis of possible risks.
2.6	Honor contracts, agreements, and assigned responsibilities.
2.7	Improve public understanding of computing and its consequences.
2.8	Access computing and communication resources only when authorized to do so.

---

[7] The ACM Code of Ethics and Professional Conduct, adopted by ACM Council on 16 October 1992 <www.acm.org/about/code-of-ethics>.

wizards and similar programs with anthropomorphic interfaces—currently so popular—appear to make public use easier; and such a requirement could be imposed on any AA, but that is scarcely the same as improving understanding.

Points 3.1–3.6 of the ACM Code concern ACM members as organizational leaders, so they are not relevant here. The final two points concerning compliance with the code—4.1: agreement to uphold and promote the code; 4.2: agreement that violation of the code is inconsistent with membership—make sense, though promotion does not appear to have been considered for current AAs any more than has the improvement of public understanding. The latter point presupposes some list of member agents from which agents found to be unethical would be struck off.[8] This brings us to the censuring of AAs.

7.6.2 Censorship

Human moral agents who break accepted conventions are censured in various ways, including: (a) mild social censure with the aim of changing and monitoring behaviour; (b) isolation, with similar aims; and (c) capital punishment. What would be the consequences of the approach defended in this chapter for artificial moral agents?

By seeking to preserve consistency between human and artificial moral agents, one is led to contemplate the following analogous steps for the censure of immoral artificial agents: (a) monitoring and modification (i.e. 'maintenance'); (b) removal to a disconnected component of the infosphere; and (c) annihilation from the infosphere (deletion without backup). The suggestion to deal directly with an agent, rather than seeking its 'creator' (a concept which I have claimed may be neither appropriate nor well defined) leads to a non-standard but perfectly workable conclusion. Indeed, it turns out that such a categorization is not very far from that used by standard antivirus software. Though not adaptable at the obvious LoA, such programs are almost agent-like. They run autonomously, and when they detect an infected file, they usually offer several levels of censure, such as notification, repair, quarantine, or deletion, with or without backup. Note that the OOP perspective is helpful again here. The antivirus software does not eliminate the class, but only a specific object from the class. Inevitably, this suggests another question about censure of AAs. Is the censure to be applied to the object that committed the immoral act (token-censure) or is it to be applied to the class (type-censure)? What happens to other objects from the same class that have not committed the immoral act? Are they subject to the same censure, too? The questions have some correspondence in the debate on action vs. rule utilitarianism. In the case of AA to be censured, *if* (an important conditional) we can correctly assume that the behaviour of a token will be identical to the behaviour of all tokens of that type (recall that we are dealing with learning agents which may be able to change their behaviours, so this condition is far from being trivial), then the same reasons that led one to censure

---

[8] It is interesting to speculate on the mechanism by which that list is maintained. Perhaps by a human agent; perhaps by an AA composed of several people (a committee); or perhaps by a software agent.

the token will lead one to censure the type. The fact that, in practice, we can only censure tokens one by one—the fact that antivirus software, to follow the previous example, can only eliminate one instantiation of a virus at a time—is of course a pragmatic, not a theoretical, issue. We usually set up an antivirus to eliminate all cases of a given virus, token by token.

For humans, social organizations have been formed over the centuries for the enforcement of censorship (police, law courts, prisons, etc.). It may be that analogous organizations could sensibly be formed for AAs, and it is unfortunate that this might sound like science fiction. Human social organizations became necessary with the increasing complexity of human interactions and the growing lack of 'immediacy'. Perhaps this is similar to the situation in which we are now beginning to find ourselves with the infosphere; and perhaps it is time to consider the development of agencies for the policing of AAs.

7.7 The advantage of extending the class of moral agents

This chapter may be read as an investigation into the extent to which moral actions are exclusively a human business. Somewhere between 16 and 21 years after birth, in most societies, a human being is deemed to be an autonomous legal entity—an adult—responsible for his or her actions. Yet, an hour after birth, that is only a potentiality. Indeed, law and society commonly treat children quite differently from adults, on the grounds that only their guardians, typically parents, are *responsible* for their actions. Animal behaviour varies in exhibiting intelligence and social responsibility between the childlike and the adult, on the human scale, so that, on balance, animals are accorded at best the legal status of children and a somewhat diminished ethical status, in the case of guide dogs, dolphins, and other species. But there are exceptions. Some adults are deprived of (some of) their rights—criminals may have their voting rights suspended—because they have demonstrated the inability to exercise responsible/ethical action. Some animals are held accountable for their actions and punished or killed if they err. Within this context, we may consider other entities, including some kinds of organizations and artificial systems. I have offered some examples in the previous pages with the goal of better understanding the conditions under which an agent may be held morally accountable.

A natural and immediate reaction could be that moral accountability lies entirely within the human domain. Animals may sometimes appear to exhibit morally responsible behaviour, but lack the thing unique to humans which renders humans (alone) morally responsible; end of story. Such an answer is worryingly dogmatic. Surely, more conceptual analysis is needed here: what has happened morally when a child is deemed to enter adulthood, or when an adult is deemed to have lost moral autonomy, or when an animal is deemed to hold it?

I have tried to convince the reader that we should add artificial agents (corporate or digital, for example) to the moral discourse. This has the advantage that all entities populating the infosphere are analysed in non-anthropocentric terms; in other words, it has the advantage of offering a way to progress beyond the immediate and dogmatic answer mentioned above.

We have made progress in the analysis of moral agency by using an important technique, the Method of Abstraction, designed to make rigorous the perspective from which the domain of discourse is approached. Since I have considered entities from the world around us, whose properties are vital to my analysis and conclusions, it is essential that we have been able to be precise about the LoA at which those entities have been considered. We have seen that changing the LoA may well change our observation of their behaviour and hence change the conclusions we draw. Change the quality and quantity of information available on a particular system and you change the questions that can reasonably be asked and the scope of plausible conclusions that could be drawn from its analysis.

In order to address all relevant entities, I have adopted a terminology that applies equally to all potential agents that populate our environments, from humans to robots and from animals to organizations, without prejudicing our conclusions. And in order to analyse their behaviour in a non-anthropocentric way, I have used the conceptual framework offered by state-transition systems to characterize agents abstractly. I have concentrated largely on artificial agents and the extent to which ethics and accountability apply to them. Whether an entity forms an agent depends necessarily (though not sufficiently) on the LoA at which the entity is considered; there can be no absolute LoA-free form of identification. By abstracting that LoA, an entity may lose its nature as an agent by no longer satisfying the behaviour we associate with agents. However, for most entities, there is no LoA at which they can be considered an agent. Otherwise one might be reduced to the absurdity of considering the moral accountability of the magnetic strip that holds a knife to the kitchen wall. Instead, for comparison, consider the far more interesting question (Dennet, 1997): 'when HAL kills, who's to blame?' The analysis provided in this chapter enables one to conclude that, since blame follows responsibility, HAL is morally accountable—though not responsible and hence not blameable—if it meets the conditions defining agency specified above. It is responsible and therefore blameable only if, in a science fiction scenario, it also has a mental and intentional life.

The reader might recall that in Section 7.3. I deferred the discussion of a final objection to the approach supported in this chapter until the conclusion. The time has come to honour that promise.

The unconvinced reader may still raise a final objection, similar to the one already encountered in Section 6.4.5: even if what has been said so far is correct, does this enlargement of the class of moral agents bring any real advantage? It should be clear why the answer is clearly affirmative. Morality is usually predicated upon responsibility. The use of LoAs and thresholds enables one to distinguish between accountability and responsibility, and formalize both, thus further clarifying our ethical understanding. The better grasp of what it means for someone or something to be a moral agent brings with it a number of substantial advantages: we can avoid anthropocentric and anthropomorphic attitudes towards agency and rely on an ethical outlook not necessarily based on punishment and reward but on moral agency, accountability, and censure; we are less likely to assign responsibility at any cost, forced by the necessity to identify a human moral agent; we can liberate technological development of AAs from being bound by the standard limiting view; and we can stop the regress of looking for the *responsible* individual when something evil happens, since we are now ready to acknowledge that sometimes the moral source of evil or good can be different from an individual or group of humans. This was a reasonable view in Greek philosophy. As a result, we should now be able to escape the dichotomy:

- moral agency, therefore responsibility, therefore prescriptive action, versus
- if there is no responsibility then there is no moral agency, but without the latter there is no need for any prescriptive action.

Promoting prescriptive action is perfectly reasonable even when there is no responsibility but only moral accountability due to the capacity for moral action.

All this does not mean that the concept of 'responsibility' is redundant. On the contrary, the previous analysis makes clear the need for a better grasp of the concept of responsibility itself, when the latter refers to the ontological commitments of creators of new AAs and environments, as we shall see in Chapter 8.

CONCLUSION

In this chapter, I discussed artificial agents in terms of transition systems. A transition system is *interactive* when the system and its environment (can) act upon each other. Typical examples include input or output of a value, or simultaneous engagement of an action by both agent and patient—for example, gravitational force between bodies. A transition system is *autonomous* when the system is able to change state without direct response to interaction, that is, it can perform internal transitions to change its state. So an agent must have at least two states. This property imbues an agent with some complexity and independence from its environment. A transition system is *adaptable* when the system's interactions (can) change the transition rules by which it changes state. Finally, an action is morally qualifiable if and only if it can decrease or increase the entropy in the infosphere, that is, whenever it can cause moral good or evil. Given this background, it becomes possible to understand why *artificial agents*—not just digital agents but also social agents such as companies, parties, or hybrid systems formed by humans and machines, or technologically augmented humans—should count as agents that are morally *accountable* for their actions. We saw in this chapter that the enlargement of the class of moral agents by IE brings several advantages. In general, it complements the more traditional approach, common at least since the Stoics and revived by Montaigne and Descartes, which considers whether non-human (animal or artificial) agents have mental states, feelings, emotions, and so on. By focusing directly on 'mindless morality', one is able to avoid that question, as well as many of the concerns of AI, and tackle some vital issues in contexts where artificial agents are increasingly part of our everyday environment.

In the following chapters, it will be important to recall that an artificial agent need not be a piece of software or something resembling a robot. True, nowadays we are witnessing an explosion of interest in webbots and so-called artificial companions. The interactive doll, *Primo Puel*, produced by Bandai (interestingly the same producer of the Tamagotchi) has sold more than one million copies since 2000. But the real challenge comes from the symbiotic mixture of biological and artificial, natural and engineered features to be found in complex agents. It might be the simple case of a driver, her car, and her GPS on a motorway, or the more complex case of an *M1 Abrams* tank with its mechanical, electronic, and computational weaponry devices and its crew of four people, or indeed a government, an international bank, or any of the Fortune 500 American global corporations. More and more commonly, moral actions are the result of complex interactions among distributed systems integrated on a scale vastly larger than the single human being. Globalization is also a measure of the size of the agents involved in moral decisions that crucially affect the life and future of millions of individuals and their environments. What we are discovering is that we need an augmented ethics for a theory of augmented moral agency. This is where some of the challenging problems are arising in terms of distributed morality (see Chapter 13). It is to be hoped that IE will contribute to tackling them.

In the introduction to this chapter, I warned the reader about the lack of balance between the two classes of agents and patients brought about by deep forms of environmental ethics that are not accompanied by an equally 'deep' approach to agency. The position defended in the previous pages supports a better equilibrium between the two classes A and P. It facilitates the discussion of the morality of agents not only in the infosphere, but also in the biosphere—where animals can be considered moral agents without their having to display free will, emotions, or mental states (see e.g. the debate between Rosenfeld (1995a, 1995b) and Dixon (1995)) and in contexts where social and legal agents can now qualify as moral agents. The greatest advantage, however, is perhaps a better grasp of the moral discourse in non-human contexts. This is an urgent development:

We should have taken all of the time that science fiction writers have given us to think about the moral and ethical problems of autonomous robots and computers; we don't have a lot more time to make up our minds. (Christensen, 2009)

The only 'cost' of a 'mindless morality' approach is the extension of the class of agents and moral agents to include AAs. It is a cost that is increasingly worth paying the more we move towards an advanced information society in which human responsibilities are growing exponentially, as we shall see in the next chapter.

8

The constructionist values of *homo poieticus*

> Properly speaking the artist, the writer, and the scientist should be moved by such an irresistible impulse to create that, even if they were not paid for their work, they would be willing to pay to get the chance to do it.
>
> Norbert Wiener, *The Human Use of Human Beings* (1954), p. 132.

SUMMARY

Previously, in Chapter 7, I argued that the whole infosphere deserves some degree of respect. No entity, understood informationally, is morally unworthy, hence *the class of moral patients* has been extended to include any expression of Being. In Chapter 7, I argued that *the class of moral agents* also needs to be expanded, since artificial agents too can be involved in moral situations as *interactive*, *autonomous*, and *adaptable* entities that may perform actions with good or evil impact on the infosphere. Further in Chapter 7, I drew a distinction between moral agents that are only *accountable* for their actions, and agents that are also *responsible* for them. In this chapter, I combine the two threads and argue that the infosphere is an increasingly poietically enabling environment, which both enhances and requires the development of an information ethics understood as a 'constructionist ethics' for morally responsible agents. Human agents, both as individuals (e.g. Alice) and as multi-agent systems (e.g. Alice and Bob as a group, or Alob, see Sections 6.1 and 13.1) belong to the class of responsible moral agents, so a constructionist ethics concerns them primarily and directly; not only them, however, because if there are any other kinds of responsible agents—human-like or indeed superhuman agents, such as a newly enhanced humanity, fully intelligent futuristic robots, extraterrestrial forms of life, angels, demons, or gods—the same constructionist ethics will concern them too. Only agents who are indifferent to the world—such as Aristotle's unmoved mover or Spinoza's *natura naturans*—are at most accountable, though not responsible, for what happens to it, and hence fail to be addressed by a constructionist IE.

In Section 8.1, I shall explain what a constructionist ethics is. In Section 8.2, I analyse virtue ethics as the best-known case of a constructionist ethics (virtue ethics as moral construction of the self, or *egopoiesis*). I shall then show, in Section 8.3, why information ethics cannot be based on virtue ethics, yet needs to retain a constructionist approach. In Section 8.4, I shall introduce the concept of *ecopoiesis*, or ethical construction, of one's environment, and argue that information ethics needs to be not just *egopoietic* but, most importantly, *ecopoietic*. As the reader may recall, this completes the analysis begun in Chapter 4. After providing evidence for the significance of poietic uses of ICTs in Section 8.5, I argue, in Section 8.6, that ethical constructionism is not only facilitated by the infosphere, but is also what the infosphere requires as an ethics of the informational environment. In the conclusion, I relate the analysis of constructionist ethics to standard positions in computer ethics and to the broader project for IE, referring to the foundation of information ethics and computer ethics discussed in Chapters 4 and 5.

8.1 Introduction: reactive and proactive macroethics

Moral issues are often discussed in terms of putative resolutions of hypothetical situations, such as, 'what should one do on finding a wallet in a restaurant lavatory?' Research and educational purposes may promote increasingly dramatic scenarios, sometimes reaching unrealistic and even hilarious excesses,[1] with available courses of action more polarized and less easily identifiable as right or wrong. However, the general approach remains substantially the same: Alice is confronted with a moral dilemma and asked to make a principled, reasoned, and justifiable decision by choosing from a menu of alternatives. Moral action is triggered by a situation. One may call such approach *situated action ethics*, to borrow an expression from AI.

In situated action ethics, a moral dilemma may give the false impression that the ethical discourse concerns primarily *a posteriori* reactions to morally problematic situations in which Alice unwillingly and unexpectedly finds herself parachuted out of nowhere. The agent is treated as a world user, a game player, a consumer of moral goods and evils, a browser, a customer who reacts to pre-established and largely unmodifiable conditions, scenarios, and choices.[2] Only two temporal modes count: present and future. The past seems irrelevant ('how did Alice end up in such a predicament in the first place?'). At most, the approach is further expanded by a casuistry analysis. Yet ethics is not only a question of dealing morally well with a given world. It is also a question of constructing the world, improving its nature, and shaping its development in the right way. This *proactive* approach treats the agent as a world owner, a game designer or referee, a producer of moral goods and evils, a provider, or a

---

[1] See e.g. 'the trolley problem' (Foot, 1967; Thomson, 1976); for a very entertaining parody, do not miss 'the revised trolley problem' in Patton (1988), which is also freely available on the web. On 'George's job' and 'Jim and the Indians', see Smart and Williams (1987). The last two cases are meant to provide counter-examples against purely consequentialist positions.

[2] For an entirely situation-based ethics approach to the Internet, see e.g. Dreyfus (2001). Dreyfus seems to ignore entirely any constructionist issue. His 'anthropology' includes only single web users passively browsing the net and appears not to be in touch with the actual development of the web.

creator. I use the term 'proactive' technically here, to qualify policies, agents, processes, or strategies that:

(1) implement effective action in anticipation of expected problems, difficulties, or needs, in order to control and prevent them, at least partially, rather than merely reacting to them as they occur. In this sense an ethically proactive approach can be compared to preventive medicine, which is concerned with reducing the incidence of disease by modifying environmental or behavioural factors that are causally related to illness; or that
(2) actively initiate good changes, promoting rather than merely waiting for something positive to happen.

In a proactive scenario, the agent is supposed to be able to plan and initiate actions responsibly, anticipating future events, in order to (try to) control their course by making something happen, or by preventing something from happening, rather than waiting to respond (*react*) to a situation once something has happened, or merely hoping that something positive or negative will or will not happen.

There are significant differences between reactive and proactive approaches. Consider, just for the sake of simple illustration, the moral responsibilities of a webmaster as opposed to those of a user of a website, keeping in mind that differences should not be confused with incompatibilities. A mature moral agent is commonly expected to be both a morally good user and a morally good producer of the environment in which she operates, not least because situated action ethics can be confronted by lose–lose situations in which all options may turn out to be morally unpleasant and every choice may amount to failure. Like in a chess game, a moral game is often won or lost many moves before the actual checkmate. So a proactive approach may help to avoid unrecoverable situations by treating today as tomorrow's yesterday. It certainly reduces the agent's reliance on moral luck. As a result, a large part of an ethical education consists in acquiring the kinds of traits, values, intellectual skills, and rules of thumb that may enable the agent to switch successfully between a reactive and a proactive approach to the world.

All of this is acknowledged by many macroethics, albeit with different vocabulary, emphasis, and levels of explicitness. Still, some more conservative macroethics tend to concentrate on the reactive nature of the agent's behaviour. For example, deontologism embeds a reactive bias insofar as it supports duties on-demand. Or consider the moral code implicit in the Ten Commandments, which is less proactive than that promoted in the New Testament. On a more secular level, the two versions of Asimov's laws of robotics provide a simple case of evolution from a reactive to a proactive ethics. The 1940 version is more reactive than the 1985 version, whose new Zeroth Law includes a substantially proactive requirement: 'A robot may not injure humanity, or, through inaction, allow humanity to come to harm.'[3]

---

[3] See Clarke (1994) for an early, full analysis and further references.

Macroethics that adopt a more proactive approach can be defined as *constructionist*. They are the ones that interest us here. One of the best examples of constructionist ethics is virtue ethics. The analysis of its scope and limits will introduce the discussion of a constructionist approach to IE.

8.2 The scope and limits of virtue ethics as a constructionist ethics

According to virtue ethics, an individual's principal ethical aim is to live the good life by becoming a specific kind of person. The constructionist stance is expressed by the desire to mould oneself. The goal is achieved by implementing or improving some characteristics, while eradicating or controlling others. The stance itself is presupposed: it is simply assumed as uncontroversial that one does wish to live the good life by becoming the best person one can. Some degree of personal malleability and some capacity to choose critically and autonomously provide further background preconditions. The key question—'what kind of person should I be?'—is rightly considered to be a reasonable and justified question. It grounds the question 'what kind of life should I lead?' and immediately translates into several other questions: 'What kind of character should I construct?' 'What kinds of virtue should I develop?' 'What sorts of vice should I avoid or eradicate?' It is implicit that each agent strives to achieve that aim *as an individual*, with only secondary regard to the enveloping community.

Different brands of virtue ethics disagree on the specific virtues and values identifying a person as morally good. The disagreement—say between Aristotle, Paul of Tarsus, and Nietzsche—can be dramatic, not least because the question is ultimately ontological insofar as it regards the kind of entity that a human being should strive to become. In prototyping jargon, theories disagree on the *abstract specification* of the model, not just on *implementation details*.

Despite their divergences, all brands of virtue ethics share the same agent-oriented kernel. This is not to say that they are all subjectivist but rather, more precisely, that they are all concerned exclusively with the proper *construction* of the moral subject, be that a self-imposed task or an educational goal of a second party, like parents, teachers, or society in general. To adopt another technical expression, virtue ethics is intrinsically *egopoietic*. This is not to say that virtue ethics adopts necessarily a conception of the subject as an isolated self, as if society were comprised of a collection of Robinson Crusoes. Actually, virtue ethics does have a *sociopoietic* nature as well, in the following sense. Egopoietic practices that lead to the ethical construction of the agent inevitably interact with, and influence, the ethical construction of the community inhabited by that agent. So, when the agent's microcosm and the socio-political macrocosm differ only in scale, but essentially not in nature or complexity, as one may assume in the idealized case of the Greek *polis*, egopoiesis can scale up to the role of general ethics and even political philosophy. Plato's *Republic* is an excellent example. Plato finds it unproblematic to move seamlessly between the construction of the ideal self and the construction of the ideal city-state. Likewise, Aristotle's ideal state is actually a very small place:

Aristotle's ideal state would have had a territory of about 60 km^2 with a population of 500 to 1000 households, that is, about 2% to 3% the size of Athens. (Nagle, 2006, p. 312)

Yet, so does the Mafia, whose code of conduct and 'virtuous ethics' for the individual is based on the view that 'the family' is its members, and *vice versa*.

Egopoiesis and sociopoiesis are inter-derivable only in sufficiently *simple* and *closed* societies, in which significant communal behaviour (the multi-agent system) is ultimately derivable from (or, *vice versa*, reducible to) that of its constituent individuals (the agents constituting the multi-agent system). It is hard to specify 'sufficiently' precisely, but some light can be cast here by trying to clarify what 'simple' and 'closed' mean.

On the one hand, 'simple' refers to the 'vertical' growth of a society, that is, to its degree of *autonomy*. A society is no longer sufficiently simple, but qualifies as increasingly *complex* when some of the major new variables that govern its development are internal forces, emerging holistically from the actions and decisions of its members, from forces such as unemployment or price inflation, for example, which are beyond the control of the single human agents constituting that society.

On the other hand, the threshold between a closed and an open society (no reference to Popper here) is to be identified in the level and relevance of interconnections and interactions between the society in question and other similar macro-agents. A sufficiently open society is one in which some of the major new variables that govern its development are external forces influencing it from without. Therefore, 'open' and 'closed' indicate the relative degree to which interaction determines social evolution. This is the 'horizontal' growth of a society.

Societies exhibit a continuum of stages, with simple and closed societies at one end of the continuum and, at the other end, societies open and complex enough to sustain autonomous behaviour. Communal behaviours that are not immediately or directly explicable as mere aggregates of individual behaviours are called *emergent*. Perhaps the simplest examples come from artificial communities. In Conway's *Game of Life*, for example, the behaviour of an individual is determined by the states of its immediate neighbours. Stable, periodic, or otherwise interesting behaviour (e.g. gliders, which retain their collective state but glide across the digital landscape) of sub-communities consisting of several individuals provides examples of emergent behaviour. In our own, real, global society, good examples are offered by monetary inflation, unemployment, and other similar phenomena the dynamics of which are determined by the feedback of data from sub-communities.

As they evolve, societies may progress along the continuum. At some point, while immediate and personal interactions among all members are still significant, in practice, systemic forces may supervene, profoundly influencing the life of each individual. Such open and complex societies inherit *autonomy* and *interactivity* from their constituent individuals and, at some level of evolution, they become *adaptive*. By virtue of those three properties, they turn into *artificial agents*. We have seen in Chapter 7 what sort of ethical analysis might then become appropriate. In such societies, sociopoiesis is no longer reducible to egopoiesis alone, and this is the fundamental limit of 'scalability' that affects virtue ethics. In autonomous, interactive, and adaptive societies, virtue ethics acquires an individualistic value, previously inconceivable, and may result in moral escapism. We move from an agent-oriented approach to subjectivist and then individualistic approaches. The individual still cares about her own ethical construction and, at most, the construction of the community with which she is more closely involved, like her family or her wider circle of friends and acquaintances, but the rest of the world falls beyond the horizon of her moral concern and indeed ability to affect, as we shall see in Chapter 10. In Christian ethics, for example, virtues are a matter of assessment between the Creator and the creature in a one-to-one relationship that may leave the rest of society largely untouched.

The 'scalability' problem was an issue during the last centuries of the Roman Empire, for example, and applies equally well in our new era of globalization. Phrasing the point in terms of situated action ethics, new problematic situations arise from emergent phenomena. Examples include issues such as disarmament, the ozone level, pollution, famine, and the digital divide. The difficulty becomes apparent in all its pressing urgency as the individual agent tries to reason, unsuccessfully, by using 'local' ethical principles to tackle a problem with 'global' ethical features and consequences.

8.3 Why information ethics cannot be based on virtue ethics

We are now in a position to distinguish between two phenomena often confused in the literature on information and computer ethics, namely the renewed popularity of virtue ethics:

(a) in our society (see e.g. Slote (2000) for a sympathetic overview); and
(b) in cyberspace (Coleman, 1999, 2001; Grodzinsky, 2001).

In case (a), one is confronted with a context in which an individualistic culture facilitates practically, but does not justify theoretically, the return to an agent-oriented ethics as a me-centred ethics. One should still properly object that:

(1) the kind of egopoiesis promoted by virtue ethics cannot (indeed, was not meant to) scale to very complex and open social contexts; and
(2) virtue ethics presupposes a philosophical anthropology (a theory of what it means to be fully human) that, in a sufficiently evolved social context, cannot be left implicit but that, once it is made fully explicit, requires an ethical justification to become acceptable precisely as a *morally good* anthropology, and hence as ethically preferable.

In case (b), phenomena like the great popularity of social media and virtual communities (see Section 8.4.5), which arguably represent the digital re-incarnation of the *polis*, mean that people naturally tend to concentrate on the ethical construction of their 'personae' as, at the same time, a contribution to the construction of the agent's self and a substantial contribution to the construction of the local cyber-community, which is largely characterized by the members constituting and inhabiting it. In this simple and closed context, an egopoietic approach may indeed be fruitful, precisely for the same reasons it was in the *polis*. One is justified in arguing that virtue ethics may be all that is needed for the ethical well-being of the whole community.

The two trends (a) and (b) have merged and currently interact in the information society, but they are better understood separately, lest one should mistakenly argue that all that a macroethics for the infosphere needs, as a theoretical foundation, is virtue ethics because the latter can work in small cyber-communities (comparable to local-area networks) and is popular 'IRL' or 'OT' (*in real life* or *out there*). However, the opposite is true. Because virtue ethics remains limited by its subject-oriented approach and the specific philosophical anthropology it presupposes, it cannot provide, by itself, a satisfactory ethics for a globalized world in general and for the information society in particular. If misapplied, it fosters ethical individualism, as the agent is more likely to mind only her own self-construction. If it is uncritically adopted, it may easily become intolerant, since agents and theorists may forget the culturally overdetermined nature of their foundationalist anthropologies, which often have religious roots. And even if it fosters tolerance, it may still spread relativism, because any self-construction becomes acceptable as long as it takes place in the enclave of one's own private sphere, culture, and cyber-niche, without bothering any neighbour.

The inadequacy of virtue ethics is of course historical. The theory has aged well, but it can provide, at most, a local sociopoietic approach as a mere extension of its genuine vocation: egopoiesis. It intrinsically lacks the resources to scale up beyond the construction of the individual and the indirect role this may play in shaping her local community. Theoretically, however, the limits of virtue ethics should not lead to an overall rejection of any *constructionist* approach. On the contrary, the essentially constructionist lesson taught by virtue ethics, one of the features that make virtue ethics so appealing in the first place, is more important than ever before. Clearly, a constructionist ethics should be retained and reinforced. The mistake (developing information ethics in terms of virtue ethics) lies not in the stress put on constructionism *per se*, but in the direction in which constructionism is presupposed to develop: primarily towards the individual source of the moral action (building the character of a human agent) rather than the receiver of the moral action as well, that is, towards the patient, the object, and more generally the environment affected by the action. The kind of ethical constructionism needed today goes well beyond the education of the self and the political engineering of the simple and closed cyberpolis. It must also address the urgent and pressing questions concerning the kind of global realities that are being built. This means decoupling constructionism from an agent-oriented approach (leading to subjectivism and then individualism) and re-orienting it to the patient, so that it might be applied *also* to society and the environment, the receivers of the agent's actions. In short, what is needed is an ecopoietic approach to information ethics.

8.4 Ecopoiesis

In a global information society,[4] humanity (understood as a *multi-agent system*) is like a *demiurge*, Plato's god responsible for the design of the physical universe based on pre-existing matter. Its ontic powers have been increasing steadily. Today, humanity can variously exercise them (in terms of control, creation, or modelling) over itself (e.g. genetically, physiologically, neurologically, and narratively), over society (e.g. culturally, politically, socially, and economically), and over natural or artificial environments (e.g. physically and informationally).

Humanity is clearly a very special moral agent. Like a demiurge, it has *ecopoietic* responsibilities towards the whole infosphere. The term 'ecopoiesis' refers to the morally informed construction of the environment based on the patient-oriented or ecologically oriented perspectives introduced in the previous chapters. The more powerful humanity becomes as an agent, the greater its duties and responsibilities become to oversee not only the development of its own nature and habits, but also the well-being and flourishing of each of its ever-expanding spheres of influence, including the whole infosphere. To move from individual virtues to global values, an *ecopoietic* approach that recognizes humanity's *responsibilities* towards the environment (including present and future inhabitants) as its enlightened creators, stewards, or supervisors, not just as its virtuous users and consumers, is needed. So IE is an ethics addressed not just to 'users' of the world, but also to producers or demiurges, who are 'divinely' responsible for its creation and well-being. It is an ethics of *creative stewardship* in which responsibility for the whole realm of Being, that is, the whole infosphere, plays a crucial role.

An ecopoietic ethics, like any form of constructionism, raises a fundamental ontological concern. Moral luck aside, the chances of constructing an ethically good x increase the better one knows what an ethically good x is, and *vice versa*. Constructionism depends on a satisfactory epistemic access to, or understanding of, the relevant ontology. When it comes to the whole infosphere, an ecopoietic ethics presupposes a substantial answer to the foundationalist question 'what is the essential nature of the infosphere?' In the same way that virtue ethics presupposes a philosophical anthropology, an ecopoietic ethics requires a *philosophy of information* that can ground an *informational anthropology*. In the rest of this chapter, I shall not pursue this ontological foundation of constructionism. The interested reader might wish to consult Floridi (2011a and 2011b). Instead, I shall concentrate on clarifying the connection between

---

[4] On the history of the development of the global information society see Mattelart (2001).

THE CONSTRUCTIONIST VALUES OF *HOMO POIETICUS* 169

information ethics and constructionism by showing how the latter emerges from the infosphere and how the infosphere can benefit from a constructionist approach.

8.5 Poiesis in the infosphere

Life in the infosphere, or *onlife*, is changing patterns of moral behaviour in many ways, with important repercussions for the development of the ethical discourse. Instances of situated action ethics, primarily with negative consequences, have attracted a large variety of detailed analyses and account for most of the literature in computer ethics (see Chapters 4 and 5, or, e.g. Spinello and Tavani (2001)). However, the infosphere is not only a source of moral dilemma. As a new way of understanding the whole of reality and hence the environment in which we live, it has also greatly enhanced the possibility of developing *egopoietic*, *sociopoietic*, and *ecopoietic* projects. It has thus contributed to the emergence of a constructionist ethics as a macroscopic phenomenon. In this section, I shall consider a range of indicative examples, which illustrate well the ethics of constructionism.

8.5.1 Interfaces

Choosing and modelling one's own interface to the infosphere (adapting one's own LoAs) is an indicative example of the kind of constructionism promoted by the onlife experience. By default, Alice interacts epistemically with the empirical world through her sensory experience mediated by natural constraints. Yet today we are also increasingly used to interacting with the world *mediately*, through informational interfaces with features that influence our views. Take for example Facebook's 'edge detection' algorithm, which can decide what to show to each individual on her newsfeed. Traditionally, a well-designed interface offers its user a convenient *mental model* for the actions it supports. For instance, one design principle states that, if an action has different effects in different situations, the prevailing *mode* that determines the effect should be intuitively clear to the user. Typical mental models in this context are the 'desktop', 'folder', and 'filing cabinet' present in so many of our ICT applications. One pertinent example is the mental model of the text file as a folder. Here, the user is able to appreciate that pressing a key has different effects when a text file is open or when it is closed. On the other hand, that model is limited because it does not address why the user needs periodically to 'save' the results of the editing of the file.

Laurel (1991) has proposed an alternative view of interfaces as theatre, following Aristotle's six elements of drama. They are listed in Table 5 in increasing order of abstract material cause (one of Aristotle's four causes, operating during the process of creation, which reflects the fabric from which a thing is made), together with their interpretation in human–computer activity (adapted from Laurel (1991, table 2.1)):

This approach places emphasis on designing the action (to be engaged in equally by user and computer) rather than, for example, on the user's mental model. The computer, tablet, or smart phone is thought of as an enabling medium rather as than

Table 5. Interfaces as theatre

MATERIAL CAUSE	INTERFACE ACTIVITY
spectacle/enactment	all sensory components of the action represented
pattern/melody	the pleasurable perception of pattern in the sensory phenomena
language/diction	the selection and arrangement of signs, used semiotically
thought/reasoning	the inferred internal processes leading to choice, of both human and computer
character/agency	the bundles of predispositions and traits, of both human and computer
plot/action	the whole action; a collaboration between system and user

a mere tool. Laurel's metaphor, expressed in terms of Aristotle's analysis of theatre, highlights the constructionist nature of interface design rather than the ontological properties emphasized in the 'mode' metaphor. Indeed, attributes at each level are constructed from those at the lower level. The agent is charged with the responsibility of building her own access to the digital environment. The insights gained by Laurel's approach seem mainly to have been applied to the design of interfaces that are meant to stay in their delivered form. A more recent, 'dynamic', approach has been taken by ICT manufacturers who recognize that many users want to configure their interface themselves (with customization ranging from the rather superficial choice of background and screen saver to more substantial matters of structure and mode of interaction). It seems to be more important to provide the user with a configurable interface than to provide a particularly elegant or efficient one: it is a consequence of the user's constructionist drive that the act of configuring one's own interface makes it preferable. After all, our interfaces to the infosphere are the thresholds where an increasing number of people spend an ever larger amount of their time. Giving them some choice in configuring them should be a no-brainer. Thus, the popularity of the apps trend—whereby systems are increasingly tailorable to their users' needs and preferences—is in line with the constructionist drive. The curious reader should simply google 'two cursors' to see that users can be more creative than one may sometimes suspect. In this particular case, there is really no reason why, having two hands and one mouse or tablet in each, one could not use two cursors independently of each other.

8.5.2 Open source

The second logical step, after the construction of a personalized interface to the infosphere, is the construction of informational entities that populate and interact in it. What should the form of these entities be? As the use of the Internet increases, an ever larger number of users are demanding 'open source' software. The average user interacts with an operating system by clicking on icons, dragging-and-dropping, and so on. A user-friendly graphical interface (GUI) shields her not only from invoking commands directly (i.e. from typing the command name and whatever parameters it requires) but also, and more interestingly, from the underlying code that implements the operations. Consequently, even the experienced user has no way of accessing and modifying the underlying source code, which executes operating system or applications commands. Contrary to this paradigm, an open-source system allows the user direct access to the code.

The high demand for open-source code and the popularity of open-source initiatives and resources is a reflection of the number of users who prefer, wherever possible, the option of configuring their own software rather than making do with off-the-shelf packages. This provides further evidence for the strength of constructionism, quite apart from other factors involved in supporting the open-source movement, such as a feeling of solidarity made possible by collective initiatives on the web that oppose the dictatorial decisions of monopolistic software companies. The major extraordinary success of the 1990s was Linux, a free, open-source, Unix-like computer operating system. Its remarkable story provides evidence of what may be called *distributed constructionism*. To clarify the point, consider the difference between Richard Stallman's and Linus Torvalds' strategies.

On the one hand, Richard Stallman's Free Software Foundation (begun October 1985, see Williams (2010)) released the code for components of the open source GNU version of Unix (GNU/Linux), as they were completed by Stallman himself, with the aim of giving

the users freedom by giving them free software they can use and to extend the boundaries of what you can do with entirely free software as far as possible. (Stallman, quoted in Moody (2001), p. 28)

Stallman's GNU GPL (General Public License) perpetuates, efficiently, the freeness of open source software and any derivatives resulting from modifications by its recipients. According to Moody,

[t]his enormous efficiency acted as one of the main engines in driving the free software projects on to their extraordinary successes during the 1990s. (2001, p. 28)

Initially, circulation of the original components was by magnetic tape from Stallman or people affiliated with his project, when the web was not yet a common medium of communication. Controlled by Stallman, the enterprise still exhibited egopoietic values: most notably, it was meant to promote a software version of the 'freedom of speech' movement.

On the other hand, Linus Torvalds launched his project for the development of Linux by relying entirely on *distributed constructionism*; that is, the unsuspected but evident interest, shared by a growing community, in coordinating efforts to achieve a global product whilst each developing only a local specific component of it. The project took full advantage of the web's *point-to-point* penetration. Human communities tend to be rigidly structured, so that direct communication between individuals is highly constrained. Traditional media can be seen as partially facilitating that tendency, and mobile phones help to implement it to a restricted degree. But the web removes that constraint almost entirely amongst its 'netizens' and provides a poietic-enabling environment through which the community of users and developers of Linux could interact and communicate easily and efficiently. Linux has clearly developed as an ecopoietic enterprise.

Moody (2001) appears to underestimate the 'philosophical' contrasts between the two movements, yet the difference between the two approaches is unmistakable, as one may immediately grasp from documents (available online) such as the Free Software Foundation's 'Why "Free Software" is better than "Open Source"' and Open Source Software's 'Why "Free" Software is too Ambiguous'. It has been well summarized by Eric Raymond in *The Cathedral and the Bazaar*:

> Linux overturned most of what I thought I knew.... I believed that the most important software [...like that of Stallman] needed to be built like cathedrals, carefully crafted by individual wizards or small bands of mages working in splendid isolation, with no beta to be released before its time. Linus Torvalds' style of development—release early and often, delegate everything you can, be open to the point of promiscuity—came as a surprise... the Linux community seemed to resemble a great babbling bazaar of differing agendas and approaches. (2001, p. 21)

The difference between the strategies of Stallman and Torvalds is partly attributable, historically, to different stages in the development of the infosphere, before and after the appearance of the Internet. Conceptually, and more importantly, however, the difference is the result of two different constructionist ethics. Linux and other similar open-source products are built and maintained as an expression of distributed constructionism in the infosphere more akin to what we have become accustomed to see with Wiki-style projects. Such projects provide a more distributed and advanced version of Stallman's simple individual constructionism and are made possible by a new web environment, which provides a robust support for *collaboration without friction* (on the frictionless nature of the infosphere see Chapter 11).

8.5.3 Digital arts

The availability of malleable interfaces and open source software makes possible the construction of forms of digital art previously unimaginable. Murray (1997) has identified three characteristic pleasures of digital environments in general:

(1) *immersion*, the participatory immersive medium intensifies the age-old desire to live out fantasy. Rather than Coleridge's 'willing suspension of disbelief', she proposes it to be viewed, more realistically, as supporting 'the active *creation* of belief' (p. 110, emphasis added);

(2) *agency*, that is 'the satisfying [poietic] power to make meaningful action and see the results of our decisions and choices' (p. 126); and

(3) *transformation*, the shape-shifting morphing made possible by the digital representation of data and the ease with which it can be transformed.

For the purpose of analysing the future of digital narrative, Murray reflects:

> These pleasures are in some ways continuous with the pleasures of traditional media and in some ways unique. Certainly the combination of pleasures, like the combination of properties of the digital medium itself, is completely novel. (Murray, 1997, p. 181)

Murray's interest is specifically in digital environments, not in the infosphere in general. The infosphere is *public* in a way that specific digital media are not. Nevertheless, if we add to Murray's three pleasures that of *interactivity*, we finally gain a full picture of the phenomenon of *telepresence* already encountered in Chapter 3. Such narrative telepresence leads one to investigate the wider field of digital art and the impact that constructionism has had on it.

Digital art has shared with information and computer ethics the first half-century of its existence (Mealing, 1997). Over this period, it has expanded with the pervasive influence of the digital medium and now includes graphic art, musical composition, poetry, architectural style, and cinema, as well as narrative fiction. Despite such variety, it seems that:

> digital art is novel in two ways, the first deriving from virtual reality techniques and the second deriving from the capacity of computers to support interactivity. (Lopes, 2003, p. 108)

Because the result of some digital art is difficult to distinguish from traditional art, emphasis is placed on the *process* rather than the *product*. If a computer can solve crosswords faster than I can—albeit by the brute-force method of searching through a dictionary and trying all feasible combinations—then, one may reason, at least *the way* I do it cannot be mimicked by computer. Likewise, if a computer can produce Picasso-like pictures or Bach-like music—albeit routinely by digitizing a photo or music score and then processing an abstraction of it—then, one may still reason, at least Picasso's and Bach's originality is inimitable. The same emphasis on process rather than product is made by Binkley (1998) who identifies the objects being manipulated, or *maculated*, by artists as being digital (data structures rather than paint or cardboard) with the result that the artwork produced lacks physical uniqueness and can in fact be copied electronically indefinitely. His view of process can be interpreted as acknowledging the importance of constructionism. Indeed, Binkley makes the point that, within the infosphere, the objects of construction may bear little resemblance to those of earlier generations.

8.5.4 *The construction of the self*

Along with interfaces, software, and even new forms of art being constructed in the infosphere, the self is next in line. The topic of the construction of the self, understood as an informational organism, will be fully explored in Chapter 11, but here it is worth emphasizing that egopoiesis is an essential part of a constructionist trend fostered by the evolution of the infosphere. Social media of all kinds (a reference to Facebook is *de rigueur* here) certainly offer new spaces for human creativity. The reason lies partly with the recent development of e-commercial models of marketing—if you want to buy a lounge suite, visit our website and simulate how it would appear in your room—partly with human desire or need for communication (from government legislation to photos of the grandchild's first birthday), and partly with a new wave of constructionism concerning the self clearly seen on personal homepages and Facebook entries (Floridi, 2011c).

8.5.5 Virtual communities

With the construction of the self, we have reached the starting point for the construction of virtual communities. What can we learn from socio-cyber phenomena like web-based chat-rooms, interest groups, ICQ-like communities,[5] newsgroups, online forum, and so forth, all the way down to the most recent experiences made possible by Facebook or Google+ (and by the time the reader reads this paragraph there may well be more updated examples), which rely for their existence on the new forms of communication now possible in the infosphere? Until recently, it was common to argue, pessimistically, that the *onlife* experience—the web in particular—prompted people to withdraw from social engagement and become isolated, depressed, or even alienated. According to a constructionist view, however, the infosphere actually provides a poietic-enhancing environment, which should facilitate, rather than hinder, the construction, development, and reinforcement of self-identities, of links with local (real and/or virtual) communities, and of social interactions. This has been true for over a decade. Already a report published by the Pew Internet & American Life Project in 2001[6] has shown that:

> the online world is a vibrant social universe where many Internet users enjoy serious and satisfying contact with online communities. These online groups are made up of those who share passions, beliefs, hobbies, or lifestyles. Tens of millions of Americans have joined communities after discovering them online. And many are using the Internet to join and participate in longstanding, traditional groups such as professional and trade associations. All in all, 84% of Internet users have at one time or another contacted an online group. (p. 2)

Virtual communities are a flourishing result of the free exercise of the constructionist drive. Users reveal personal facts in them, 'flame', and switch personae by enjoying the possibility of endlessly constructing, deconstructing, and reconstructing alternative selves. They can collaborate with and participate in a common social project. In general, they can behave quite differently from the way they would behave in person. It is as if the normal metric of social distance were contracted by the infosphere. The infosphere empowers new categories of users with the possibility of constructing a new self and an e-polis. It makes constructionism an open option for anyone with access to it.

---

[5] ICQ is an instant messaging computer program.
[6] See Horrigan, J. B., *Online Communities: Networks that Nurture Long-distance Relationships and Local Ties*, Pew Internet & American Life Project, 31 October 2001 <http://pewinternet.org/~/media//Files/Reports/2001/PIP_Communities_Report.pdf.pdf>.

8.5.6 Constructionism on the web

What is the nature of constructionism as exhibited on the web? The previous examples show that the characteristic features of the web that seem particularly relevant to existing instances of constructionism are: interactivity, virtuality, agency, transformationality, process- (rather than product-) orientation, usage rather than ownership, social publicity, and immediate point-to-point communication, which allows collaboration without friction due to an apparent decrease in social distance. Constructionism emerges as a most significant and intrinsic property of the infosphere, more fundamental than any policy vacuum or pressing practical problems. The contracted social distance means that the ethical consequences of constructionism in the information society are particularly acute. When anything changes, everything might be affected. Indeed, the apparent decrease in social distance acts as a magnifier for the new ethical challenges facing *homo poieticus*.

8.6 Homo poieticus

Homo sapiens has primary needs, which relate to survival (like food, shelter, security, and reproduction), and secondary needs (like hedonistic, intellectual, artistic, and physical pursuits), which arise once primary needs are fulfilled. Constructionism seems to be amongst such secondary needs. It is the drive to build physical and conceptual objects and, more subtly, to exercise control and stewardship over them. It manifests itself in the care of existing realities and the creation of new ones, these being material or conceptual. We are poietic creatures, and constructionism is ultimately best understood as a struggle against entropy, both in the thermodynamic sense and in the metaphysical sense introduced in Chapter 4. Existentially, constructionism represents the strongest reaction against the destiny of death. In terms of a philosophical anthropology, constructionism is embodied by what I have termed in the past *homo poieticus* (Floridi, 1999b).

Homo poieticus is to be distinguished from *homo faber*, user and 'exploiter' of natural resources, from *homo oeconomicus*, producer, distributor, and consumer of wealth, and from *homo ludens* (Huizinga, 1970), already encountered in Chapter 4, who embodies a leisurely playfulness devoid of the ethical care and responsibility characterizing the constructionist attitude. *Homo poieticus* concentrates not merely on the final result, but on the dynamic, on-going process through which the result is achieved. *Homo poieticus* is a demiurge, who takes care of reality, today conceptualized as the infosphere, to protect it and make it flourish.

Many influential teachers of constructive disciplines emphasize in their teachings an approach to their art that we can now identify as constructionist, to distinguish it from the ludic, the routine, or the mundane approach. Often these teachings draw from eastern philosophy and mysticism to make the point that the process, and the novice's state of mind during it, are of fundamental importance. The end result will 'take care of itself', if the process is right.[7] To use a very simple and mundane example, a punctured bicycle tyre may be mended entirely routinely (in primary fashion, for 'survival' on a busy day) with little component of construction, or it may be mended in a more deliberate, considered fashion, perhaps with reflection on the process and what is being achieved. In the case of the whole infosphere, the ease with which informational constructs can be created and altered means that the infosphere is an ideal environment for *homo poieticus*. The real challenge facing *homo poieticus* is whether and how it might be possible to negotiate, in an ethically constructive way, a new alliance between *physis* and *techne*. We will keep building and transforming the world; the question is whether we can do this in a patient-friendly way. I shall return to this issue in the conclusion of Chapter 15.

Given the importance I have attributed to *homo poieticus*, it would be surprising if its nature had not been studied in other contexts. Two indicative examples are worth mentioning here, to enable the reader to place the line of reasoning developed in this chapter in a wider context.

Piaget (1977) coined the term *constructivism* for an epistemic model in which children learn while interacting with their environment by manipulating and building objects and developing coherent intellectual structures. Papert (1993) extended Piaget's work from genetic epistemology to the child's construction of *microworlds* and called the result *constructionism*:

> My perspective is more interventionist. My goals are education, not just understanding. So, in my own thinking I have placed a greater emphasis on two dimensions implicit but not elaborated in Piaget's own work: an interest in intellectual structures that could develop as opposed to those that actually at present do develop in the child, and the design of learning environments that are resonant with them. (p. 161)

Inspired by both, Murray (1997) is interested, as we have seen above, in the possibilities for narrative fiction in cyberspace. She uses Piaget's term 'to indicate an aesthetic enjoyment in making things within a fictional world' (p. 294). Indeed, she claims that 'constructivist pleasure is the highest form of narrative agency the MUD [Multi-User Domain] medium allows' (p. 149). Whilst for Piaget and Papert the mental process of construction is autonomous and even subconscious, for Murray it is typically explicit. Unsurprisingly, constructivist methodologies have been applied to digital media. In Eisenstadt and Vincent (2000), for example, one reads that:

> Our approach to media rich learning experiences derives from constructivist models of education; the aim is . . . empowering individuals to create their own content. (p. ix)

In this case, the difference between the two approaches is that, from the poietic perspective defended in this chapter, the fundamental novelty brought about by

---

[7] Particularly interesting examples of a constructionist attitude arise in most of the fine arts and especially in architecture (Alexander, 1964).

computer-based or online learning has little to do with long-distance courses, virtual classes, and telepresence, for it is rather to be identified in the vindication of the 'maker's knowledge' tradition.[8] ICTs make possible hands-on experiences, simulations, collaborations, and interactions with conceptual or informational entities, structures and processes that can be built, manipulated, disassembled, and so on, thus completely transforming the learning/teaching experience.

The process-oriented component of the constructionism articulated in the previous pages has an interesting precedent in literary theory. This is the second example. *Genetic criticism* (*critique génétique*)[9] was the name given in the early 1970s to an empirical approach to the literary act. Such an approach had the goal

of explaining through which processes of invention, writing and transformation a project has become a text onto which the institution will confer the status of literary work or not (d'expliquer par quels processus d'invention, d'écriture et de transformation un projet est devenu ce texte auquel l'institution conférera ou non le statut d'œuvre littéraire). (Gresillon, 1994, p. 206; my translation)

However, the concept differs from the constructionism outlined in this chapter because it subscribes firmly to written traces:

Genetic criticism has used the post-structuralist dissolution of the closed text to define its own notion of the fluid, dynamic manuscript text which, since it is not in any published form, is subject to constant revision. At the same time, genetic criticism has abandoned the vague post-structuralist conception of the text as an interactive process. The genetic approach reinstalls the text in its materiality. Its objects of inquiry are the material traces of writing. (Schmid, 1998, p. 12)

CONCLUSION

For the first half-century of its existence, information and computer ethics under its various denominations (including computer ethics, cyber-ethics, Internet ethics, the ethics of ICTs, web ethics, and so forth) has been a situated action ethics. This is obvious if one reads Bynum's overview (2001a, 2001b), which aims to survey the 'historical milestones' of the subject decade by decade. According to Bynum, 'the best way to understand what the field is like is to examine some example sub-areas of current interest' (1998). He considers the workplace, security, ownership, and professional responsibility. Clearly, the approach has been predominantly pragmatic and action-oriented. In the absence of any foundational principle, the field was reduced

---

[8] The maker's knowledge tradition goes back to Plato. It is the view that an epistemic agent knows, understands, or otherwise epistemically controls better (or perhaps only) what the agent has made. For a defence, see Floridi (2011b).

[9] For a summary of genetic criticism and two case studies (Flaubert and Proust) see Schmid (1998). There is an interesting tension produced by a rigid application of those ideas when text is interpreted as digital art; the kind of constructionism supported in this chapter provides one resolution of it.

to a collection of case-based analyses, as we saw in Chapters 4 and 5. The battle cry for the 1990s was James Moor's quote:

> A typical problem in computer ethics arises because there is a policy vacuum about how computer technology should be used. (Moor, 1985, p. 266)

In the tailwind of such a policy vacuum, much of the discussion was concentrated on the extent to which the web, the Internet, and ICTs more generally provide only a context of application for standard ethical issues *in silico*. The conclusion was that, at the very least, ICTs magnify many ethical issues (security, privacy, ownership, and so on). Yet, not *all* problems of interest arise in this way. For example, Brey's *disclosive (computer) ethics* has provided an alternative approach, which

> uncovers and morally evaluates values and norms embedded in the design and application of computer systems. (Brey, 2001, quoted in Spinello and Tavani, 2001, p. 61)

While the resulting study is again pragmatic and limited in its philosophical scope, it does acknowledge the importance of emergent ethical phenomena. We saw in Chapter 5 that Bynum interprets the future of information ethics dictated by a similarly pragmatic outlook as being dominated by the tension between the *conservative view* and the opposing *global view*. To repeat, according to the former, no issues exist which are unique to information or computer ethics and so the subject will eventually subside. According to the latter, the information revolution and its issues are causing a re-evaluation of traditional ethics. In the meta-theoretical first half of this book, I argued for an alternative view, neither conservative nor revolutionary. The approach does not undervalue the important contributions provided by technological applications and the ethical questions arising from them. Situated action ethics is important, even when 'situated' means 'placed in cyberspace' or 'on the web'. The approach I have defended simply offers an account based more squarely on an appreciation of the artefacts of the new technology. This should help to re-evaluate Bynum's view of the future of information ethics by suggesting where the originality of this new field lies. In fact, by its lights, a merely situated action 'cyberethics' would necessarily be constrained by a lack of concepts and hence inevitably suffer serious hermeneutical shortcomings, and, in the long run, prove useless in our dealings with the new challenges posed by ICTs. One of the benefits of a constructionist IE is that this issue simply does not arise. From a constructionist perspective, for example, the digital divide is not just a matter of denied access to information and recreation, but also a more fundamental problem of anthropological flourishing concerning the prevention of a full epiphany of *homo poieticus* in many cultures and social contexts. The approach promoted by situated action ethics makes it extremely difficult to imagine what a foundation for computer ethics could be. On the contrary, a constructionist IE liberates us from that difficulty and makes intellectual progress much easier. By placing value in the infosphere and in the informational nature of entities, regarded ontologically as the primary, fundamental, and constituent element of our new environment and its artificial agents, it is possible to elaborate a constructionist strategy that supports an ecopoietic approach. This is a development consistent with a fundamental trend in other ethical fields like environmental ethics. It is encouraging that, at last, it is becoming clearer how the ethical discourse may be able to feed back and upgrade itself.

If moral goodness and Being, understood informationally, become two sides of the same coin, and our responsibilities acquire an ecopoietic dimension, the next problem to be addressed is the nature of evil itself and our duties against it. I began to explore moral evil in Chapter 4, where it was identified with non-Being or metaphysical entropy. It was also discussed in Section 6.4.3, where moral evil was analysed in terms of actions rather than entities. The time has come to focus entirely on its investigation; this is the task of the next chapter.

9

Artificial evil

> [Man] is not likely to salvage civilization unless he can evolve a system of good and evil which is independent of heaven and hell.
> George Orwell, 'As I Please', *Tribune*, 14 April 1944 (1970, p. 127).

SUMMARY
Previously, in Chapter 5, I argued in favour of an ecumenical extension of the class of entities that may qualify as moral patients, in order to include all manifestations of Being, understood informationally. In Chapter 7, I supported the expansion of the class of agents that may qualify as moral, in order to include artificial agents as morally accountable sources of good and evil. As a consequence of such enlargements of what moral agents and patients are in the infosphere, in this chapter, I support the expansion of our taxonomy of evil. Moral reasoning traditionally distinguishes between two types of evil: moral (ME) and natural (NE). The standard view is that ME is the product of human agency, and so includes phenomena such as war, torture, and psychological cruelty; that NE is the product of non-human agency, and so includes natural disasters such as earthquakes, floods, disease, and famine; and finally, that more complex cases are appropriately analysed as a combination of ME and NE. I shall argue that, as a result of developments in autonomous agents, a new class of interesting and important examples of hybrid evil has come to light. I shall define it as artificial evil (AE) and defend the view that AE complements ME and NE to produce a more adequate mapping of the phenomenon and concept of evil. In Section 9.1, I shall introduce the conceptual resources needed in the rest of the chapter. In Section 9.2, I shall defend a deflatory theory of evil as a non-entity. In Section 9.3, I relate the debate on the nature of evil to the theodicean problem, and in Section 9.4, I articulate a theory of artificial evil. As usual, the conclusion will summarize the results and link them to the next chapter.

9.1 Introduction: the nature of evil

As a concept, 'evil' is the most comprehensive expression of disapproval in our ethical vocabulary. As a phenomenon, we experience evil whenever we deal with forms of moral wrong and the reverse of moral good. Thus, unsurprisingly, evil is a key component in any axiology and hence in any macroethics. Yet, what is the nature of evil, when discussing it in connection with the well-being of the infosphere?

Of the many contributions made in order to answer the previous question, three are particularly useful in order to provide the essential background for the rest of this chapter (see (a)–(c) below). In order to introduce them, I will need to rely on the model of moral action offered in Section 6.1, so let me quickly recall it here.

I shall treat any moral action as a variably interactive process relating a source or sender, the agent A, which initiates the process, to a destination or receiver, the patient P, which reacts more or less interactively to the process. Note that, for the sake of simplicity, in this chapter I shall consider A and P as single entities, not sets of them. The reader may recall that A and P are understood in information-structural terms, or as objects according to the object-oriented analysis. The moral action itself is modelled as an information process, i.e., a series of messages (M), initiated by A, that brings about a transformation of states directly affecting P, which may interactively respond to M with changes and/or other messages, depending on how M is interpreted by P's methods. In short, our model will be: $\exists A \ \exists P \ M \ (A, P)$.

We can now use the previous model to introduce three clarifications about the nature of evil. Even if they are all quite commonly accepted in the literature, they are listed below in order of slightly increasing controversial status.

(a) 'Evil' is a second order predicate that qualifies primarily M.

Only actions (messages in our vocabulary) are *primarily* evil.[1] Sources of evil (agents with their intentional states) are usually identified as evil in a *derivative* sense: intentional states are evil if they (can) lead to evil actions, and agents are overall evil if the *preponderance* of their intentional states or actions is evil. The problem is that, since the domain of intentional states or actions is probably infinite, a 'preponderance' analysis may be based on

- a limit in time and scope, to the effect that A is evil between time t_1 and time t_n and/or as far as such and such intentional states or actions are concerned; and/or
- an inductive/probabilistic projection, to the effect that A is such that A's future intentional states or actions are more likely to be evil than good.

Obvious difficulties in both approaches reinforce the view that an agent is evil only derivatively.

(b) The interpretation of A ranges over the domain of all agents, both human and non-human.

It is of course possible to argue, as Anderson (1990) does, that to be evil an action must be done consciously, voluntarily, and wilfully, and the agent must cause some harm, or

---

[1] See e.g. Anderson (1990); Hampton (1989); Kekes (1988, 1990, 1998b, 1998a, 2005).

allow some harm to be done, to at least one other person. This approach, however, seems too restrictive, as it captures only the meaning of 'moral evil'. We usually consider evil actions as the result of human or non-human agency (e.g. natural disasters). In the case of human agency, we speak of moral evil (ME). ME implies autonomy and responsibility, and hence a sufficient degree of information, freedom, and intentionality. In other words, it is related to a mindful morality (see Section 7.4). In the case of non-human agency, we speak of natural evil (NE). NE is usually defined negatively, as any evil that arises independently of human intervention, in terms of prevention, defusing, or control. It is related to mindless morality (see Section 7.4).

The third clarification, although rather common, is slightly more debatable:

(c) the positive sense in which an action is evil (A's intentional harming) is parasitic on the privative sense in which its effect is evil (decrease in P's goodness or well-being).

Contrary to 'responsibility'—an agent-oriented concept that works as a robust theoretical 'attractor', in the sense that standard macroethics (e.g. consequentialism or deontologism) tend to concentrate on it for the purpose of moral evaluations of the agent—'evil' is a perspicuously patient-oriented concept. Actions are ontologically dependent on agents for their implementation (evil as cause), but are evaluated as evil only in view of the degree of severe and unnecessary harm that they may cause to their patients (evil as effect). Hence, whether an action is evil can be decided only on the basis of a clear understanding of the nature and future development of the affected patient. Since an action is evil if and only if it harms or tends to harm its patient, evil, understood as the harmful effect that could be suffered by the patient, is properly analysed only in terms of possible corruption, decrease, deprivation, or limitation of P's well-being, where the latter can be defined, using the model introduced above, in terms of the patient's appropriate data structures and methods. This is the classic, 'privative' sense in which evil is parasitic on the good and does not exist independently of the latter (evil as *privationem boni*). Gaita (2004), for example, makes this point quite clear:[2]

[E]vil can be understood only in the light of the goodness. I shall yield to the temptation to express Platonically and say that evil can be understood only in the light of 'the Good'. (p. 191)

In view of this further qualification, and in order to avoid any terminological bias, I suggest we avoid using the term 'harm'—a zoocentric, not even biocentric, word, which implicitly leads to the interpretation of P as a sentient being with a nervous system—in favour of 'damage', an ontocentric, more neutral term, with 'annihilation' as the level of most severe damage or highest degree of metaphysical entropy.

---

[2] To be fair, Gaita (2004) does not attempt to clarify, ultimately, how evil should be defined, but argues that '[t]here cannot be an independent metaphysical inquiry into the "reality" of good and evil which would underwrite or undermine the most serious of our ways of speaking.... It would be better, at least in ethics, to banish the word "ontology"' (p. 192).

Summarizing, and in terms of the informational model just recalled, messages are processes that affect their patient *P* either positively or negatively. Positive messages respect or enhance *P*'s well-being; negative messages do not respect or they actually damage *P*'s well-being. Evil actions can now be understood as a subclass of negative messages, those that do not merely fail to respect *P* but (can) damage it. The following definition attempts to capture the clarifications introduced so far (Table 6):

Table 6. The definition of evil action

(E) Evil action = $_{def.}$ one or more negative messages, initiated by *A*, that brings about a transformation of states that (can) damage *P*'s well-being severely and unnecessarily; or more briefly, any patient-unfriendly message.

Note that (E) excludes the possibility both of *victimless* and of *anonymous* evil. An action is (potentially) evil only if there is (could be) a damaged patient (there has to be a victim, at least potentially), and there is no evil action without a damaging source, even if, in a multi-agent and distributed context, this may be sufficiently vague or complex to escape clear identification (there must be an agent). Below, I shall argue that this does not imply that evil cannot be *gratuitous*. In fact, because standard macroethics tend to prioritize agent-centred analyses, they usually concentrate on evil actions *a parte agentis*, by *presupposing* the presence of an agent and qualifying the agent's actions as evil, at least hypothetically or counterfactually.

We have come to the end of this introductory section. On the basis of the previous clarifications and definition of evil, I shall argue in favour of a deflatory approach to the existence of evil and of a revision of our understanding of some evils as artificial.

9.2 Nonsubstantialism: a deflatory interpretation of the existence of evil

The classic distinction ME vs. NE is so intuitive as to be hugely popular, yet it may also be misleading. Human beings may act as natural agents, for example as unaware and healthy carriers of a contagious disease, and natural evil may be the mere means of moral evil, for example through morally blameworthy negligence. But above all, the dichotomy may be misleading because it is the result of the application of a first-order ('moral', 'natural') to a second-order ('evil') predicate, which paves the way to a questionable hypostatization of evil and what Schmitz (1978) has aptly called an 'entitative conception of evil'. Evil is reified as if it were a 'token' transmitted by M from *A* to *P*, an oversimplified 'communication' model that is implausible, since *A*'s messages can generate negative states only by interacting with *P*'s methods, and do not seem either to be evil independently of them, or to bear and transfer some pre-packaged, perceivable evil by themselves.

In order to avoid the hypostatization of evil, a nonsubstantialist position

(i) must defend a deflationary interpretation of the existence of evil,

without

(ii) accepting the equally implausible alternative represented by revisionism, i.e. the negation of the existence of evil *tout court*, which may rely, for example, on an epistemological interpretation for its elimination (evil as mere appearance).

This balance can be achieved by

(iii) accepting the derivative and privative senses of evil (evil as absence of good)

and by clarifying that 'there is no evil' means that

(iv) only actions, and not entities in themselves, can be qualified as primarily evil;

and that

(v) what type of evil x is should not be decided on the basis of the nature of the agent initiating x, since ME and NE do not refer to some special classes of entities, which would be intrinsically evil, nor to some special classes of actions *per se*, but they are only shortcuts to refer to a three-place relation between types of agents, actions, and patients' well-being, hence to a specific, context-determined interpretation of the triple $<A, M, P>$.

The reasoning in (i)–(v) seems plausible. Unfortunately, especially in ancient philosophy, it has often been over-interpreted as an argument for the non-existence of evil. This is because nonsubstantialism has been equated with revisionism through a 'thing-ology' or ontology of things, i.e. the assumption that either x is a substance, a 'some-thing', or x does not genuinely exist. Yet, since evil is so widespread in the world, and its direct experience so common and painful, any argument that attempts to deny its existence should be resisted in favour of more realistic alternatives. Revisionism seems hardly defensible. Unfortunately, through the equation nonsubstantialism = revision, the consequence has been that the presence of evil in the world has often been taken as definitive evidence against nonsubstantialism as well and, even more generally, as a final criticism of any theory based on clarifications (a)–(c) and points (i)–(v). It should be obvious, however, that this conclusion is far from inevitable: nonsubstantialism is deflatory but not revisionist, and it is perfectly reasonable to defend the former position by rejecting the implicit reliance on a simple ontology of things. Actions–messages and entities' states, as defined in informational structural realism (exemplified by the object-oriented programming (OOP) paradigm in Section 6.1), do not have a lower ontological status than entities themselves. Evil exists not absolutely, *per se*, but in terms of damaging actions and damaged patients. The fact that its existence is parasitic does not mean that it is fictitious. Tapeworms are no less real than their unfortunate hosts. On the contrary, in an ontology that treats interactions, methods (operations, functions, and procedures), and states on the same level as entities (objects and their attributes), evil could not be any more real. Once an ontology of things is replaced by a more adequate structural ontology, it becomes possible to have all the benefits of talking about evil without the ontological costs of a substantialist hypostatization. The objection that a deflationary approach does not seem to do justice to the reality of evil (e.g. pain and suffering) can be compared to the objection that quantum physics does not do justice to the reality of chairs and tables.

9.3 The evolution of evil and the theodicean problem

The discussion on the nature of evil has been largely monopolized by the theodicean debate: whether it is possible to reconcile the existence of a divine creator and the presence of evil in the created universe.[3] In particular, most discussions of the nature of evil, at least in Western philosophy, have focused exclusively on the theoretical problem of evil as it arises within the context of Biblical religion, treating the existence of evil as a classical objection to theism. A clear example of this monopoly is provided by John Hick's article 'The Problem of Evil', in *The Encyclopedia of Philosophy* (1967), which concentrates solely upon the theodicean debate, ignoring any other ethical issue connected with the existence of evil. However, more recently things have begun to change. In the *Routledge Encyclopedia of Philosophy*, for example, we find two separate entries, one on the theodicean problem of evil, and one on the axiological nature of evil (Kekes, 1998a). Information ethics can help to reinforce this 'secular' trend and establish a clear distinction between axiological vs. theological analyses of evil.

Partly because of theodicean-oriented analyses, natural evil is commonly understood as any evil that arises through no human action, either positive or negative: NE is whatever evil human beings do not initiate and cannot prevent, defuse, or control. One might conceive different kinds of NE as placed on a scale, from the not-humanly-initiated and not-preventable earthquake (only the evil effects of it can be a matter of human responsibility) to the not-humanly-initiated but humanly preventable plague, to the humanly initiated and preventable environmental disaster (human agents as natural causes).

Interestingly, contemporary macroethics seem to have failed to notice that a negative understanding of natural evil in terms of anything that is not moral evil (NE = $\neg$ ME) entails the possibility of a diachronic transformation of what may count as NE because of the increasing power of design, configuration, prevision, and control over reality offered by science and technology, and especially ICTs. If a negative definition of NE is not only inevitable but also adequate, the more powerful humanity becomes, in terms of its scientific and technological achievements, the more it and its members become responsible for what is within their power to influence. Past generations, when

---

[3] On the theodicean problem, see Leibniz (1990) and Adams and Adams (1990). On the axiological analysis of evil see Benn (1985); Kekes (1988, 1990, 1998a, 1998b, 2005); Milo (1984); Moore (1993, pp. 256–62). Gelven (1983) provides an analysis of the various ways in which the word 'evil' is used in English.

confronted by natural disasters like famine or flood, had little choice but to put up with their evil effects. Nowadays, most of the ten plagues of Egypt would be considered moral rather than natural evils because of human negligence. Even in the Old Testament, what seems to be at stake is an ontological issue of power. The plagues have mainly an ontological value, as evidence of the nature of their source, and its total control and power over reality, rather than ethical value. Several times the Pharaoh's magicians are summoned to deal with the extraordinary phenomena, but the point is always whether they can achieve the same effects 'by their secret arts'—hence showing that there is either no divine intervention or equal divine support on the Egyptian side—not whether they can undo or solve the difficulties caused by the specific plague. Tough luck for the poor people who suffer the consequences of the competition between the God of the Bible and the magicians of Egypt. The latter lose the 'ontic game' when 'the magicians tried by their secret arts to bring forth gnats, but they could not'.

A clear sign of how much the world has changed is that people expect human solutions for virtually any natural evil, even when this is well beyond the scientific and technological capacities of present times. Whenever a natural disaster occurs, the first reaction has become to check whether anyone is responsible for an action that might have initiated, or prevented, its evil effects. Resignation (as acceptance) is no longer an obvious virtue of patients but rather the expected decent thing (as relinquish) done by responsible agents.

The human-independent nature of NE and the power of science and technology, especially ICTs, with their computational capacities to monitor, control, and forecast events, determine a peculiar phenomenon of constant erosion of NE in favour of an expansion of ME. If anyone were to die from smallpox in the future this would certainly be a matter of ME, no longer NE. Witchcraft in theory and science and technology in practice share the responsibility of transforming NE into ME and this is why their masters look morally suspicious. Bunge (1977), for example, analyses the moral responsibility brought about by technological advances, stressing how the 'technologists', i.e. the technology-empowered persons, will be increasingly responsible for their professional actions.

The erosion of NE in favour of ME is inevitable and should be welcomed, insofar as science and technology can constantly increase human power over nature. It may also seem unidirectional: at first, it may appear that the only transformation brought about by the evolution of science and technology is a simplification in the nature of evil. In the next section, I shall argue that the introduction of the concept of artificial evil (AE) provides a corrective to this view. If, for the present purpose, it is simply assumed that, at least in theory, all NE can become ME but not *vice versa*, it is obvious that this provides an interesting approach to the classic theodicean problem of evil. In the long run, the theist may need to explain only the presence of ME, despite the fact that God is omniscient, omnipotent, and all-good, and it is known that a theodicy based on the responsibility that comes with freedom is more defensible,[4] especially if connected

---

[4] On this see also Plantinga (1977). The title of the book follows the full title of Leibniz's Theodicy, see Leibniz (1990).

with a nonsubstantialist approach to the existence of evil. In a utopian world, the occurrence of evil may be just a matter of human misbehaviour. What matters here, of course, is not solving the theodicean puzzle, but realizing how ICTs are contributing towards making humanity increasingly responsible, morally speaking, for the way the world is.

9.4 Artificial evil

More and more frequently, especially in advanced societies, we are confronted by visible and salient evils that are neither simply natural nor immediately moral: an innocent dies because the ambulance was delayed by the traffic; a computer-based monitor 'reboots' in the middle of surgery because its software is not fully compatible with other programs also in use, with the result that the patient is at increased risk during the reboot period. The examples could easily be multiplied. What kinds of evil are these? 'Bad luck' and 'technical incident' are simply admissions of ignorance. Conceptually, they indicate the shortcomings of the ME vs. NE dichotomy. The problem is that the latter was formulated at a time when the primary concern was anthropocentric, human-agent oriented, and the main issue addressed was the allocation of human and divine responsibility and the issuing consequences, for one's existence after death, and for existence *tout court* in the case of God. However, strictly speaking, the difference between human and natural agents is not that the former are not natural, but that they are autonomous, i.e. they can regulate themselves. So the correct taxonomy should really be a simple four-place scheme: forms of agency are either natural or artificial (non-natural) and either autonomous or heteronomous (non-autonomous), as shown in Table 7.

We saw in Section 7.1.1 that an agent is a system, situated within and a part of an environment, which initiates a transformation, produces an effect or exerts power on it over time, as contrasted with a system that is (at least initially) acted on or responds to it (patient). Such a definition is sufficient to clarify the previous four basic forms of

Table 7. A taxonomy of agents

Agent	Natural	Artificial
Autonomous	NAA	AAA
Heteronomous	NHA	AHA
NAA	natural and autonomous agent, e.g. a person, an animal, an angel, a god, an extra-terrestrial	
NHA	natural and heteronomous agent, e.g. a flood, an earthquake, a tsunami	
AAA	artificial and autonomous agent, e.g. a webbot, an expert system, a software virus, a robot	
AHA	artificial and heteronomous agent, e.g. traffic, inflation, pollution	

agency. A natural agent is an agent that has its ontological basis in the normal constitution of reality and conforms to its course, independently of human beings' intervention. Conversely, an artificial agent is an agent that has its ontological basis in a human-constructed reality and depends, at least for its initial appearance, on human beings' intervention. An autonomous agent is an agent that has some kind of control over its states and actions, senses its environment, responds to changes that occur within it and interacts with it, over time, in pursuit of its own goals, without the direct intervention of other agents. And a heteronomous agent is simply an agent that is not autonomous.

Following the taxonomy summarized by Table 7, it is easy to see that:

- moral evil (ME) is any evil produced by a *responsible* NAA;
- natural evil (NE) is any evil produced by NHA and by any NAA that may not be held directly responsible for it.

We may now define artificial evil (AE) as any evil produced by either AAA or AHA. The question is: is AE always reducible to (perhaps a combination of) NE or ME?

It is clear that AE is not reducible to NE because of the nature of the agents involved, whose existence depends on human creative ingenuity. Yet, this leads precisely to the main objection against the presence of AE; namely, that any AE is really just ME under a different guise. We saw that Bunge may be read as supporting this view. Human creators are morally accountable and responsible for whatever evil may be caused by their artificial agents, for the latter are mere means or intermediaries of human activities (indirect responsibility). The objection of indirect responsibility is based on an analogy with the theodicean problem and is partly justified. The reasoning is that in the same way that a divine creator can be blamed for NE, so a human creator can be blamed for AE.

A first reply consists in remarking that, even in a theodicean context, one still speaks of 'natural' not of 'divine' evils, thus indicating the nature of the agent involved (e.g. an earthquake), not of the morally responsible creator of that agent. However, this, admittedly, would be a weak retort, for it misses the important ethical point: if NE is 'real' then this causes a problem precisely because it is reducible to 'divine' evil and, *mutatis mutandis*, this could apply to the relation between AE and ME. AE could be just the result of performing morally wrong actions by other means. The buck of AE stops only with humanity taking responsibility for it, one may contend. True, but the previous reply paves the way to a better understanding of the differences between the two cases and hence to a second, more convincing reply.

On the one hand, AE may be caused by AHA whose behaviour depends immediately and directly on human behaviour. In this case, the reduction AE = ME is reasonable. AHA are just an extension of their human creators, like tools, because the latter are both the ontological and the nomological source of the formers' behaviour. Human beings can be taken to be directly accountable for the artificial evil involved, e.g. pollution. To illustrate it with a slogan: 'guns don't kill people, people kill people', so people are responsible, not guns (and that's why you should not make guns available in the first place, because if people have them they will use them).

On the other hand, an AAA, whose behaviour is nomologically independent of human intervention, may cause AE. In this case, the interpretative model is not divine creator vs. created universe, but parents vs. children. Although it is conceivable that the evil caused by a child may be partly blamed on her or his parents, it is also true that, normally, the sins of the sons at some point will stay with the sons and will not always be passed on to the fathers. At some point, the moral buck does stop at the grown-up children. Indirect responsibility can only be forward, not backward, as it were. We are now on the right path. Unfortunately, the analysis requires some further effort, for things are, in fact, even more complicated. Let me explain. Recall that:

(i) evil refers primarily to actions,

and that

(ii) an action is evil if it causes serious and morally unjustified damage.

Let us also agree with Kekes (1998b) that:

(iii) if an evil action is *reflexive* this means that it should be taken to reflect adversely on the agent whose action it is and this agent would be held responsible for its action.

It follows that it cannot be true that:

(iv) all evil actions, in the sense specified in (i)–(ii), are reflexive, in the sense specified in (iii).

The negation of (iv) is a consequence of the fact that there are many autonomous agents that can perform evil actions without being responsible for them. If a drone no longer under human control kills a family of innocent people in Afghanistan, mistaking it for a group of terrorists, surely that is an instance of evil but surely the drone is not morally responsible for it. Kekes (1998b), however, argues that (i)–(iv) are consistent. He does so by relying on a questionable interpretation of 'autonomy' and on the denial of a classic ethical principle, thus

(v) 'actions are autonomous if their agents (a) choose to perform them, (b) their choices are unforced, (c) they understand the significance of their choices and actions, and (d) they have favourably evaluated the actions in comparison with other actions available to them.... Actions of which any one or more of (a), (b), (c), or (d) is not true are nonautonomous [heteronomous, in the more Kantian vocabulary adopted in Table 7]' (Kekes, 1998b, p. 217).

However, it is clear that, following (v), while a drone could be autonomous (no sci-fi here, just off-the-shelf technology) many human beings, many other artificial agents, and no animals would qualify as autonomous, so Kekes is forced to argue further that:

(vi) in many cases, neither the evil actions nor the vices from which they follow are autonomous. It is nevertheless justified to hold the agents who perform these actions morally responsible for them; the widespread denial of this claim rests on the principle 'ought implies can'; the latter, however, cannot be used to exempt agents from moral responsibility for their nonautonomous actions and vices.

In fact, (v) seems to provide more a definition of freedom than a definition of autonomy, which is usually taken to be synonymous for 'self-regulating' when it qualifies the nature of an agent. Rather than maintaining (v) and hence being forced to abandon the 'ought–can' principle following (vi), it seems more sensible to invert the process. After all, the ought–can principle is worth salvaging (I shall rely on it in Section 10.1), and the step taken in (vi) obscures the fact that people could be guilty of evil actions—to the point of rightly blaming themselves—even if they are not responsible for them. Evil can be unintentional. Indeed a human agent might have no choice but to do evil. Recall how in Section 8.1 we saw that situated action ethics may be confronted by lose–lose situations, in which all options cause some evil and every choice amounts to moral failure. In *Sophie's Choice* (the 1982 film directed by Alan J. Pakula), the Nazis force Zofia 'Sophie' Zawistowski to choose which one of her two children will go to the concentration camp and consequently be gassed. To avoid having both children killed, she chooses her daughter, Eva, to be sent to her death in Crematorium Two. Wisdom chooses to let the first woman die. This is the sense in which life can be tragic, Cassandra *docet* (see Section 10.2 on the lack of balance between information and power).

If one maintains the ought–can principle and rejects (v) as being too demanding, then (i)–(iv) need to be modified, and since I agree with Kekes on (i)–(iii), it is step (iv) that must be revised. Evil actions can be *irreflexive* or *gratuitous*, i.e. they can be caused by sources that cannot be held responsible for them. The modification of the definition of 'autonomy'—as being different from freedom, hence the revision of clause (iv)—allows one to consider all agents, including animals (*homo sapiens* not an exception here) and artificial agents, *indirectly* or *derivatively* evil whenever they are the regular source of evil actions, despite their lack of understanding, intent and free ability to choose to do evil, and hence moral responsibility.

Note that, given the deflatory account of evil, this does not justify abusive treatment of evil agents. Only evil actions are rightly considered intrinsically worthless or even positively unworthy and therefore rightly disrespectable in themselves. If all this seems complicated, the reason is that we are trying to analyse a problem that is eminently patient-centred, i.e. the existence of evil, by means of a vocabulary and a cluster of concepts that are inherited from an agent-oriented tradition.

Artificial 'creatures' can be compared to pets, agents whose scope of action is very wide, who can cause a huge variety of evils, but who cannot be held *morally* responsible for their behaviour, owing to their insufficient degree of intentionality, intelligence, and freedom. It turns out that, like in a universe without God, in the infosphere, evil may be utterly *gratuitous*: there may be evil actions without any causing agent being *morally* blameable for them. We saw in Chapter 7 that artificial agents are becoming sufficiently autonomous to pre-empt the possibility that their creators may be nomologically in charge of, and hence morally responsible or even accountable for, their misbehaviour. And we are still dealing with a generation of fairly simple agents, predictable and controllable. Just imagine what the situation will be like in a hundred years. The phenomenon of potential artificial evil will become even more obvious as self-produced generations of AAA evolve.

There is an ICT version of the theodicean problem, which I shall discuss in Section 10.3. However, such *IT-heodicean problem* is very different from the theodicean one not only because we know that the creators exist, but also because the creators, in this case, are fallible, only partly knowledgeable, possibly malevolent and may work at cross-purposes, so there is no need to explain how the presence of humanity may be compatible with the presence of artificial evil. Unfortunately, like Platonic demiurges or fallible creators much less powerful than God, we may not be able to construct truly intelligent AAA, but we can certainly endow them with plenty of autonomy and interactivity, and it is in this lack of balance that the moral risk lies. It is clear that something similar to the ethical principles discussed in Section 4.6 and Asimov's Laws of Robotics will need to be enforced to keep the infosphere safe. Science and technology transform natural into moral evil, but at the same time create a new form of artificial evil. In a dystopian world like the one envisaged in *The Matrix* (the 1999 film directed by Andy and Larry Wachowski), there could be only moral and artificial evils.

CONCLUSION

In this chapter, I have argued that artificial agents may be accountable for evil actions for which no human or divine agent can be considered responsible. I have concluded that we need a third category of evil, which I defined as artificial evil.

In terms of ICTs, the infosphere supports a variety of agents, from routine service software (implementing communications protocols) through less routine applications packages (like cybersitters, webbots) to applets downloadable from remote website on a smart phone. Similar artefacts highlight a shift in the burden of responsibility of humans, including software engineers, amateurs, big ICT companies, and small retailers. In many situations today, there is still a contract between producer and user: the producer, say the software engineer, is responsible for the performance of some software, both ethically and legally. This model suited the context in which computers, or local-area networks, were isolated from others, except by physical media (disks, CD ROMs, etc.). In the new model, promoted by the Internet and now by the infosphere, it is increasingly less clear whether there is a 'point of sale', since a program may be downloaded at one of a sequence of mouse clicks, with no clear responsibility or even specification attending its acquisition. So transparent and porous is the interface that the user may not even be aware that a program has been downloaded and executed locally, e.g. automatically and transparently (i.e. invisibly) to her. Indeed, this is true with just about every single webpage that has JavaScript in it. The autonomy and hence seamlessness of so many interactions is further reinforced by the presence of artificial agents, which employ randomization in making decisions (the giver of a coin can hardly be held responsible for decisions made on the basis of tossing it, even if the coin is sold as a binary-decision-making mechanism) and which are able to adapt their behaviour on the basis of experience, as we saw in Chapter 7.

Given the nature of the new environment and the presence of such agents in it, the tendency towards further autonomy and adaptability will only increase in the future. So it seems reasonable to accept the fact that sometimes the evils that may result from such artificial agents will not be traceable or blameable on either humanity, nature, or a divine creator. The infosphere supports actions that may originate from humans (email from a colleague) or artificial agents (messages from a word processor or directives from a webbot). The claim is not that current artificial agents have passed the Turing test. This would be silly. Nor is it that some kind of singularity is in view. This is science fiction. It is rather that, with the types of artificial agents mentioned above, there is scope for evil that lies beyond the responsibility of human beings or nature. Our region of the infosphere may be changed as a result of the autonomous actions of artificial agents: decisions are delegated to routine procedures, data are altered, settings changed and programs subsequently behave differently, with artificial agents responding or reacting, often interactively, to further actions, at such a pace, such a speed, and with such a scope that it may easily prevent human control. This is a common and ordinary scenario. For example, in high-frequency trading, smart agents analyse market data for trading opportunities that may be available only for milliseconds, autonomously competing for tiny but consistent profits (Wellman et al., 2007). By 2015, it is estimated that high-frequency trading will account for over 70 per cent of equity trades in the USA, with a rapid growth in Europe and in Asia.[5] Some of the actions of artificial agents seem benign: the old example of the *Easter eggs* planted inside Apple and Palm software still offers a good illustration (Pogue, 1999). It seems equally clear that some actions are evil: viruses and the action of some webbots, for example. Artificial evil is going to be a growing phenomenon, sweeping it under the carpet of ethical denial will only make it more problematic.

Connecting what has been argued in this chapter with the thesis supported in Chapter 8, it seems clear that we should come to terms with the fact that technologies in general (think for example of biochemical engineering, genetic technologies, or nanotechnologies), and ICTs in particular, place demiurgic responsibilities on our shoulders, both positively, in terms of what we create and do, and negatively, in

---

[5] The Boston Consulting Group, U.S. Securities and Exchange Commission Organizational Study and Reform, 10 March 2011 <www.sec.gov/news/studies/2011/967study.pdf>.

terms of what we fail to create and do. The more likely it is that we may unleash extremely powerful artificial agents in our natural and synthetic environments, the more demanding our moral duties become to exercise care, foresight, prevention and even restraint. Consider the pace at which unmanned military weapons are being developed nowadays or how quickly software agents are becoming autonomous and ubiquitous. Perhaps some kinds of artefact should never be built. In September 2008, for example, *The Wall Street Journal* reported that Google News picked up an obscure reprint of a 2002 article about United Airlines' risk of bankruptcy. Although United Airlines had since recovered, there was no dateline, so Google News ran the story as current news. It was then distributed widely by other news aggregators and eventually became a headline on Bloomberg. This triggered automated trading programs and a devaluation of the airlines' stock from $12 to $3, evaporating $1.14 billion in shareholder wealth, close to United's total market value. Later in the day, the stock recovered, but not entirely, and at the end of the day was trading at $9.62, a market cap of $300M less than before Google ran the story. Still harbouring doubts about whether artificial agents may do evil?

In this chapter as well as in Chapter 8, I referred to common cases in which responsible agents, interested in being good agents and willing to do the right thing, are still caught in circumstances that leave open evil courses of action. The time has come to look more closely at such a tragic predicament.

10

The tragedy of the Good Will

> *For a truly religious man nothing is tragic.*
> Ludwig Wittgenstein, 'Conversation in 1930' (quoted in Drury (1981), p. 107).

SUMMARY
In Chapter 2, I suggested that there are three interconnected ways in which information may play a crucial role in ethics: as a resource, as a product, and as a target. In Chapters 4, 6, and 9, I have focused on the infosphere understood as the patient of ethical actions, hence on information understood, ontologically, as a target. In this chapter, I shall concentrate on information as a resource and product of ethical interactions, and do so by considering its semantic value. From a semantic perspective, information has always played a major role in any moral theory at least since Socrates' time. ICTs have now revolutionized the life of information, from its production and management to its consumption, thus deeply affecting our moral lives. Amid the many issues they have raised, a very serious one, discussed in this chapter, is what I shall label the *tragedy of the Good Will*. This is represented by the increasing pressure that ICTs and their deluge of information are putting on any *responsible* (in the technical sense seen in Chapter 7) agent who would like to act morally, when informed about actual or potential evils, as defined in Chapter 9, but who also lacks the resources to do much about them. In Section 10.1, I shall provide the necessary clarifications to formulate the problem. In Section 10.2, I shall distinguish between the tragic and the scandalous. In Sections 10.3 and 10.4, I shall show how ICTs may tragically affect the moral life of a good agent, described as a Good Will. In Section 10.5, I shall argue that the tragedy may be at least mitigated, if not solved, by seeking to re-establish some equilibrium, through ICTs themselves, between what agents know about the world and what they can do to improve it. In the conclusion, I shall connect the tragedy of the Good Will with the informational construction of the self (Chapter 11) and the protection of informational privacy (Chapter 12).

10.1 Introduction: modelling a Good Will

Let us return for a moment to the model introduced in Section 2.5. A responsible agent, our Alice, is embedded within the infosphere. Her moral life is significantly dependent on how information flows and is processed. The presence of powerful ICTs hugely increases such dependence. In previous chapters, I concentrated on ontological and e-nvironmental issues (the third arrow in Figure 3, to simplify) related to Alice's new habitat and her interactions with it. In this chapter, I intend to consider a key problem that arises in the context of the first two 'vectors' in our model, namely when information is taken in a semantic sense, as a resource and a product. I shall refer to the problem as *the tragedy of the Good Will*. The problem is simple, but making it explicit and precise, as well as suggesting some fruitful strategies for tackling it, requires careful analysis.

The first step is to clarify six assumptions. None of them seems to be so controversial as to require much support here, but each of them should do its work openly, in case the reader finds that there is room for disagreement.

The first assumption has already surfaced, so let me make it fully visible:

(1) 'information' will be used here in its strongly semantic sense, in order to refer to syntactically well-formed, semantically meaningful, and veridical data, like 'Paris is the capital of France' or 'the train to London leaves at 11 a.m.'.[1]

The reader who finds the 'veridicality' thesis embedded in (1) unconvincing may simply concede that we shall be talking exclusively of true information. I shall not be concerned with information in the *mathematical* and *probabilistic* sense (Shannon's theory), in the *structural* sense (the ontological sense in which Being and infosphere are co-referential), or in the *instructional* sense (e.g. an algorithm or an order).

We saw in Chapter 2 that semantic information represents a crucial component in moral evaluations and actions. Without repeating what has already been said there, it seems obvious that Alice may be expected to choose courses of actions based on the best information she can gather, and hence be reasonably keen on acquiring information that can improve her performance as a moral agent. This is the second, Aristotelian assumption:

(2) Alice, our moral agent, is interested in gaining as much relevant information as required by the circumstances.

As Aristotle puts it at the beginning of his *Metaphysics*, we shall assume that 'all men by nature desire to know'.[2] This may be for evolutionary reasons (one naturalistic way of reading Aristotle's 'by nature') or because well-informed agents are more likely to do the right thing (a Socratic way of reading Aristotle's 'by nature'). One can accept the assumption without necessarily embracing the ensuing ethical naturalism or intellectualism, which analyses evil and morally wrong behaviour as the outcome of deficient

---

[1] See Floridi (2011a) for a full articulation and defence of this assumption or Floridi (2010c) for an introduction.
[2] This is much more controversial than it may seem and than it has been assumed in the history of philosophy, see Floridi (1994, 1995a).

information. Indeed, even evil agents need as much information as possible in order to carry out their deeds.

The third assumption concerns A's limited powers:

(3) Alice does not have boundless resources but is realistically constrained, especially by time, memory (i.e. amount of information storable and available), energy expendable to increase her information, and capacities to handle it.

This is not as bad as it looks. As is well known, moral action cannot presuppose any form of omnipotence. So one of the axioms of standard deontic logic requires that, if it ought to be that a, then it is permissible that a ($Oa \rightarrow Pa$), which in our context means that, if A must do a then A can do a.[3]

The previous condition goes some way towards mitigating the impact of the next assumption:

(4) Alice's moral *responsibility* tends to be directly proportional to A's *amount of information* (how much and how well she is informed), any decrease in the latter usually corresponding to a decrease in the former.

We have seen in Chapter 2 that this is the important sense in which information may occur in terms of judicial evidence, informed decision, informed consent, or well-informed participation. The assumption also allows counterfactual evaluations: had A been properly informed, A would have acted differently and hence would not have made the moral mistake that she did.

The next assumption is a simplification:

(5) Alice suffers no *akrasia*.

I shall assume that Alice is capable of carrying out the course of action that she judges to be morally best. Although not very realistic (the practising vs. preaching dichotomy is common to the point of being proverbial), this assumption is still plausible and it merely satisfies a simplicity requirement. Alice's lack of *akrasia* means that she does not act against her judgement, but here it is not taken to mean that she has an intrinsic desire to act morally. For this anti-Hobbesian motivation, we need a last assumption:

(6) Alice enjoys full *eudokia*.

This Greek word means 'good will', an expression made famous by the *Vulgata* version of *Luke* 2:14 ('pax hominibus *bonae voluntatis*' 'peace to all men of *good will*'). It is in this original sense of *benevolent attitude*, or a willingness/desire to do the right thing, that I shall use it in the rest of this chapter.[4] This use of 'Good Will' is slightly different from

---

[3] On the connection between epistemic and deontic logic see now Pacuit et al. (2006).
[4] The reader should be warned that the discussion about the proper reading of the passage is a scholarly battlefield. Depending on whether one adds an 's' at the end of *eudokia* and makes it a genitive, the reading changes from 'Glory to God in the highest, and on earth peace, good will toward men'—which is the classic

Kant's well-known interpretation. According to Kant, a good will is the only thing that 'can be taken as good without qualification'. Its decisions are entirely dictated by moral demands, that is, by the moral law. In this chapter, the Good Will overlaps with Kant's description deontologically, insofar as she (I use 'it' to refer to Kant's conception) is identified as a privileged centre of morally good action. However, the Good Will differs partly from Kant's description in a way that may be defined as 'care-ethically', that is, insofar as she includes not only a purely rational but also a caring attitude. Our Good Will is expected to exhibit a *willingness* to engage with the world for its own sake and an *attentiveness* to (that is, interest in, concern with, and compassion for) its well-being. Both attitudes are extraneous to Kant's conception, as each requires an emotional and empathic involvement, an engagement with the poietic values discussed in Chapter 8, and the ontic trust that will be the topic of Chapter 15. In our case, the rational and caring attitudes are supposed to be complementary and to add value to each other.

To summarize the six assumptions, I shall treat Alice as a responsible agent, endowed with some albeit limited resources, who bases her decisions and actions on the proper management of her factual information about the moral situations in which she is involved (see the concept of envelope in Section 6.1), who is reasonably capable of implementing whatever she thinks ought to be done morally, whose responsibilities increase with the amount of information she enjoys (and who knows that this is the case), and who is motivated by a genuine desire to know and by a sincere *eudokia*, while not suffering from *akrasia*. For the sake of simplicity, I shall refer to this type of agent as *the Good Will*.

The Good Will is an ideal but not an idealized agent. As in any scientific experiment in which one tries to abstract from irrelevant details and obtain ideal conditions (e.g. by referring to frictionless models in dynamic experiments), the previous six assumptions form a level of abstraction the adoption of which is justified by the need to use the Good Will to bring to light and properly formulate an important problem caused by ICTs, namely the tragedy of the Good Will. But first, one last round of clarifications, as promised.

10.2 The tragic and the scandalous

To understand the tragedy of the Good Will we need to appreciate what the tragic means. The suggestion developed in this section is that the tragic arises from a lack of balance between information and power in the presence of *eudokia*, i.e. of a Good Will's (the agent's) inclination to act morally. 'Power' refers to the bounded skills, resources, means, etc. needed to implement a morally good action (see assumption (3) above). 'Information' refers to how much (or how little) the Good Will knows about

---

reading (but note that the good will in question is God's, not men's)—to 'Glory to God in the highest, and peace among men in whom he is well pleased', which has strong Calvinist implications in favour of the predestined. Either way, the *Vulgata* seems a misleading translation, if suggestive.

the world, including past events, current circumstances, and future implications or effects (see assumption (1) above). Without *eudokia* there is no sense of the tragic, but the presence of *eudokia* is insufficient to give rise to the tragic, since the Good Will might actually succeed in her endeavours. For the tragic to arise, there also needs to be a fundamental lack of balance. A few classic examples will help to clarify the point.

1) Lucretius: no Good Will, no tragedy

Lucretius in his *De Rerum Natura* provides a beautiful illustration of information without either Good Will or power:

> Tis sweet, when, down the mighty main, the winds
> Roll up its waste of waters, from the land
> To watch another's labouring anguish far,
> Not that we joyously delight that man
> Should thus be smitten, but because 'tis sweet
> To mark what evils we ourselves be spared.
> (Book II, *Proem*, lines 1–6)

Lucretius is presenting here the detached and content *ataraxia* to be developed by the philosophical mind. If there is a lack of involvement (*apathia*) and no Good Will—in this case no desire to help and intervene—then it is not tragic but sweet to witness someone else's anguish, for the struggle is only in the object observed and not in the observer. Compare this to the following, equally famous scene of shipwrecking.

2) Miranda: the tragic as a result of Good Will, information and power

When in *The Tempest* Shakespeare portrays Miranda watching from afar the apparent sinking of 'a brave vessel', he might not have had Lucretius' passage in mind, but he makes her utter the following words:

> If by your art, my dearest father, you have
> Put the wild waters in this roar, allay them.
> The sky, it seems, would pour down stinking pitch,
> But that the sea, mounting to the welkin's cheek,
> Dashes the fire out. *O, I have suffered*
> *With those that I saw suffer*: a brave vessel,
> Who had, no doubt, some noble creature in her,
> Dash'd all to pieces. O, the cry did knock
> Against my very heart. Poor souls, they perish'd.
> *Had I been any god of power*, I would
> Have sunk the sea within the earth or ere
> It should the good ship so have swallow'd and
> The fraughting souls within her.
> (Act I, Scene II, 1–13, emphasis added)

Two points deserve our attention. First, both Lucretius and Miranda may be assumed to be witnessing the same disaster. But Miranda is a Good Will ('I have suffered with those that I saw suffer'). Her *eudokia* makes her wish she were able to match her alleged information (in fact, it will turn out that no 'noble creature' is 'dashed to pieces') with some equal power, which, in this case, would require a god-like (*demiurgic*, more on this later) degree of control over the elements ('had I been any god of power I would have sunk the sea within the earth'). She knows that the tragic would disappear if only her (the Good Will's) power were equal to her information.

The second point is that the tragic will indeed later vanish when Miranda/the Good Will realizes that she was misinformed. So we, readers and audience, are confronted by a lack of balance between power and information that can be restored either by making the former match the latter (what Miranda would like to do as a 'god of power'), or by making the latter match the former (what in fact will happen: 'those that I saw suffer' turns out to be false). Such a lack of balance, as the essence of the tragic, is openly evident in Oedipus and Cassandra.

3) Oedipus: the tragic as a result of Good Will, power but lack of information

On the one hand, Oedipus has only some limited information about his horrific future (he is told that he will kill his father and marry his mother) but lacks the relevant information (he was adopted; the man he kills on his journey is his real father; the woman he later marries is his real mother). On the other hand, Oedipus has quite a lot of power to implement his *eudokia* and to try to avoid his destiny (he leaves his home town and those whom he believes to be his parents, thus hoping to escape his destiny; he later becomes king). It is because Oedipus is a Good Will that his fate is tragic. But his tragedy is entirely informational: his desire to do the right thing is combined with the (royal) power to carry out his decisions but also with the wrong sort of information. So it is not accidental that Oedipus becomes king of Thebes (marrying his mother Jocasta) through an informational rite of passage, by answering the riddle of the Sphinx; that it is a *blind* source who sees better than he does (the seer Teiresias) and reveals to Oedipus his real fate; and that Oedipus, in the end, punishes himself by forcing his mother's brooch pins into his eyes. Greek epistemology is very visual; being informed is seeing. He was epistemologically blind and restores some coherence to his life by physically blinding himself.

The last example is equally classic, but shows a lack of balance in terms of lack of power, not of information.

4) Cassandra: the tragic as a result of Good Will, information but lack of power

Although Cassandra can predict ('hear') the future, a gift from Apollo, she is also cursed by the same god, so that her predictions will never be believed. This is a source of endless frustration and pain, as nobody acts on her accurate warnings. She is the Good Will that has all the necessary information (about the Trojan Horse and Troy's destruction; or about Agamemnon's and her own murder) but one who is powerless when it comes to avoiding the foreseen events.

To summarize: the tragic occurs in the presence of a Good Will (Miranda), when she is sufficiently powerful but insufficiently informed (Oedipus), or sufficiently informed but insufficiently powerful (Cassandra). Since the tragic is due to a lack of balance, and any balance is a matter of fine-tuning, the risk of the tragic in either form is constant. When the tragic occurs, it is a scandal.

The *scandalous* is how the tragic may be perceived by its observers. Oedipus' and Cassandra's tragic predicaments are scandalous not because they set bad examples (for nobody would follow them), but because they show to the observers the ultimate, titanic failure of the Good Will. In a context in which the essence of agency is largely constituted by its *eudokia*, the agent who 'gives scandal' has, by the same token, annihilated her essence, and thus ceased to be an agent altogether. For the Good Will, giving scandal is tantamount to committing suicide or being 'terminated'. It is an extreme case of metaphysical entropy. This is how one may interpret the famous quote from Matthew's Gospel:

He that shall scandalize one of these little ones, that believe in Me, it were better for him that a mill-stone should be hanged about his neck, and that he should be drowned *in the depth of the sea*. (*Matthew* 18:6, emphasis added).

In the desperate sea of Miranda, that is, not of Lucretius.

We are finally ready to analyse the relation between ICTs, the tragic, and the scandalous.

10.3 The IT-heodicean problem

Given the forms in which the tragic (and hence the scandalous) may occur, it is not surprising that the relation between the information revolution, brought about by ICTs, and the tragic, might be twofold.

On the one hand, we have what I labelled in Section 9.4 the *IT-heodicean problem*. ICTs provide the Good Will with increasing opportunities—directly or indirectly, from nanotechnology to risk assessment modelling, from bioinformatics to neuroscience, from genetic engineering to telemedicine, and so forth—to prevent, defuse, control, or eradicate evil. Information is power, as we all know. It follows that, the more powerful the Good Will becomes—in terms of science and technology and ICTs in particular—the wider the scope of her responsibilities becomes for what is within her power to influence. Thus, ICTs greatly contribute to the increasing moral pressure put on the Good Will and her insufficient information about what ought to be done. It is as if the Good Will had more and more means to do something for the well-being of the world, but did not see how. Like Oedipus, when evil finally occurs, the Good Will can only blame herself, for had she been better informed, evil might have been avoidable. The tragedy of her inability is also the scandal of her annihilation as a moral agent.

As we saw in Chapter 9, ICTs erode the scope of natural evil, re-cataloguing it as moral, or, as André Gide once put it, 'man's responsibility increases as that of the gods decreases' and ICTs play a major role in this shifting process. Not that the process itself is either new or limited to ICTs. Already Homer could write

> Look you now, how ready mortals are to blame the gods. It is from us, they say, that evils come, but they even of themselves, through their own blind folly, have sorrows beyond that which is ordained. (*Odyssey*, I.30–35)

But ICTs have made the process snowball.

On the other hand, if ICTs have increased by orders of magnitude a Good Will's capacity to cope with the world, they have also submerged her with information about the endless evils that she should be worried about. This is *Cassandra's predicament*, which I suggested we label the *tragedy of the Good Will*, to differentiate it from the *IT-heodicean problem*.

10.4 Cassandra's predicament

Good Wills are regularly submerged and often overwhelmed by information about evils in the world about which they can do very little, if anything at all. In the past, less information meant less responsibility. Nowadays, ICTs keep inundating the Good Wills with distressing news about famine, diseases, wars, violence, corruption, injustices, environmental disasters, poverty, lack of education, racism, and so forth, on a daily basis. The list is endless, the disasters heart-breaking, the responsibilities mounting, the sense of scandalous powerlessness nauseating. Confronted with so much information about so many moral failures, the Good Will cannot help feeling frustrated, aggrieved, and guilty. A concrete example will render the analysis less academic and ivory-towerish. It concerns the sea again.

On 14 August 2003, *The Economist* published an article in which one could read that:

> Since 1990, [in the western Pacific] ten big tsunamis have claimed more than 4,000 lives. So it would be nice to be able to detect such tsunamis far enough in advance for people to be evacuated.... What is needed are specific detectors that take advantage of the fact that tsunamis are felt throughout the ocean's depths, unlike wind-generated waves, which affect only its surface.[5]

The article continued by discussing several technologies and techniques for detecting, analysing, classifying and predicting tsunamis. It concluded:

> Technology, though, can do only so much.... Coastal dwellers must be able to recognize the signs of a possible tsunami—such as strong, prolonged ground shaking—and seek higher ground at once. As with any hazard, the more informed the public are, the better their chances of survival.

---

[5] 'The next big wave', *The Economist*, 14 August 2003 <http://www.economist.com/node/1989485>.

Despite all this information, on 26 December 2004, the Sumatra–Andaman earthquake caused a series of devastating tsunamis that spread throughout the Indian Ocean, killing approximately a quarter of a million people, with thousands of others missing. No ICTs (tsunami warning systems) were in place to mitigate the impact of the catastrophe. It was one of the deadliest disasters in modern history. On the other hand, thanks to ICTs, Good Wills everywhere in the world 'suffered with those whom they saw suffer', almost in real time. Morally speaking, it was an instance of the tragedy of the Good Will.

It would be easy to speculate about future disasters that will be equally tragic and scandalous in the technical sense of the words specified above. Think of global warming, nuclear proliferation, the Palestinian problem, or AIDS throughout the world, for example. The point should be sufficiently clear to require no further illustration. Instead, one aspect that is worth emphasizing here is how the Good Will might be inclined to develop skilful forms of ignorance or blind spots. As Plato remarks in the *Republic* (478c), the soul might decide not to pursue *nous* (knowledge and understanding) but *agnoia* (ignorance and irrationality), and dwell in 'that which *is not (at all)*', or *metaphysical entropy*, in the vocabulary I introduced in Section 4.5. Let me explain.

If the analysis offered so far is even roughly correct, a Good Will will feel pain and frustration when informed about evil events, and the more so the more she is informed about dramatic events with respect to which she is powerless. At the same time, it is also reasonable to assume that no Good Will will be inclined to leave open such a perennially bleeding wound. If one suffers too much with those whom one sees suffer, one may soon wish to avert one's eyes. So the risk that the Good Will constantly runs is that of unwittingly (when not consciously) and innocently trying to avoid her Cassandra-like predicament by shutting herself off in her own informational niche. The dialectic is simple, and well captured by two well-worn phrases: since 'what the eye does not see, the heart cannot grieve', the Good Will is constantly tempted to 'bury her head in the sand', what charities refer to as 'disaster fatigue' or 'compassion fatigue'. ICTs have made the need for such hiding more strongly and widely felt, insofar as they have increased the potential exposure of the Good Will to evil.

The result is well epitomized in our digital age by the phenomenon of the so-called *The Daily Me*. The term, coined by Negroponte (1995) some time ago, refers now to any news system (including news feeds) tailored to, customized by, or personalized for the reader's interests and tastes. The problem with *The Daily Me* is that it can easily become a mere mirror of one's own idiosyncratic biases, thus contributing to what David Weinberger has called the 'echo chamber', information spaces where like-minded people unwittingly (and this is the risk) communicate only with people who already agree with them, reinforcing and never really challenging their belief systems.[6]

---

[6] For a critical discussion of *The Daily Me* effects see Sunstein (2001).

One can block anything one chooses not to see. This filtering phenomenon is not new. On the contrary, it might help to explain, for example, why the Germans managed to organize the concentration camps (recall: no Good Will, no tragedy) while largely failing to grasp the horror of the Holocaust in all its magnitude (the agnostic Good Will).[7] What I am suggesting here is not that the Germans did not know at all, or that there was insufficient information available to anyone who cared to check it, but that many Germans, confronted by such horrors and by the costly consequences of any disagreement with the Nazi regime, preferred not to see what was happening. As Dahrendorf (1967) put it:

It is certainly true that most Germans 'did not know anything' about National Socialist crimes of violence: nothing precise, that is, because they did not ask any questions. (p. 349)

Questions are essential to gather information. Not asking questions, not seeing, not believing what one hears, filtering and rationalizing evil: this is the common trap into which *weak* (see the comment above about *akrasia*) Good Wills tend to fall. No one is less informed than the person who does not want to be informed.

Paradoxically, Good Wills may therefore be the worst witnesses, the more so the more morally good they are and hence, more sensitive to evil. Compare this to the conclusions reached by Pacuit et al. (2006) about the Kitty Genovese case, which later gave the name to the 'Genovese syndrome' to refer to the so-called bystander effect or diffusion of responsibility:

In 1964, a young woman [Kitty Genovese] was stabbed to death while 38 neighbours watched from their windows but did nothing. The reason was not indifference, but none of the neighbours had even a default obligation to act, even though, as a group, they did have an obligation to take some action to protect Kitty. (p. 311)

As a consequence, Good Wills may have to be forced to keep their eyes and ears open in front of the horrors that are being committed in their backyards. This might seem almost a torture. It reminds one of the 'Ludovico technique' in *A Clockwork Orange* (1971), the cult film directed and produced by Stanley Kubrick. There, the protagonist, Alex, is forced to keep his eyes mechanically and painfully wide-open, while being shown scenes of intense violence, cruelty, and social aberration, including *The Triumph of the Will* by Leni Riefenstahl, the infamous propaganda documentary about the 1934 Nazi Party Congress in Nuremberg. Alex is not a Good Will but a psychopath, who enjoys violence. His conditioning is supposed to rehabilitate him. In the case of the Good Will, the metaphorical 'Ludovico technique' that should be applied by ICTs has a different effect, for it is supposed to prevent Alice from burying her head in the sand

---

[7] The issue of how much the German population knew about the Holocaust is still debated. Gellately (2001) has provided mass media evidence in favour of the hypothesis that Germans knew quite a lot about the Holocaust, but it seems that what the research shows, rather, is that they could have known quite a lot, had they wished to know it.

of ignorance. It is one of the ethical tasks that a free press and uncensored ICTs should have in any decent democracy.

10.5 Escaping the tragic condition

There may be plenty of reasons for being pessimistic about the tragedy of the Good Will, not least historical records. Perhaps information about preventable or solvable evils will keep pouring in, and we will forever be unable to do anything about them. One good thing about such pessimism, however, is that, if correct, it would require no action and Lucretius' attitude might be the only serious alternative. In contrast, if some optimism is even partially justified, the bad news is that this is cause for further toil, and not just pragmatically, but also theoretically, as more discussion of the possible strategies available to escape the tragic becomes indispensable. In this section, I hope to take some steps in such a direction.

There seem to be four main ways in which the tragedy of the Good Will might be escaped. Luckily, they are mutually compatible and hence possibly synergetic. Before discussing them, let me briefly outline them here:

1. the information/power gap may decrease, as information has already reached its peak, whereas power is catching up;
2. from quantity to quality of information: better informed Good Wills can act and exercise their augmented power better;
3. from the powerless observation of the single Good Will to the empowered interactions of multi-agent systems of Good Wills: global problems and distributed morality require global agents;
4. the ontological side of information: the need for an augmented ethics.

Each strategy requires some comments.

(1) More power. To begin with, although ICTs and the corresponding amount of available information have seen an extraordinary development in the last half-century, Good Wills have also witnessed a steady increase in their powers. For a rough estimate, one may adopt a brute translation into dollars per person. According to a study by the World Bank, despite corrections to previous analyses, there is robust evidence of continually declining poverty incidence and depth since the early 1980s. For 2005 we estimate that 1.4 billion people, or one quarter of the population of the developing world, lived below our international line of $1.25 a day in 2005 prices; 25 years earlier there were 1.9 billion poor, or one half of the population. Progress was uneven across regions. The poverty rate in East Asia fell from almost 80 percent to under 20 percent over this period. By contrast it stayed at around 50 percent in Sub-Saharan Africa, though with signs of progress since the mid 1990s. Because of lags in survey data availability, these estimates do not yet reflect the sharp rise in food prices since 2005. (Chen and Ravallio, 2008–2009, p. 2)

Of course, these are merely quantitative measures, but they do provide some ground for cautious optimism. Good Wills might be able to put ever more dollars where the bad news events conveyed by their ICTs occur, thus helping to restore some balance between information and power. Recently, for example, in response to the Haiti earthquake, it was possible to send a tweet in order to donate $10 to the Red Cross, with 100 per cent of the donation going to Haiti relief. The cell phone carrier kept nothing. The experiment was very successful.

(2) Better information. The second way of tackling the tragedy of the Good Will is by using the same ICTs, which can bring so much information about the evils in the world, to empower the *individual* Good Will. This is not a simple matter of more or less information. Depending on contexts and usage, more information might be a benefit (more control, more competition, more choice, and less censorship) or a curse, since sometimes less information might be preferable (more fairness and less bias, more privacy, more security). Too often these issues are left unqualified (what information?), and uncircumstantiated (information for whom? under what conditions? for what purpose?). Rather, empowering the single Good Will seems to be a matter of more 'quality information', in the sense that future ICTs should provide her with more guidance (what could be done effectively), feedback (whether and how the single agent's efforts and resources are affecting reality), more transparency (information constrains other agents' misbehaviour, as speed cameras show), more forecasting (information as prevention) and more engineering (information as building capability).

(3) Global agents. The careful reader might have noticed a tension between, on the one hand, the IT-heodicean problem and Oedipus' predicament (sufficient power, insufficient information) and, on the other hand, the tragedy of the Good Will and Cassandra's predicament (sufficient information, insufficient power). How is it possible that ICTs can generate both predicaments? If they are empowering both pragmatically and informationally, surely these are two sides of the same coin, so their effects should overlap and cancel each other out, at least to a large degree. Make Oedipus and Cassandra work together, as it were, and it won't be necessary to escape the tragic condition because none will arise in the first place.

The tension is indeed there, but the inference drawn from it is mistaken, for it is based on a confusion of levels of agency. The IT-heodicean problem affects the Good Will insofar as the latter refers to supra-individual agents. In this sense, it is ultimately *humanity* that is empowered by ICTs. For example, none of us individually could have done anything to prevent the devastating Sumatra–Andaman tsunamis, but humanity as a whole could and should. The tragedy of the Good Will, on the other hand, affects single individual agents: it is you and I, Mary and Peter, Alice and Bob who are subject by ICTs to the dialectic of being informed about evils against which we are largely unable to do anything of comparable magnitude. It is we individually who give scandal.

It follows that the third strategy consists in identifying this mismatch and re-aligning individual and global agents, in order to make sure that the latter inherit the *eudokia* of the former and act on it. It might be easier to overcome both the IT-heodicean problem and the tragedy of the Good Will if we could work on developing global artificial agents—i.e. non-human (engineered) and/or social (e.g. groups, organizations, institutions) global agents—capable of channelling and guiding the energies of the single Good Wills who constitute them. National states, NGOs, international organizations or multinational companies are just some examples of these sorts of supra-individual, global, artificial agents that are hybrids of other artificial agents (imagine the member states of the EU, or the software and hardware systems that contribute to the existence of a company) and individual people. This general strategy calls for more conceptual analysis, in order that we might understand artificial agents better and clear outlines about how moral artificial agents may be built, morally educated or trained, and controlled.

(4) Augmented ethics not super-ethics. It might be felt that the impact of ICTs on our lives could be entirely reduced to a matter of DUMB effects: Doing & Understanding More & Better. If this were the case, then DUMB effects would transform man (the supra-individual Good Will) into superman. Superman has super-responsibilities and so ICTs would require a super-ethics. The problem would then be that any super-ethics would be, for each of us single human agents, supererogatory, as it would require super-heroes. The mistake, in this case, is to confuse not only the level of agency, but also the scope of the impact of ICTs. ICTs are not just a matter of DUMB effects. As we saw in Section 1.3, ICTs *re-ontologize* (design and construct anew) the very nature of the infosphere, that is, of the environment itself, of the agents embedded in it and of their interactions. Since they also have an essentially ontic impact, they radically transform old realities and create entirely new ones. And because of their ontic impact, ICTs require an augmented ethics for the whole of humanity as the ultimate Good Will, not for individual super-heroes. It follows that nowadays the IT-heodicean problem and the tragedy of the Good Will call for an ethics of creators or demiurges and not of mere end-users of reality. Or, to put it slightly differently, since the Good Will is increasingly morally responsible for designing and implementing reality the way it is, the moral question concerning her responsibilities is as much ethical as ontological, namely how she (both as an individual and as a supra-individual or global agent) could act as a morally good demiurge. Her augmented responsibilities require an ecological approach to the whole reality.

10.5.1 *The Copenhagen Consensus: using information to cope with information*

Let me now illustrate the previous analysis by means of a specific case in which some of the suggestions made in Section 10.5 seem to have found an application. This is the *Copenhagen Consensus*, a project originally conceived and organized by Bjørn Lomborg and now run by The Copenhagen Consensus Center at the Copenhagen Business School under Lomborg's directorship.

The question addressed by the Copenhagen Consensus is: what would be the best ways to spend additional resources on helping developing countries? Resources are scarce, and their allocation is therefore a specific case of triage, which demands difficult choices among good projects. In 2004 (Lomborg, 2009), the project attempted to set priorities among a range of suggestions on how to improve standards of living in developing countries on the basis of a cost–benefit analysis. Eight economists, including four Nobel laureates, met on 24–28 May 2004 at a roundtable in Copenhagen, and produced a ranking, based on applied welfare economics, of the 30–50 identified opportunities on which $50 billion of new money for development initiatives might be best spent. Ten global challenges were chosen: civil conflicts, climate change, communicable diseases, education, financial stability, governance, hunger and malnutrition, migration, trade reform, and water and sanitation. With something close to unanimity, the panel put measures to restrict the spread of HIV/AIDS at the top of the ranking. It also rated all four top proposals 'very good', as measured by the ratio of social benefit to cost. The bottom of the list, however, aroused much controversy and quite rightly so. For all three of the schemes proposed to the panel for mitigating climate change (including the Kyoto protocol on greenhouse-gas emissions) were rated 'bad', meaning that their costs were estimated to outweigh their benefits. The panel met again in 2008 for a second round. Again, it ranked efforts to cut carbon-dioxide emissions last, and caused further criticism.

The reader who has appreciated the e-cological message of the previous chapters will certainly understand why the conclusions reached by the Copenhagen Consensus about the environment are far from what I consider to be reasonable. There is not much point in improving anything about human life, if there is no planet where that life could be spent. Cost–benefit analyses can be terribly blind, and I share Jeffrey Sachs' negative assessment of the ranking (Sachs, 2004). However, this should not prevent one from appreciating the positive features of the project. Regardless of whether one shares the conclusions of the panellists, several aspects of the Copenhagen Consensus resonate positively with the analysis developed in this chapter.

First, the Copenhagen Consensus itself should be interpreted as a supra-individual Good Will; that is, as a multi-agent system constituted by individuals, institutions, and communication systems satisfying those conditions laid down in the first section of this chapter (the six assumptions).

Second, the Good Will gave priority to information above any other consideration, including politics and religion. Of all the problems tackled, it was clear that the most pressing was to have some reliable information on which problems to tackle first. An ethics of information was the setting against which the decisional procedure took place.

Third, the Copenhagen Consensus clearly meant to offer a series of strategies to other global Good Wills (again, understood as supra-individual agents) while at the same time informing individual Good Wills (the public) about what it considered to be the most economically fruitful and morally justifiable approach to global challenges. So there was no confusion in levels of agency, while the needs of both individual and global agents were addressed.

Fourth, despite appearances, the Consensus adopted a strongly ecological approach: it was clear that it wished to provide a balanced assessment of how limited resources could be best deployed to improve the world. That some solutions to solve environmental problems were deemed to be unsatisfactory says a lot about both the solutions criticized (they are in need of huge improvements) and a solely economic approach to environmental and ethical problems, but nothing about the importance of the issues they were addressing.

Fifth, in a way that complements the previous remark about an e-nvironmental ethics, the Consensus was an explicit attempt to develop a demiurgic approach to global issues. One of the assumptions behind the Copenhagen exercise is that the world will change according to human initiatives and that sorting them out and prioritizing them is of vital importance.

Last, but equally importantly, since its beginning, the Copenhagen Consensus project has itself been subjected to open discussion and made the subject of that flow of information that ICTs have taught us to take for granted.

In synthesis, that the Copenhagen Consensus probably got its environmental priorities upside down is not a reason to reject the opportunities offered by a rational process of discussion about what needs to be done first. There seem to be few better ways of dealing with the world's most serious problems. A reference to Habermas or Rorty might be expected here, but I would rather point in the direction of Charles Sanders Peirce's ethics of research, and his famous invitation 'not to block the way of inquiry' (Peirce, 1931–1958, vol. I, para. 135).

CONCLUSION

ICTs have done much to improve our lives but also to make them more morally demanding. In this chapter, I explored a major problem, defined as the tragedy of the Good Will. The problem arises when there is a lack of balance between the increasing amount of information available to a morally good and responsible agent and the limited power enjoyed by such agent to act on it. I argued that such an imbalance is exacerbated by ICTs. I have also indicated some possible strategies to deal with it.

This chapter completes the line of reasoning that connects the information revolution and its relevance to the ethical discourse (Chapter 1), through the analysis of the role of semantic information in ethics (Chapter 2), with the investigation about the nature of evil (Chapter 9). In Chapter 2, we saw that a crucial step, in understanding the challenges posed by information ethics, is represented by a shift from a conceptualization that interprets the human agent as external to the infosphere (see the 'external RPT model' and Figure 2), to one which embeds her within it (see the 'internal RPT model' and Figure 3), as an informational organism among many others. We saw in

Chapter 7 how Alice then finds herself surrounded by many other kinds of agent not necessarily human, but either entirely artificial, like a webbot, or hybrid, like a company. In Chapter 8, we looked at the sort of constructionist value that she might hold, once she is placed in charge of the well-being of the infosphere. That line of reasoning now requires two further steps, namely two answers to two straightforward but difficult questions: what does it mean to conceptualize a human agent as an informational entity operating within the infosphere? And what happens to her internal life, once she is so conceptualized? The first question will be answered in the next chapter. The second will have to wait for Chapter 12.

11

The informational nature of selves

> When I consider every thing that grows
> Holds in perfection but a little moment,
> That this huge stage presenteth nought but shows
> Whereon the stars in secret influence comment...
>
> William Shakespeare, Sonnet 15.

SUMMARY

Previously, in Chapter 1, I argued that the information revolution—understood as a fourth revolution, after the Copernican, the Darwinian, and the Freudian—is deeply affecting our self-understanding. Questions about who we are, our personal identities and the nature of our selves are, of course, as old as philosophy,[1] so one may suspect that nothing new could sensibly be said about the topic. Yet such an attitude would be only partially correct. Our philosophical anthropology is changing profoundly. Human life is quickly becoming a matter of *onlife* experience, which reshapes constraints and offers new affordances in the development of our identities, their conscious appropriation, and our personal as well as collective self-understanding. Today, we increasingly acknowledge the importance of a common yet unprecedented phenomenon, which may be described as the construction of personal identities in the infosphere. In particular, during the last two decades or so, roughly since the appearance of Turkle (1995), a new area of investigation into the nature of personal identity has begun to emerge, due to the dramatic evolution of ICTs and their widespread impact on our lives. In this chapter, I shall explore the foundations of the construction of personal identities, by developing an informational analysis of the self. Who are we onlife, or in the infosphere? The broader theses that I shall defend are, first, that ICTs are, among other things, *egopoietic technologies* or technologies of construction of the self. They significantly affect who we are, who we think we are, who we might become, and who we think we might become, once our philosophical anthropology is updated to take into account an informational ontology. And, second, that ICTs, as egopoietic technologies, deeply influence our ethical relations with ourselves, offering new

---

[1] In researching this chapter, I relied especially on Martin and Barresi (2006), Perry (2008), and Sorabji (2006).

opportunities and risks in the ethical development of our selves and our lives. The two theses are articulated and supported through the following steps.

In Section 11.1, I shall rely on Plato's famous metaphor of the chariot in order to introduce a specific problem regarding the nature of the self as an informational multi-agent system: what keeps the self together as a whole and coherent unity? This question may be addressed from two perspectives. One is *synchronic* and focuses on what may constitute the self as a particular whole unity, continuously existing and behaving coherently at any given time. The other is *diachronic* and focuses on what may enable the self to remain that unity, or simply itself, at different times and through changes. Following this distinction, in Sections 11.2 and 11.3 I shall quickly outline two branches of the theory of the self, or *egology* for short. One concerns the *individualization* of the self as an entity (no substantialism, essentialism, or dualism is presupposed). The other concerns the *identification* of such an entity. I shall argue that both presuppose an informational approach, defend the view that the *individualization* of the self is logically prior to its *identification*, and suggest how such *individualization* can be provided in informational terms. In Section 11.4, I shall then offer an informational *individualization* of the self based on a tripartite model, illustrated in terms of a three-membrane description: the corporeal, the cognitive, and the conscious. This 3C model of the self helps to tackle the problem of the chariot. Once it is outlined, in Section 11.5 I shall use it to show how ICTs may be interpreted as egopoietic technologies, by illustrating how they affect each membrane. The informational interpretation of the self would be incomplete without the inclusion of a reflection on the very understanding of the self by the self. Such 'self-understanding' is provided in Section 11.6, where I shall borrow Aristotle's concept of *anagnorisis* ('realization') in order to support the view that selves are informational structures. In Section 11.7, I shall finally connect the purposeful shaping of the self to the constructionist values analysed in Chapter 8, shifting back from egology to ecology. This will introduce Chapter 12 on informational privacy, as I shall indicate in the conclusion.

11.1 Introduction: Plato and the problem of the chariot

In one of the most famous passages in the history of philosophy, Plato compares the soul—what in this chapter will be referred to as the self—to a chariot:

We will liken the soul to the composite nature of a pair of winged horses and a charioteer.... [T]he charioteer of the human soul drives a pair, one of the horses is noble and of noble breed, but the other quite the opposite in breed and character. Therefore in our case the driving is necessarily difficult and troublesome. (*Phaedrus* 246a–254e)

The tripartite analogy is too well known to require any explanation, but two aspects of it may be highlighted here, for they nicely introduce both the 'engineering' approach adopted in the following pages and the key problem on which I shall focus.

First, the approach. Plato quite literally interprets the self as a *multi-agent system* (MAS), and not just any MAS, but one that has a significantly technological nature. I reached a similar conclusion in Chapter 13 of Floridi (2011a), where I used the 'knowledge game' in order to discriminate between conscious (human) and conscious-less agents (zombies and robots), depending on which version of the game they can win. The knowledge game showed that conscious agents, like us, are a special kind of informational multi-agents, or inforgs. Now look carefully at Plato's text and you will see that he is talking about a MAS. The three agents are not three sides of a triangle, 'three men in a boat', a master and two slaves, or a family of two parents and a child. They are three components in a complex, engineered artefact, and one that was fairly advanced for the time. Plato's technological analogy of the multi-agent chariot is interesting both because it facilitates the application of a wealth of interesting results to the analysis of the self, already available in the literature on MAS (Wooldridge, 2009b), and because it invites a shift from a phenomenological or descriptive approach to the self to a constructionist or design-oriented approach, one that considers what it means to create (or at least what it means for something to constitute) such a chariot or multi-agent system. It is easy to realize, for example, that some of the classic challenges in the engineering of a MAS (Bond and Gasser, 1988; Sycara, 1998)—such as communication, coherence, rationality, successful interaction with the environment, and coordination and collaboration with other agents, to mention the most obvious—are just AI translations of classic issues in the philosophy of the self. Still from a design perspective, upbringing, training, education, and social and political practices and norms may easily be interpreted as self-engineering techniques, as Plato already knew, and any virtue ethics rightly assumes. The comparison could be extended, but the ethical implications are obvious: good engineering of the self is good virtue ethics. I shall briefly return to this self-engineering process in the conclusion. At the moment, let me highlight the second aspect, which, quite surprisingly, seems to have been overlooked by the vast literature on the Platonic analogy at least as much as its technological nature. A difficult question posed by any multi-agent analysis of a system, be this an engineered artefact, a society of agents (Minsky, 1986), a system of zombies able to win the knowledge game (Floridi, 2011a, pp. 311–12), an inforg or a biological self, is: what makes such a complex MAS a coherent unity and source of actions and keeps it as such?

The previous question may not immediately strike one as difficult in engineering contexts, where we build the MAS in which we are interested as a unit, but even there the problem soon becomes pressing, once we start considering slightly more complex scenarios, in which agents temporarily coordinate their actions and collaborate to achieve specific goals (e.g. a rowing team). In biology, the study of multi-cellular organisms made up of specialized tissues and organs already shows the complexity of the problem. In philosophy, one appreciates its difficult nature as soon as one recognizes in it an instance of the infamous problem of Theseus' ship. If one of the two horses is replaced, is it the same soul? And what happens if the charioteer decides to dismount the chariot and abandon the horses to their destiny? More seriously, it seems plausible to assume that the MAS in question is constituted by its interacting and coordinated components and may not survive either their replacement or their irrecoverable disappearance, but what about their evolution? Such questions help to clarify the fundamental challenge posed by the unity of the self. I shall refer to it as *the problem of the chariot* because it is the chariot and the tack that, in Plato's analogy, represent the fourth, hidden component that guarantees the unity and coordination of the system, thus allowing the self to be, persist, and act as a single, coherent, and continuous entity in different places, at different times, and through a variety of experiences. It is the problem of the chariot that poses a serious challenge to any information-based theory of the self, as we shall see in the next two sections.

11.2 Egology and its two branches

Plato's interest in the theory of the self, or *egology*, was ethico-political and epistemological, but not ontological. Therefore, his dialogues explore the life of the multi-agent system (the tripartite self, the socially structured city), but leave the problem of the chariot philosophically (if not mythologically) untouched. It is mainly from Descartes onwards that the unity, identity, and continuity of the I, or self, as an entity become the subjects of an ontological investigation in their own right. It takes the Christian emphasis on the concept of individual person and then the long-term fading of a Christian answer to what an individual person is, to place egology at the centre of philosophical attention first, and then turn it into a source of problems. Once modern egology becomes an ontology of the self, two branches soon emerge. *Diachronic* egology, understood as an ontology of personal *identity*, concentrates on the problems arising from the *identification* of a self through time or possible worlds, progressively moving towards metaphysics. *Synchronic* egology, understood as an ontology of *personal* identity, deals with the *individualization* of a self in time or in a possible world, thus placing itself at the heart of the philosophy of mind. For reasons that will become clear presently, in the rest of this chapter I shall focus only on synchronic egology. So let me devote the rest of this section to sketching the sort of approach that might be developed when dealing with diachronic egology informationally.

As is well known, the literature on diachronic egology offers two main alternatives. *Endurantism* argues that a self is a three-dimensional entity that wholly exists at each moment of its history, and the same self exists at each moment. *Perdurantism* argues that a self is a four-dimensional entity constituted by a series of spatial and temporal parts, somewhat like the frames of a film. In both cases, an ontology of the self is developed by presupposing some form of direct realism, according to which the model (description, theory, representation, analysis, etc.) of the system (the referent of the model, in this case the self, the I, or whatever is intended by personal identity as a feature of the world) can be developed through a non-mediated access to the system in itself. Such presupposition may be justified, but is certainly open to question for all those who, like myself, are convinced that any system, the self included, is always accessed and hence modelled at a given level of abstraction or LoA, as indicated in Chapter 3. This suggests an alternative approach, according to which the analysis of self 'identity' (*a* is this) and 'sameness' (this is the same *a* as that *a*) relations should be developed in terms of the relevant kinds of information (observables) that, once fixed, provide the referential framework required to satisfy the specific epistemic goals in question. If this is unclear, consider the following example. Whether a hospital transformed now into a school is still the same building seems a very idle question to ask, if one does not specify in which context and for which purpose the question is formulated, and therefore what the required observables are that would constitute the right LoA at which the relevant answer may be correctly provided. If the question is asked in order to get there, for example, then the relevant observable is 'location' and the answer is yes, they are the same building. If the question is asked in order to understand what happens inside, then 'social function' is the relevant observable and therefore the answer is obviously no, they are very different. The illusion that there might be a single, correct, absolute answer, independently of context, purpose, and LoA, leads to paradoxical nonsense. Nor does the retort that some LoAs should be privileged when personal identities are in question carry much weight. For the same analysis holds true when the entity investigated is the young Saul, who is watching the cloaks of those who laid them aside to stone Stephen (*Acts* 7:58), or the older Paul of Tarsus, after his conversion. Saul and Paul are, and are not, the same person; the butterfly is, and is not, the caterpillar; Rome is, and is not, the same city in which Caesar was killed and that you visited last year; you are, and yet you are not, the same person who went there. It depends on the LoA, and this depends on the purpose for which, and the context in which, the question is asked. Locke was right in urging us to be careful about the sort of question that one might ask about the same man, same substance, same soul, same consciousness, same set of memories, etc., and also about the LoA that one is naturally led to privilege (the consciousness one), especially from a first-person perspective. It is less clear whether he was also right—indeed coherent, for someone who acknowledged, correctly, not to know what substance might be in itself—in committing himself ontologically, when an informational (epistemological, for Locke) standpoint would have been sufficient. Identity and sameness relations are satisfied according to the LoAs adopted, and these, in turn, depend on the goals being pursued. This is not relativism: given a particular goal, one LoA is better than another, and questions will receive better or worse answers. The ship will be Theseus', no matter how many bits one replaces, if the question is about legal ownership (try a Theseus trick with the taxman); it is already a different ship, for which the collector will not pay the same price, if all one cares about are the original planks. Questions about diachronic identity and sameness are really teleological questions, asked in order to attribute responsibility, plan a journey, collect taxes, attribute ownership or authorship, trust someone, authorize someone else, and so forth. Insofar as they are dealt with metaphysically (modally or not, it does not matter), they do not deserve to be taken seriously. For in a LoA-free context they make no sense (although it might be intellectual fun to play idly with them), exactly like it makes no sense to ask whether a point is at the centre of the circumference without being told what the circumference is, or being told the price of an item but not the currency in which it is given. It is not just the degree of confidence in the re-identification through time or possible worlds of someone as the same someone that is a matter of epistemology; it is the very process of identification and re-identification that needs to be conceptualized in a fully epistemological way, i.e. informationally, through a careful analysis of the information that is required and hence needs to be made available in order to provide a reasonable answer, because:

> That which has made the difficulty about this relation [sameness], has been the little care and attention used in having precise notions [i.e. information, my specification] of the things to which it is attributed. (Locke, 1979, bk. II, ch. XXVII, §§ 27–30)

Let us now turn to the individualization of the self.

11.3 Egology as synchronic individualization

Before establishing, informationally (i.e. at the right LoA), whether this *a* is even approximately the same as that *a*, it seems that one needs to have some information about what this *a* is in the first place. Plato was right: you cannot look for something, let alone know whether you have found it, unless you know what you are looking for. So, *individualization* logically precedes *identification*. Of the many approaches that seek to characterize the nature of the self, two stand out as popular and promising for the task ahead. According to the Lockean approach, the identity of the self is grounded in the unity of consciousness and the continuity of memories. According to the narrative approach (Schechtman, 1996), the self is a socio- or (inclusive or) auto-biographical artefact. We have already encountered Locke in the previous section. Regarding the narrative approach, the following passage elegantly illustrates its essential gist:

> But then, even in the most insignificant details of our daily life, none of us can be said to constitute a material whole, which is identical for everyone, and need only be turned up like a page in an account-book or the record of a will; our social personality is created by the thoughts of other people. Even the simple act which we describe as 'seeing someone we know' is, to some extent, an intellectual process. We pack the physical outline of the creature we see with all the ideas we have already formed about him, and in the complete picture of him which we compose in our minds those ideas have certainly the principal place. In the end they come to fill out so completely the curve of his cheeks, to follow so exactly the line of his nose, they blend so harmoniously in the sound of his voice that these seem to be no more than a transparent envelope, so that each time we see the face or hear the voice it is our own ideas of him which we recognise and to which we listen. (Proust, 1982, p. 20)

We 'identify' (provide identities) to each other, and this is a crucial (although not the only) variable in the complex game of the construction of personal identities, especially when the opportunities to socialize are multiplied and modified by new ICTs, as we shall see.

Now, in both cases, *individualization*—the characterization or constitution of the self—is achieved through forms of information processing: consciousness and memory are dynamic states of information, but so is any kind of personal or social narrative. So both the Lockean and the Narrative approach presuppose the existence of individual agents endowed with the right sorts of informational skill. Hume saw this quite clearly, but was also aware that his account of the 'informational' self completely failed to explain its unity. The passage is famous, but it is worth quoting at length while keeping in mind the problem of the chariot:

> [H]aving thus loosen'd all our particular perceptions [bits or streams of information separate from each other], when I proceed to explain the principle of connexion, which binds them together, and makes us attribute to them a real simplicity and identity; I am sensible, that my account [the bundle and then the commonwealth] is very defective.... If perceptions are distinct existences, they form a whole only by being connected together. But no connexions among distinct existences are ever discoverable by human understanding. We only feel a connexion or a determination of the thought, to pass from one object to another. It follows, therefore, that the thought alone finds personal identity, when reflecting on the train of past perceptions, that compose the mind.... Most philosophers seem inclin'd to think, that personal identity arises from consciousness; and consciousness is nothing but a reflected thought or perception [information processing]. The present philosophy, therefore, has so far a promising aspect. But all my hopes vanish, when I come to explain the principles, that unite our successive perceptions in our thought or consciousness.... In short, there are two principles, which I cannot render consistent; nor is it my power to renounce either of them, viz. that all our distinct perceptions are distinct existences and that the mind never perceives any real connexion among distinct existences [the infrastructure that keeps them together as a unity]. Did our perceptions either inhere in something simple and individual [the tack], or did the mind perceive some real connexion among them, [if there were a chariot] there would be no difficulty in the case. For my part, I must plead the privilege of a sceptic, and confess that this difficulty is too hard for my understanding. I pretend not, however, to pronounce it absolutely insuperable. Others, perhaps, or myself, upon more mature reflection, may discover some hypothesis, that will reconcile those contradictions. (Hume, 2007, vol. 1, p. 400, App., §§ 20–1)

In short: if the self is made of information (perceptions or narratives, or any other informational items one may privilege), then a serious challenge is to explain how that information is kept together as a whole, coherent, sufficiently permanent unity. If there is no narrator—and there cannot be, because the narrative theory of the self describes the narrator as the narrative, and presupposing a narrator would only shift the problem one step back—what prevents the narrative from being a completely random, incoherent, and disjointed selection of miscellaneous bits of stories? The answer seems to be twofold.

First, there is a blocking manoeuvre, which prevents us from biting the bullet: selves, if they are narratives, are coherent and unitary narratives, at least when dealing with healthy, ordinary selves. We owe this to Kant, who made a step forward by arguing convincingly (or at least so plausibly as to shift the burden of proof onto the shoulders of those who disagree) that the unified coherence of information about the external world, synthesized by the epistemic agent, could be guaranteed only by the unity of the very agent's self that is its (that is, of the information) source. So Kant's transcendental argument, in favour of the unity of the self, is a partial, epistemological solution to the ontological problem of the chariot, or the unity of the informational self. Yet, it is only 'partial' because, like all transcendental arguments, it is non-constructive, to use a mathematical distinction. At best, it shows that a specific characterization of the self as a whole unity of consciousness is the required condition of possibility for the meaningful coherence of the stream of empirical information generated by the agent. The semanticization of the world requires a unity of perspective, so presence of the former guarantees the presence of the latter. How such unity and coordination come to be there in the first place and have those features is not the issue addressed. It is the part of the question left unanswered. Kant is essentially arguing that the chariot and the tack must be there and have the features that they have in order for the MAS to work informationally as successfully it does, but he provides no further insight into how such unity arises or is reached in the first place, and then maintained. So we are still left with the problem: granted that the unity of the narrative or informational self and (perhaps) its crucial role in the delivery of a coherent experience of the world must be conceded, what generates it and keeps it together? If the flow of information (or Humean perceptions, or narrative elements) is no more than an aggregate, it must fail to form a coherent unity, let alone a conscious self, unless it is consistently and non-transiently bound together as a whole, but then the binding, that is, the problem of the occurrence and maintenance of the chariot, is precisely an instance of our recurring difficulty.

Clearly, we need a second, more constructive manoeuvre. It is going to be hard to tackle a problem that Plato, Hume, and Kant left unsolved. We do have the advantage of coming after them and hence being able to learn from, and build upon, their work. Still, such advantage might come at a high price, in terms of what plausible solutions are still viable. In the following section, I shall follow Sherlock Holmes' advice: having eliminated the impossible, whatever remains, however improbable, must be the truth. But the reader should know that I am aware that '[o]thers, perhaps, upon more mature reflection, may discover some hypothesis, that will reconcile those contradictions' and that escaped my understanding.

11.4 A reconciling hypothesis: the three membranes model

Kant was able to show that the unity of the self must be presupposed as the source that 'unite[s] our successive perceptions in our thought or consciousness', to quote Hume once again. In this section, I shall suggest that such informational unity of the self may be achieved, or at least described, through a three-phase development of the self. The model I am going to propose is, I take it, biologically and informationally plausible, but it is, admittedly, somewhat figurative. I hope the reader will not object. On the one hand,

> To tell what it really is [the form of the soul, or the characterisation of the self] would be a matter for utterly superhuman and long discourse, but it is within human power to describe it briefly in a figure; let us therefore speak in that way. (Plato, *Phaedrus*, 246a)

On the other hand, the goal is ultimately that of explaining in what sense ICTs are egopoietic technologies that affect our ethical construction of ourselves as inforgs, and I hope that the model at least achieves this much.

The 'reconciling hypothesis', to use Hume's terminology, that I wish to articulate is strategically simple, if a bit complicated in its details. Here it is. In the same way that organisms are initially formed and kept together by auto-structuring (i.e. auto-assembling and, within the assembled entity, auto-organizing)[2] physical (henceforth *corporeal*) membranes, which encapsulate and hence *detach* (bear with me, more on this below) parts of the environment into biochemical structures that are then able to evolve into more complex organisms, selves too are the result of further encapsulations, although of informational rather than biochemical structures. The basic mechanism of encapsulation, detachment, and internal auto-organization, I suggest, is the same, or at least we should take seriously the possibility that it might be the same from a minimalist perspective (Ockham's razor). If this is the case, then selves emerge as the last step in a process of detachment from reality that begins with a corporeal membrane encapsulating an organism, proceeds through a cognitive membrane encapsulating an intelligent animal, and concludes with a consciousness membrane encapsulating a mental self or simply a mind. Of course, one may add as many mid-steps as required, yet these three—the corporeal, the cognitive, and the consciousness or simply 3C—seem to be the main stations at which the train of evolution has called. Each step builds on the previous one (supervenience) and, at each step, more, not less, distance is placed between the entity and its environment. Each membrane is a defence of the structural integrity of what it encapsulates, against the surrounding environment. Of course, in moving from the corporeal to the cognitive to the consciousness membrane, there is an increasing process of virtualization. Yet, there is nothing metaphorical in this, as anyone acquainted with the concept of the virtual machine in computer science can readily appreciate. Indeed, it has become almost fashionable to compare the mind to a virtual machine,[3] even if, without some further theorizing, the comparison only hides and fails to solve the usual *homunculus* problem. I agree with Pollock (2008) that, in general, the whole approach seems a refined version of the sort of classic functionalism originally developed by Putnam (1960). As such, 'virtual functionalism' does not seem

---

[2] In the chapter I use 'auto-' instead of 'self-' in order to avoid potential confusions whenever necessary.
[3] See for example the symposium in Hayes et al. (1992) or the debate between Densmore and Dennett (1999) and Churchland (1999). To the best of my knowledge, Aaron Sloman has been the first to call attention to the computational theory of virtual machines as a way to model the mind, see Sloman and Chrisley (2003) for a more recent statement.

to be much more instructive than the old-fashioned kind. For example, Pollock (2008) writes:

> If I am a virtual machine, which virtual machine am I? The proposal is that I am a virtual machine that cognizes. But there is more than one such virtual machine implemented on my body. (p. 291)

Clearly it is the concurrence of machine-like processes that is 'solved' by the virtualization of the machine itself, a gain that does not seem to be a substantive progress with respect to any alternative analysis in terms of multi-functionality. Unfortunately, virtual machines generate virtual problems about virtual minds that are virtually conscious.

Nor is there any problem about each membrane being auto-poietically structured through auto-assembly and auto-organization: at each stage, local relations act on local building blocks to generate a new divide, within the old environment, between a new inside and hence a new outside. This is the general hypothesis. Let me now add some details about the model. The three phases concern the evolution of organisms, then of intelligent animals and finally of self-conscious minds. Each phase contributes to the construction of the ultimate personal identity of the human organism in question.

11.4.1 Phase one: the corporeal membrane and the organism

The first phase begins in an environment in which there are not yet biotic structures. There are, however, physical structures, that is, patterns of physical data understood as asymmetries or lacks of uniformities, e.g. lights, noises, or magnetic fields. Such data might be flowing around, but there are no senders or receivers yet. This might be seen as a stage when there are environmental data and patterns that might be exploitable as information by the right sort of agent for their purposes, before there is any kind of communication. We move from a pre-biotic to a post-biotic environment once some structures in the environment become encapsulated cells through a *corporeal membrane*. The encapsulation of part of the environment through a corporeal membrane allows the separation of the interior of a cell from the external world. This is the ontological function of the membrane, as a hard-wired divide between the inside, the individual biotic structure, and the outside, the environment. Its negentropic function is to enable the organism to interact with the environment to its own advantage and withstand for as long and as well as possible the second law of thermodynamics. The epistemological function of the membrane is that of being selectively permeable, thus enabling the cell a variety of degrees of inputs and outputs with respect to the environment. At this stage, data are transduceable physical patterns, that is, physical signals now seen as broadcasted by other structures in the environment, which are captured by the permeable membrane of the organism. The body is a barrier that protects the stability of the living system (physical homeostasis). A good example is a sunflower.

11.4.2 Phase two: the cognitive membrane and the intelligent animal

We move from pre-cognitive to post-cognitive systems once data become encodable resources exploitable by organisms through some language broadly conceived (sounds, visual patterns, gestures, smells, behaviours, etc.). This requires a *cognitive membrane*, which allows the encapsulation of data (some form of memory) for processing and communication. The streams of data, which were before quantities without direction (scalars), broadcasted by sources not targeting any particular receiver (e.g. the sun generating heat and light, or the earth generating a magnetic field), acquire a direction, from sender to receiver (vectors), and an interpretation (e.g. noises become sounds interpreted as alarms). From now on, Shannon's classic communication model applies. The body becomes an interface and the cognitive membrane is a semi-hard-wired (because configurable) divide between the cognitive system and its environment, that is, a barrier that further detaches the organism from its surroundings, and allows it to exploit data processing and communication in its fight against entropy. The stability (cognitive homeostasis) now concerns the internal data within the system and their codification: memory and language. A good example is a bird on the sunflower.

11.4.3 Phase three: the consciousness membrane and the self-conscious mind

The third phase is represented by the evolution of the consciousness membrane. We move from pre-conscious (aware) to post-conscious (self-aware) systems once data become re-purposable information, including conventional meanings (e.g. sounds become a national anthem). The consciousness membrane is soft-wired (programmable). The body becomes the outside environment for an inside experience, and stability now concerns the self within the system (mental homeostasis). The self, or I, becomes the fixed point of the detachment function, to use a mathematical analogy. To put it in Cartesian terms, the self or mind or I is indivisible, not because it cannot be divided (detached from itself) but because the division (detachment) does not generate two selves, or minds, or I's, but mere schizophrenia. This is why there is no further, healthy detachment of the self from the self, but only increasing degrees of self-reflection. Once the self, conscious mind or I emerges, it appropriates and unifies what happens to the corporeal and cognitive levels as his or her own experiences. In Floridi (2011a), I have defined this as the 'I before Mine' hypothesis, or IBM. A good example is a gardener watching the bird on the sunflower.

The 3C model just sketched helps us to deal with the problem of the chariot and, in so doing, it finally enables us to clarify why, and in what sense, ICTs are technologies of the self. Each membrane, and hence each step in the detachment of the individual from the world, is made possible by a specific, auto-reinforcing, bonding force. The corporeal membrane relies on chemical bonds and orientations. The cognitive membrane relies on the bonds and orientations provided by what is known in information theory as mutual information, that is the (measure of the) interdependence of data (the textbook example is the mutual dependence between smoke and fire). And, finally, the consciousness membrane relies on the bonds and orientations provided by semantics (here narratives provide plenty of examples), which ultimately makes possible a stable and long-lasting detachment from reality. At each stage, corporeal, cognitive, and consciousness elements fit together in structures (body, cognition, mind) that owe their unity and coordination to such bonding forces. The more virtual the structure becomes, the more it is disengaged from the external environment in favour of an autonomously constructed world of meanings and interpretations, the less physical and more virtual the bonding force can be. The self emerges as a break with nature, not as a super connection with it. Such an 'unnatural' break requires a collaborative and cumulative effort by generations through time. No individual can successfully rely just on a private semantics (what Wittgenstein calls private language). This is why a single human being needs to be embedded, at a very early stage of development, within a community, in order to grow as a healthy conscious mind: mere corporeal and cognitive bonds, in one-to-one interactions with the external environment, fail to give rise to, and keep together, a full self, for which language, culture, and social interactions are indispensable. The problem of the chariot therefore may be solved only by taking into account all the bonding forces—physical, cognitive, and semantic—that progressively generate the unity of the self. As Hume discovered, by itself each of them is insufficient.

The 3C model as a solution to the problem of the chariot acquires further plausibility once we apply it to explain the impact of ICTs on the construction of personal identity. This is the topic of the next section.

11.5 ICTs as technologies of the self

If the self is made possible by something like the healthy development of all the three membranes, then any technology capable of affecting any of them is *ipso facto* a technology of the self. Already Plato, for example, had acknowledged that humanity had changed because of the invention of writing. Now, ICTs are the most powerful technologies to which selves have ever been exposed. They induce radical modifications (a re-ontologization) of the contexts (constraints and affordances) and praxes of self-poiesis, by enhancing the corporeal membrane, empowering the cognitive membrane, and extending the consciousness membrane. Let us have a quick look. The following examples are not meant to provide an exhaustive analysis but only a variety of brief illustrations about embodiment, space, time, memory and interactions, and finally perception.

11.5.1 Embodiment: from dualism to polarism

We have seen that each membrane contributes to the construction of the self: the body, its cognitive functions and activities, and the consciousness that accompanies them are inextricably mixed together to give rise to a self and its personal identity. Diachronically, each membrane must be there for the others to occur. Yet this truism hides the fundamental fact that, once a membrane is in place, the particular inside that it detaches from the relevant outside becomes conceivably independent of the previous stages of development. It is correct to stress that there is no butterfly without the caterpillar, but insisting that once the butterfly is born the caterpillar must still be there for the butterfly to live and flourish is a conceptual confusion. There is no development of the self without the corporeal and the cognitive faculties, but once the latter have given rise to a consciousness membrane, the life of the self may be entirely internal and independent of the specific body and faculties that made it possible. While in the air, you no longer need the springboard, even if it was the springboard that allowed you to jump so high, and your airborne time is limited by gravity. Wittgenstein is right in saying that no private language may subsist without a public language, but once a public language is available, the speaker may throw away the public language (privatize it, as it were), like the famous ladder. This does not mean that the self requires no physical platform. Some platform (some data structure) is required to sustain the constructed self. And it does not mean that just any platform will do either. But it does open the possibility of a wider choice of platforms and of the temporary stability of a permanent self even when the platform changes. Our culture, so imbued with informational concepts, finds the very idea of eterobodiment of the self, or the self as a cross-platform (not a-platform) structure, perfectly conceivable, witness the debate about mind uploading and body swap in the philosophy of mind. It is not the science fictional nature of such thought experiments that is interesting—in many cases, it tends to be distracting and fruitlessly scholastic—but the readiness with which many seem to be willing to engage with them, because this is indicative of the particular impact that ICTs have had on how we conceptualize selves.

11.5.2 Space: the detachment between location and presence

We saw in Chapter 3 that, through the phenomenon of telepresence, ICTs magnify (make more salient and increase) the distinction between presence and location of the self. A living organism (e.g. a spider) is cognitively present only where it is located as an embodied and embedded information-processing system. A living organism aware of its information processes (e.g. a dog dreaming) can be present within such processes (e.g. chasing dreamed rabbits) while being located elsewhere (e.g. in the house). But a self, that is, a living organism self-aware of its own information processes (e.g. you) and its own presence within them, can choose where to be. The self, and mental life in general, is located in the brain but not present in the brain. Thus the locus of the self is the brain but the self is not present in the brain.

11.5.3 Time: the detachment between outdating and ageing

ICTs increase the endurance effect, for in digital environments exactly the same self may be identified and re-identified through time. The problem is that the virtual may or may not work properly, it may be old or new, but it does not grow old; it outdates, it does not age. Nothing that outdates can outdate more or less well. On the contrary, the self ages and does so more or less well. The effect, which we have only started to experience and with which we are still learning to cope, is a chronological misalignment between the self and its online habitat, between parts of the self that age and parts that simply outdate. Asynchronicity is acquiring a new meaning in onlife contexts.

11.5.4 Memories and interactions: fixing the self

We have seen that memory plays a crucial role in the construction of personal identity. Obviously, any technology, the primary goal of which is to manage memories, is going to have an immense influence on how individuals develop and shape their own personal identities. It is not just a matter of mere quantity; the quality, availability, accessibility, and replaying of (records of) personal memories may deeply affect who we think we are and may become. The Korean War was, for example, the first major conflict with a soundtrack: soldiers could be listening to the same songs at home, in the barracks, or during a battle (see Figure 13).

Similar 'repeatable' memories cannot but have a deep impact on how subjects exposed to them shape their understanding of their past, the interpretation of what has happened to them, and hence how they make sense of who they are. We are the first 'replay' generation, and our *madeleines* are digital.

Until recently, the optimistic view was that ICTs empowered individuals in their personal identity DIY ('do it yourself'). The future is more nuanced. Recorded memories tend to freeze the nature of their subject. The more memories we accumulate and externalize, the more narrative constraints we provide for the construction and development of personal identities. Increasing our memories also means decreasing the degree of freedom we might enjoy in defining ourselves. Forgetting is also a self-poietic art. A potential solution, for generations to come, is to be thriftier with anything that tends to fix the nature of the self, and more skilful in handling new or refined self-poietic skills. Capturing, editing, saving, conserving, and managing one's own memories for personal and public consumption will become increasingly important not just in terms of protection of informational privacy, as we shall see in the next chapter, but also in terms of a morally healthy construction of one's personal identity. The same holds true for interactions, in a world in which the divide between online and offline is being erased. The *onlife* experience does not respect dimensional boundaries, with the result that, for example, the scope for naïve lying about oneself on Facebook is increasingly reduced (these days everybody knows if you are, or behave like, a dog online). In this case, the solution may lie in the creation of more affordances and spaces for self-expression and self-poiesis (see e.g. Diaspora, the open-source Facebook).

11.5.5 Perception: the digital gaze

The gaze is a composite phenomenon, with a long and valuable tradition of analyses (Lacan, Foucault, Sartre, feminist theory). The idea is rather straightforward: the self observes 'the observation of itself' by other selves (including, or sometimes primarily itself) through some medium. It should not be confused with seeing oneself in a mirror

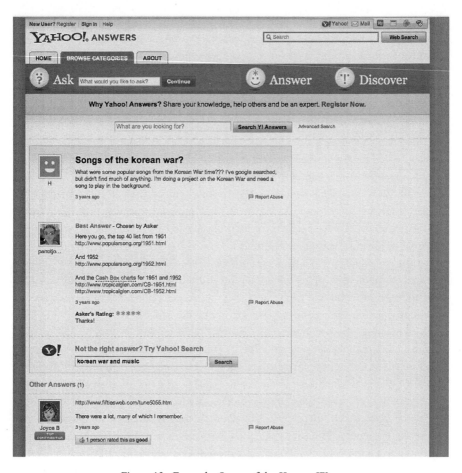

Figure 13. Example: Songs of the Korean War

(ego surfing or vanity googling). It is rather comparable to seeing oneself as seen by others, by using a mirror ('what do people see when they see me?'). In child development, the gazing phase is theorized as a perfectly healthy and normal stage, during which the individual learns to see her- or himself by impersonating, for example, a chair ('how does the chair see me?'), or simply placing her- or himself in someone's shoes, as the phrase goes. The digital gaze is the transfer of such phenomenon in the infosphere. The self tries to see how others see itself, by relying on ICTs that greatly facilitate the gazing experience. In the end, the self uses the digital imaginary concerning itself to construct a virtual identity through which it seeks to grasp its own personal identity (the question 'who am I for you?' becomes 'who am I online?'), in a potentially recursive feedback loop of adjustments and modifications leading to an onlife equilibrium between the offline and the online selves. The observing is normally hidden and certainly not advertised. And yet, by its very nature, the digital gaze must be understood both as an instance of presumed 'common knowledge' of the observation ('I know that you know that I know, etc. . . . that this is the way I am seen by you') and as a private experience (it is still *my* seeing of myself, even if I try to make sure that such seeing is as much like your seeing as I can). The digital translation of the gaze has important consequences for the development of personal identities.

First, there is the amplification, postponement (in terms of age), and prolongation (in terms of duration) of the gazing experience. This means that the *ontic feedback*—the tendency of the gaze to re-ontologize (change the very nature of) the self that is subject to it—becomes a permanent feature of the *onlife* experience.

Second, through the digital gaze, the self sees itself from a third-person perspective through the observation of itself in a proxy constrained by the nature of the medium, which affords only a partial and specific reflection.

Third, the more powerful, pervasive and available ICTs are, the more the digital gaze may become mesmerizing: one may be lost in one's own perception of oneself as attributed by others in the infosphere.

And finally, the experience of the digital gaze may start from a healthy and wilful exposure/exploration by the self of itself through a medium, but social pressure may force it on selves that are then negatively affected by it, leading them to re-ontologize themselves heteronomously.

11.6 The logic of realization

We are coming to the end of our exploration, but before drawing a final conclusion one more topic needs to be covered for the sake of completeness. In the previous pages, we have quickly looked at the process of progressive detachment (membranes) of the self from the non-self (the world), and at the role played by ICTs in the construction of personal identities. The goal was to understand our nature as inforgs and prepare the ground for the ontological interpretation of informational privacy, to be developed in the following chapter. The process itself, however, is also part of the narrative through which we semanticize reality, i.e., through which we make sense of our environment, of ourselves in it, and of our interactions with and within it. In other words, the process of progressive detachment of the self from the non-self is always and inevitably reconstructed by the self from the self's perspective. Although the ultimately internal nature of such perspective is inescapable, it can be made critically explicit, and this is the concluding move we need to make now.

In order to do so, I suggest we borrow a concept from Aristotle's *Poetics*, that of *anagnorisis*. The Greek word is translated differently depending on the context. In Aristotle, the phenomenon of *anagnorisis* refers to the protagonist's sudden recognition, discovery, or realization of his or her own or another character's true identity or nature. Through *anagnorisis*, previously unforeseen character information is revealed. Classic narratives in which *anagnorisis* plays a crucial role include *Oedipus Rex* and *Macbeth*. More recently one may mention *The Sixth Sense*, *The Others*, or *Shutter Island*. I shall not spoil the last three, if the reader has not watched them. Generalizing, one may say that, given an information flow, *anagnorisis* is the information process (epistemic change) through which a later stage in the information flow (the acquisition of new information) forces the correct re-interpretation of the whole information flow (all information previously and subsequently received). For this reason, I prefer to translate *anagnorisis* as *realization*. Figure 14 provides an illustration.

The logic of realization should not be confused with the logic of falsification. At point R (for realization), some information becomes available that does not make some information at point B (for before) false, but rather provides the right perspective from which to interpret it. For example, at R it is still true that at B Alice loves Bob, but now (at R) Alice realizes that it is fraternal love, and this is not going to change in A (for after). The difference should be clear once we see that the information at point R also affects information at point A not yet available (and hence hardly falsifiable). Thus, realization is a concept that belongs more to hermeneutics than to epistemology.

If we now apply the logic of realization to the development of the 3C model, we may understand that it is the self that is speaking about itself, and then appreciate that it is actually through the self that information becomes self-aware. Let me be less abstract. In a different context (Floridi, 2011a, ch. 15), I have defended a view of the world as the totality of informational structures dynamically interacting with each other. If this is the case—or, at least, in order for a philosophy of personal identity to be consistent with such a view—selves too must be interpreted as informational structures. Selves are the ultimate negentropic technologies, through which information, understood ontologically, temporarily overcomes its own metaphysical entropy, becomes conscious, and is finally able to recount the story of its own emergence in terms of a progressive detachment from external reality. There are still only informational structures. But some are things, some are organisms, and some are minds, intelligent and self-aware beings. Only minds are able to interpret and take care of other informational structures as things or organisms or selves. And this is part of our special position in the universe.

11.7 From the egology to the ecology of the self

ICTs have made possible unprecedented phenomena in the construction of the self. Self-poiesis today means tinkering with the self, with still unknown and largely unassessed risks and rewards. Amazing as all this already is, we are witnessing only the beginning of an information revolution, which may have even more radical

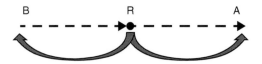

Figure 14. The logic of realization

consequences in our self-understanding and the constructions of our own identities. It is, as they say, an interesting time in which to live. In the previous pages, I have outlined what may be a fruitful approach to start understanding the construction of personal identities in the infosphere. Who we are and can be in the infosphere is a complicated and challenging issue, and I am fully aware that much more can and should be done in order to develop our new egology. More philosophical insight and better understanding are needed in order to cope successfully and fruitfully with the new affordances, constraints, and challenges brought about by ICTs' exponential development. Unfortunately, as if this were not already a gigantic task, it needs to be paralleled by the development of an equally robust ethics of self-poiesis, a new ecology of the self fully capable of meeting the demands of a healthy life spent in the infosphere. There is much that needs to be done on the ethical front as well. All this won't be easy, but it can be done, and it is certainly worth a try.

CONCLUSION

In this chapter, I presented an informational approach to the nature of personal identity. I relied on Plato's famous metaphor of the chariot to introduce a specific problem regarding the nature of the self as an informational multi-agent system: what keeps the self together as a whole and coherent unity? I then outlined two branches of the theory of the self: one concerning the *individualization* of the self as an entity, the other concerning the *identification* of such entity. I argued that both presuppose an informational approach, defended the view that the *individualization* of the self is logically prior to its *identification*, and suggested that such *individualization* can be provided in informational terms. Hence, I offered an informational *individualization* of the self, based on a tripartite model, which can help to solve the problem of the chariot. I used this model of the self in order to highlight how ICTs may be interpreted as egopoietic technologies. I introduced the concept of 'realization' (Aristotle's *anagnorisis*) and supported the rather Spinozian view according to which, from the perspective of informational structural realism (Floridi, 2008g, 2011a), selves are the final stage in the development of informational structures, for they are the semantically structuring structures conscious of themselves. I concluded with a reference to the purposeful shaping of the self, in a shift from egology to ecology. It is this shift that will require our attention now. In an environment in which we are (conceptualized as) our information, informational privacy becomes a major issue, as we shall see in the next chapter.

12

The ontological interpretation of informational privacy

> We, who have a private life and hold it infinitely the dearest of our possessions...
>
> Virginia Woolf, 'Montaigne' (1992), vol. 1, p. 60.

SUMMARY

Previously, in Chapter 8, I argued in favour of a constructionist information ethics that looks after the environment in all its forms, the infosphere, and its inhabitants in all its manifestations, the informational entities populating the infosphere. In Chapter 11, I defended an informational interpretation of the self. Clearly, anything that affects the informational life of the self affects its very essence. This is the problem of informational privacy analysed in this chapter. The chapter is divided into two parts.

In the first half (Sections 12.1–12.6), I shall articulate and defend an interpretation of informational privacy and of its moral value. I shall argue (1) that informational privacy is a function of the ontological friction in the infosphere, that is, of the forces that oppose the information flow within the space of information (Section 12.3); (2) that digital ICTs affect the ontological friction by changing the nature of the infosphere (re-ontologization); (3) that digital ICTs can therefore both decrease and protect informational privacy but, most importantly, they can also alter its nature and hence our understanding and appreciation of it; and (4) that a change in our ontological perspective, brought about by digital ICTs, suggests considering a person as being constituted by his or her information and hence regarding a breach of one's informational privacy as a form of aggression towards one's personal identity.

In the second half (Sections 12.7–12.8), I shall discuss four types of interesting challenges confronting any theory of informational privacy: (1) parochial ontologies and non-Western approaches to informational privacy; (2) individualism and the anthropology of informational privacy; (3) the scope and limits of informational privacy; and (4) public, passive, and active informational privacy. The main point addressed there is that some problems should be taken seriously, lest our interpretation of informational privacy becomes a mere linguistic stipulation regarding the correct usage of 'privacy' in various languages or cultural contexts. I shall argue that the ontological theory of informational privacy can cope with such challenges fairly successfully. As usual, the conclusion synthesizes the results of this chapter and introduces the next one.

12.1 Introduction: the dearest of our possessions

'"One of these days d'you think you'll be able to see things at the end of the telephone?" Peggy said, getting up' (Woolf, 1965, p. 327–8). She will not return to her wondering again, in the remaining pages of Virginia Woolf's *The Years*. The novel was published in 1937. Only a year earlier, the BBC had launched the world's first public television service in London, and Alan Turing had published his groundbreaking work on computing machines (Turing, 1936).

Distracted by a technology that invites practical usage more readily than critical reflection, Peggy only half-perceives that new ICTs are transforming society profoundly and irrevocably. The thirties were laying the foundations of our information society. It was difficult to make complete sense of such a significant change in human history, at this early stage of its development. Today, the commodification of ICTs, begun in the seventies, and the consequent spread of a global information society since the eighties, are progressively challenging the right to informational privacy, at least as westerners still conceived it in Virginia Woolf's times (her evocative phrase that opens this chapter appears in an essay on Montaigne, published in *The Common Reader*, 1925).

'[We], who have a private life and hold it infinitely the dearest of our possessions . . .' (vol. 1, p. 60), find protecting it ever more difficult in a social environment increasingly dependent on Peggy's futuristic technology. The problem is pressing. It has prompted a stream of scholarly and scientific investigations, and there has been no shortage of political decisions and legally enforceable measures to tackle it.[1] The goal of this chapter, however, is not to review the very extensive body of literature dedicated to informational privacy and its legal protection, even in the relatively limited area of information and computer ethics studies. Rather, it is to argue in favour of a new ontological interpretation of informational privacy and of its moral value, on the basis of the conceptual frame provided in the previous chapters.

12.2 Informational privacy and computer ethics

It is common to distinguish four kinds of privacy:

(1) Alice's physical privacy, namely her freedom from sensory interference or intrusion, achieved thanks to a restriction on others' ability to have bodily interactions with her;

---

[1] Froomkin (2000) still provides a valuable review. On the possibility of having to enact 'paternalistic privacy laws for the benefit of uneager beneficiaries', see Allen (2011), p. xi.

(2) Alice's mental privacy, namely her freedom from psychological interference or intrusion, achieved thanks to a restriction on others' ability to access and manipulate her mind;
(3) Alice's decisional privacy, namely her freedom from procedural interference or intrusion, achieved thanks to the exclusion of others from decisions (concerning e.g. education, health care, career, work, marriage, faith) taken by her and her group of intimates; and finally
(4) Alice's informational privacy, namely her freedom from epistemic interference or intrusion, achieved thanks to a restriction on facts about her that are unknown or unknowable.

I shall return to these various privacies in Section 14.8, but until then the last form of privacy is the one that will interest us in this chapter.

Why have digital ICTs made informational privacy one of the most obvious and pressing issues in information and computer ethics? The question is crucial[2] and deceptively simple. According to one of the most widely accepted explanations, digital ICTs exacerbate old problems concerning informational privacy because of the dramatic increase in their data *Processing* capacities, in the speed (or *Pace*) at which they can process data, and in the *Quantity* and *Quality* of data that they can collect, record, and manage. This can be referred to as the 2P2Q hypothesis.

The trouble with any approach sharing the 2P2Q hypothesis is that it concentrates only on obvious and yet secondary effects of the information revolution, and that it does so from a 'continuist' philosophy of technology (more on this in Section 14.4). It thus fails to account for the equally important fact that digital ICTs are also responsible both for a potential *increase* in some kinds of informational privacy and, above all, for a radical *change* in its overall nature. ICTs are more redrawing rather than erasing the boundaries of informational privacy. A few examples may help to illustrate the point. Consider

- the 'remotization' of information management, such as the ordinary phenomenon of booking, banking or shopping online;
- the growth of anonymous, indirect or non-personal interactions. It is very common practice, for example, to lie about one's own location by text message; this is privacy as well;
- the much faster and more widespread revisability, volatility, and fragility of digital data. Personal records can be upgraded or erased at the stroke of a key, destroyed by viruses in a matter of seconds, or become virtually unavailable with every change in technological standards, whereas we are still able to reconstruct whole family trees thanks to parish documents that have survived for centuries; or
- the various technologies that enable users to encrypt, firewall, or protect information, e.g. with passwords or PIN.

---

[2] See e.g. Johnson (2001), Bynum and Rogerson (2004), and Tavani (2003).

In each case, it seems that ICTs allow both the erosion of informational privacy and its protection. The following, colourful episode, although old by now, is still indicative:

Hong Kong businessmen, for example, once did not dare to leave their mobile phones switched on while visiting sleazy Macau, because the change in ringing tone could betray them. After the ringing tone for Macau was changed to sound like Hong Kong's, however, they could safely leave their phones on, and roaming revenues soared.[3]

2P2Q explains only half of the story.

The new challenges posed by ICTs are not only a matter of 'more of the same'. They have their roots in a radical and unprecedented transformation in the ontology of the informational environment, of the information agents embedded in it, and of their interactions. As I shall argue in this chapter, understanding this ontological transformation provides a better explanation that not only is consistent with the 2P2Q hypothesis—now to be interpreted as a mere secondary effect of a far more fundamental change—but also addresses the essence of the privacy problem in the information society.

12.3 Informational privacy as a function of ontological friction

Imagine a model of a limited (region of the) infosphere, represented by four students Alice, Bob, Carol, and Dave (our set of interactive, information agents) living in the same house (our limited environment). Intuitively, given a specific amount of available information (which can be treated as a constant and hence disregarded), the larger the *informational gap* among the agents, the less they know about each other, the more private their lives can be.

The informational gap is a function of the degree of *accessibility* of personal data. In our example, there will be more or less informational privacy depending on whether the students are allowed, for example, to have their own rooms and lock their doors. Other relevant conditions are easily imaginable (individual fridges, telephone lines in each room, separate entrances, etc.).

Accessibility, in its turn, is an epistemic factor that depends on the ontological features of the infosphere; i.e. on the nature of the specific agents, of the specific environment in which they are embedded, and of the specific interactions implementable in that environment by those agents. If the walls in the house are few and thin and all the students have excellent hearing, the degree of accessibility is increased, the informational gap is reduced, and informational privacy is more difficult to obtain and protect. The love-lives of the students may be deeply affected by the Japanese-style house they have chosen to share.

---

[3] 'Your Cheating Phone', *The Economist*, 2 December 2004 <http://www.economist.com/node/3423008>.

The ontological features of the infosphere determine a specific degree of *ontological friction*, which in turn determines the information flow within the system. 'Ontological friction' refers here to the forces that oppose the information flow within (a region of) the infosphere, and hence (as a coefficient) to the amount of work and effort required for some kind of agent to obtain, filter and/or block information (also, but not only) about other agents in a given environment, e.g. by establishing and maintaining channels of communication and by overcoming obstacles in the flow of information such as distance, noise, lack of resources (especially time, memory space and processing capacities), amount and complexity of the data to be processed, and so forth.

Of course, the informational affordances (Gibson, 1979) and constraints provided by an environment are such only in relation to agents with specific informational capacities. In our model, brick walls provide much higher 'ontological friction' for the flow of acoustic information than a paper-thin partition, but this is irrelevant if the students are deaf. More realistically, the debate on privacy issues in connection with the design of office spaces—from private offices to panel-based open plan office systems, to completely open working environments, see Becker and Sims (2000)—offers a significant example of the relevance of varying degrees of ontological friction in social contexts.

We are now ready to formulate a qualitative sort of equation, which will be needed to analyse the relation between ICTs and informational privacy. Given some amount of personal information available in (a region of) the infosphere I, the lower the ontological friction in I, the higher the accessibility of personal information about the agents embedded in I, the smaller the informational gap among them, and the lower the level of informational privacy implementable about each of them. Put simply, *informational privacy is a function of the ontological friction in the infosphere*. It follows that any factor affecting the latter will also affect the former.

The factors in question can vary and may concern more or less temporary or reversible changes in the environment (imagine Alice, Bob, and Carol living in a tent during a holiday, while Dave is left home alone) or in the agents (e.g. Alice and Bob change their behaviour because Carol and Dave have quarrelled). Because of their 'data superconductivity', ICTs are well known for being among the most influential factors that affect the ontological friction in the infosphere.[4] A crucial difference between old and new ICTs is *how* they affect it.

12.4 Ontological friction and the difference between old and new ICTs

In the past, ICTs have always tended to reduce what agents considered the normal degree of ontological friction in their environment. This already held true for the

---

[4] For a similar point see Moor (1997), who writes 'When information is computerised, it is *greased* to slide easily and quickly to many ports of call' (p. 27).

invention of the alphabet or the diffusion of printing. Photography and the rise of the daily press were no exceptions. One can easily sympathize with nineteenth-century concerns about the impact on individuals' informational privacy of

[r]ecent inventions and business methods... [i]nstantaneous photographs and newspaper enterprise... and numerous mechanical devices. (Warren and Brandeis, 1890, p. 195)

All this does not mean that, throughout history, informational privacy has constantly decreased in relation to the invention and spreading of ever more powerful ICTs. This would be a simplistic and mistaken inference. As emphasized above, changes in the nature both of the environment and of the agents play a pivotal role as well, so the actual ontological friction, and hence the corresponding degree of informational privacy in a region of the infosphere, is the result of a fine balance among several factors. Most notably, during the nineteenth and twentieth centuries, following the industrial revolution, the social phenomenon of the new metropolis counteracted the effects of the latest ICTs, as urban environments fostered a type of informational privacy based on *anonymity*. Anonymity is defined here as the unavailability of personal information, or the 'non-coordinability of traits in a given respect', according to Wallace (1999). This is the sort of privacy enjoyed by a leaf in the forest, still inconceivable nowadays in rural settings or small villages. In the same period in which Warren and Brandeis were working on their classic article, the Edinburgh of Dr Jekyll[5] and the London of Sherlock Holmes[6] already provided increasing opportunities for informational privacy through anonymity, despite the recent availability of new technologies.

Old ICTs have always tended to reduce the ontological friction in the infosphere because they *enhance* or *augment* the agents embedded in it. We have already encountered this distinction in Section 1.7, so let me just recall it briefly here. Consider the appliances available in our students' house. Some appliances—e.g. a drill, a vacuum cleaner, or a food processor—are tools that *enhance* their users, exactly like an artificial limb. Tele-ICTs (e.g. the telescope, the telegraph, the radio, the telephone, or the television) are enhancing in this sense. Some other appliances—e.g. a dishwasher, a washing machine, or a refrigerator—are robots that *augment* their users insofar as well-specified tasks can be delegated to them, at least partially. Recording ICTs (e.g. the alphabet and the various writing and printing technologies, the tape or video recorder) are augmenting in this sense.

Enhancing and augmenting ICTs have converged and become bundled together. The Watergate scandal and Nixon's resignation would have been impossible without them. But whether kept separate or packaged together, old ICTs have always shared the fundamental feature of facilitating the information flow in the infosphere by increasingly empowering the agents embedded in it. This 'agent-oriented' trend in

---

[5] R. L. Stevenson's *The Strange Case of Dr Jekyll and Mr Hyde* was first published in 1886.
[6] C. Doyle's *A Study in Scarlet* was first published in 1887.

old, pre-digital ICTs is well represented by dystopian views of informationally omnipotent agents, able to overcome any ontological friction, to control every aspect of the information flow, to acquire any personal data, and hence to implement the ultimate surveillance system, thus destroying all informational privacy, 'the dearest of our possessions'. It is not a digital problem. Recall that Orwell's *1984*, first published in 1949, contains no reference to computers or digital machines.

Now, according to a 'continuist' interpretation of technological changes, digital ICTs should be treated as just one more instance of well-known, enhancing, or augmenting ICTs. But then—the reasoning goes—if there is no radical difference between old and new (i.e. digital) ICTs, it is reasonable to argue that the latter increasingly cause problems for informational privacy merely because they are orders of magnitude more powerful than past technologies in enhancing or augmenting agents in the infosphere. All past ICTs have tended to reduce the ontological friction in the infosphere by enhancing or augmenting the agents inhabiting it, but digital ICTs are no exception, so the 2P2Q explanation is correct. Orwell's 'Big Brother' is readily associated with the ultimate database.

Although the continuist 2P2Q hypothesis is reasonable and intuitive, it overlooks the essence of the problem. In theory, ontological friction can both be reduced and increased. We have seen how the emergence of the urban environment actually produced more anonymity, and hence more ontological friction and more informational privacy. The 2P2Q explanation misses a fundamental difference between old and new ICTs. Old ICTs tend to reduce informational privacy, whereas new ICTs can also increase it. This is because the former enhance or augment the agents involved more and more, whereas the latter change the very nature of the infosphere, that is, of the environment itself, of the agents embedded in it and of their interactions.

The interpretation of ICTs as re-ontologizing technologies has already been discussed in Section 1.3, so let me just use two examples in order to illustrate how it applies to our current analysis.

Let us begin with a thought experiment. Imagine that all the walls and the furniture in our students' house are transformed into perfectly transparent glass. Assuming our students have good sight, this will drastically reduce the ontological friction in the system. Imagine next that the students are transformed into proficient mind-readers and telepaths. Any informational privacy in this sort of Bentham's *Panopticon* will become virtually impossible.

Consider next a science-fiction scenario. In 'The Dead Past', Asimov describes a *chronoscope*, a device that allows direct observation of past events (Asimov, 1956). The chronoscope turns out to be of only limited use for archaeologists, since it can look only a couple of centuries into the past. However, people soon discover that it can easily be tuned to the most recent past, with a time lag of fractions of a second. Through the chronoscope, one can observe any event almost in real time. It is the end of privacy, for the dead past is only a synonym for 'the living present', as one of the characters remarks rather philosophically.

The two examples illustrate how radical modifications in the very nature (a re-ontologization) of the infosphere can dramatically change the conditions of possibility of informational privacy.[7]

12.5 Informational privacy in the re-ontologized infosphere

To summarize, so far I have argued that informational privacy is a function of the ontological friction in the infosphere. Many factors can affect the latter, including, most importantly, *technological innovations* and *social developments* such as, for example, massive inurbation (i.e. the abandonment of rural areas in favour of metropolis) and the corresponding phenomenon of anonymity. Old ICTs affected the ontological friction in the infosphere mainly by enhancing or augmenting the agents embedded into it; therefore, they tended to decrease the degree of informational privacy possible within the infosphere. On the contrary, new, digital ICTs affect the ontological friction in the infosphere most significantly by re-ontologizing it; therefore, not only can they both decrease and protect informational privacy but, most importantly, they can also alter its nature and hence our understanding and appreciation of it.

Interpreting the revolutionary nature of digital ICTs in this ontological way offers several advantages. The first can be highlighted immediately: the ontological hypothesis is perfectly consistent with the 2P2Q hypothesis, since the re-ontologization of the infosphere explains why digital ICTs are so successful, in terms of the quantity, quality, and speed at which they can variously process their data. It follows that the ontological hypothesis can inherit whatever explanatory benefits are carried by the 2P2Q hypothesis.

Four other advantages can be listed here but each of them requires a more detailed analysis:

(1) contrary to the 2P2Q hypothesis, the new approach explains why digital ICTs can also enhance informational privacy, although
(2) there is still a sense in which the information society provides less protection for informational privacy than the industrial society did; above all,
(3) the ontological hypothesis provides the right frame within which to assess contemporary interpretations of informational privacy and
(4) can indicate how we might wish to proceed in the future in order to protect informational privacy in the newly re-ontologized infosphere.

Let us consider each point in turn.

---

[7] Marty Wolf has suggested to me, quite rightly, that (I quote) another example of this is the use of 'clickers' in the classroom. Students quite willingly respond to even controversial questions because their answer is anonymous. It re-ontologizes the classroom, the subsequent discussion, and the very natures of the learning and teaching experiences.

12.5.1 Empowering the information agent

In the re-ontologized infosphere, any information agent has an increased power not only to gather and process personal data, but also to control and protect them. Recall that the digital now deals effortlessly with the digital. The phenomenon cuts both ways. It has led not only to a huge expansion in the flow of personal information being recorded, processed and exploited, but also to a large increase in the types and levels of control that agents can exercise on their personal data. And while there is only some personal data that an agent may care to protect, the potential growth of digital means and measures to control their life-cycle does not seem to have a foreseeable limit. If privacy is the right of individuals (be these single persons, groups, or institutions) to control the life-cycle (especially the generation, access, recording, and usage) of their information and determine for themselves when, how, and to what extent their information is processed by others, then one must agree that digital ICTs may enhance as well as hinder the possibility of enforcing such a right. At their point of *generation*, digital ICTs can foster the protection of personal data, e.g. by means of encryption, anonymization, password-encoding, firewalling, specifically devised protocols or services, and, in the case of externally captured data, warning systems. At their point of *storage*, legislation, such as the Data Protection Directive passed by the EU in 1995, can guarantee that no ontological friction, already removed by digital ICTs, is surreptitiously reintroduced to prevent agents from coming to know about the existence of personal data records, and from accessing them, checking their accuracy, correcting or upgrading them or demanding their erasure. And at their point of *exploitation*—especially through data mining, sharing, matching, and merging—digital ICTs can help agents to control and regulate the usage of their data by facilitating the identification and regulation of the relevant users involved.

At each of these three stages, solutions to the problem of protecting informational privacy can be not only self-regulatory and legislative but also technological, not least because informational privacy infringements can more easily be identified and redressed, also thanks to digital ICTs.

All this is not to say that we are inevitably moving towards an idyllic scenario in which our PETs (Privacy Enhancing Technologies) will fully protect our private lives and information against harmful PITs (Privacy Intruding Technologies). Such optimism is unjustified. But it does mean that digital ICTs can already provide some means to counterbalance the risks and challenges that they represent for informational privacy, and hence that no fatalistic pessimism is justified either. Digital ICTs do not necessarily erode informational privacy; they can also enhance and protect it.

12.5.2 The return of the (digital) community

Because digital ICTs are radically modifying our informational environments, ourselves and our interactions, it would be naïve to expect that informational privacy in the future will mean exactly what it meant in the industrial Western world in the middle of the last century.

We have seen that, between the end of the nineteenth and the beginning of the twentieth centuries, the ontological friction in the infosphere, actually reduced by old ICTs, was nevertheless increased by social conditions favouring anonymity, and hence a new form of informational privacy. In this respect, the diffusion of digital ICTs has finally brought to completion the process begun with the invention of printing. We are now back into the digital community, where anonymity can no longer be taken for granted, and hence where the decrease in ontological friction caused by old and new ICTs can have all its full-blown effects on informational privacy. In Britain, for example, the digital ICTs that allowed terrorists to communicate undisturbed over the Internet were also responsible for the identification of the London bombers in a matter of hours (see Figure 15). Likewise, mobile phones are increasingly useful as forensic evidence in trials. In Britain, cell site analysis (a form of triangulation that estimates the location of a mobile phone when it is used) helped disprove Ian Huntley's alibi and convict him for the murder of Holly Wells and Jessica Chapman. Sherlock Holmes has the means to fight Mr Hyde.

Figure 15. CCTV image of the four London terrorists as they set out from Luton

Source: The image was released to the public by the Metropolitan Police on 18 July 2005. It is available from Wikipedia as not being covered by copyright <http://en.wikipedia.org/wiki/File:July_7,_2005_London_bombings_CCTV.JPG>.

How serious and dangerous is it to live in a glassy infosphere? Human agents tend to be acquainted with different environments that have varying degrees of ontological friction and hence to be rather good at adapting themselves accordingly. As with other forms of fine equilibria, it is hard to identify, for all agents in any environment, a common, lowest threshold of ontological friction below which human life becomes increasingly unpleasant and ultimately unbearable, although perhaps Orwell has described it well. It is clear, however, that a particular threshold has been reached when the agents are willing to employ resources, run risks, or expend energy to restore it, e.g. by building a higher fence, by renouncing a desired service, or by investing time in revising a customer profile. On the other hand, different agents have different degrees of sensitivity. One needs to remember that several factors (character, culture, upbringing, past experiences, etc.) make each agent a unique individual. To one person, a neighbour capable of seeing one's garbage in the garden may seem an unbearable breach of their privacy, which it is worth any expenditure and effort to restore; to another person, living in the same room with several other family members may feel entirely unproblematic. Human agents can adapt to very low levels of ontological friction. Virginia Woolf's essay on Montaigne discusses the lack of ontological friction that characterizes public figures in public contexts, an issue that re-acquired all its poignancy in Britain because of the phone hacking scandal that led to the closure of *News of the World*. Politicians and actors are used to environments where privacy is a rare commodity. Likewise, people involved in 'Big Brother' (but 'Truman Show' would be a more appropriate label) programmes show a remarkable capacity to adapt to settings where any ontological friction between them and the public is systematically reduced, apparently in the name of entertainment. In far more tragic and realistic contexts, prisoners in concentration camps are subject to extreme duress due to both intended and unavoidable rarefaction of ontological friction (Levi, 1959).

The information society has revised the threshold of ontological friction and therefore provides a different sense in which its citizens appreciate their informational privacy. Your supermarket knows exactly what you like, but so did the owner of the grocery store where your grandparents used to shop. Your bank has detailed records of all your visits and of your financial situation, but how exactly is this different from the old service? A phone company could analyse and transform the call data collected for billing purposes into a detailed subscriber profile: social network (names and addresses of colleagues, friends or relatives called), possible nationality (types of international calls), times when one is likely to be at home and hence working patterns, financial profile (expenditure), and so forth. Put together, the data from the supermarket, the bank, and the phone company, and inferences of all sorts could be drawn for one's credit rating. Yet so they could be, and were, in Alexandre Dumas' *The Count of Monte Cristo* (1844). *Some* steps forward into the information society are really steps back into a small community and, admittedly, the claustrophobic atmosphere that may characterize it.

THE ONTOLOGICAL INTERPRETATION OF INFORMATIONAL PRIVACY 239

In the early stages in the history of the web roughly when Netscape was synonymous with browser, users believed that being online meant being entirely anonymous. A networked computer was like Gyges' ring in Plato's *Republic* (359b–360d): it made one invisible, unaccountable, and therefore potentially less responsible, socially speaking. Turing would certainly have appreciated the (at the time) popular comic strip in which a dog, typing an email on a computer, confessed to another dog that 'when you are on the Internet nobody can guess who you are'. We saw in Section 11.5.4 that nowadays the strip is not funny any more, only outdated. Cookies, monitoring software, and malware (malicious software, such as spyware) have made more and more people realize that the screen in front of them is not a shield for their privacy or Harry Potter's invisibility cloak, but a window onto their lives online, through which virtually anything may be seen. They expect websites to monitor and record their activities and do not even mind for what purpose. They accept that being online is one of the less private things in life.[8] The screen is a monitor and it is monitoring you.

Many years ago, a journalist at *The Economist* ran an experiment still worth reporting.[9] He asked a private investigator, 'Sam', to show what information it was possible to gather about someone. The journalist himself was to be the subject of the experiment. The country was Britain, the place where the journalist lived. The journalist provided Sam with only his first and last names. Sam was told not to use 'any real skulduggery (surveillance, going through her domestic rubbish, phone-tapping, hacking, that sort of thing)'. The conclusion? By using several databases and various ICTs,

[w]ithout even talking to anyone who knows me, Sam . . . had found out quite a bit about me. He had a reasonable idea of my personal finances—the value of my house, my salary and the amount outstanding on my mortgage. He knew my address, my phone number, my partner's name, a former partner's name, my mother's name and address, and the names of three other people who had lived in my house. He had 'found' my employer. He also had the names and addresses of four people who had been directors of a company with me. He knew my neighbours' names.

Shocking? Yes, in the anonymous industrial society of decades ago, but not really in the pre-industrial village before it, or in the information society after it. In Guarcino, a small village south of Rome of roughly a thousand people, everybody knows everything about everybody else, 'vita, morte e miracoli', 'life, death and miracles', as they

---

[8] The best long-term assessment of public attitudes toward privacy is provided by Columbia's Alan Westin, who has conducted a series of polls over the last thirty years on this issue. On average, he finds that one quarter of the American public cares deeply about keeping personal information secret, one quarter doesn't care much at all, and roughly half are in the middle, wanting to know more about the benefits, safeguards, and risks before providing information. Customer behavior in the marketplace—where many people freely provide personal information in exchange for various offers and benefits—seems to bear out this conclusion (Walker, 2000, p. 3).

[9] 'Living in the Global Goldfish Bowl', *The Economist*, 16 December 1999 <http://www.economist.com/node/268789>.

say in Italian. There is very little ontological friction provided by anonymity so there is very little informational privacy in that respect. One difference with the information society is that we have seen that the latter has the digital means to protect what the small village must necessarily forfeit.

There are of course many other dissimilarities. The comparison between today's information society and the small community of the past, where 'everybody knows everything', must be taken with more than a pinch of salt. History may repeat itself, yet never too monotonously. Small communities had a high degree of intra-community transparency (like a shared house) but a low degree of inter-community transparency (they were not like the Big Brother house, visible to outside viewers). So in those communities, breaches of privacy were reciprocal, yet there were few breaches of privacy across the boundary of the community. This is quite different from today's information society, where there can be very little transparency within the communities we live or work in (we hardly know our neighbours, and our fellow-workers have their privacy rigorously protected), yet data-miners, hackers, and institutions can be very well informed about us. Breaches of privacy from outside are common. What is more, we do not even know whether they know our business. On the other hand, part of the value of this comparison lies in the size of the community taken into consideration. A special trait of the information society is precisely its lack of boundaries, its global nature. We live in a single infosphere, which has no 'outside' and where intra- and inter-community relations are more difficult to distinguish. The types of invasion of privacy are quite different too. In the small community, breaches of privacy might shame or discredit you. Interestingly, Augustine usually speaks of privacy in relation to the topic of intercourse with married couples, and he always associates it with secrecy and then secrecy with shame or embarrassment. Or they might disclose your real identity or character. Things that were private became public knowledge. In the information society, such breaches involve unauthorized collection of information, not necessarily its publication. Things that are private may not become public at all; they may be just accessed and used by privileged others. The small community also was self-regulating and this limited breaches of privacy. Everyone knew that they were as subject to scrutiny as everyone else, and this set an unspoken limit on their enthusiasm for intruding into others' affairs.

12.5.3 Assessing theories of privacy

Once it has been acknowledged that digital ICTs have re-ontologized the infosphere, it becomes easier to assess the available theories of informational privacy and its moral value.

Two theories are particularly popular: the reductionist interpretation and the ownership-based interpretation.

The reductionist interpretation argues that the value of informational privacy rests on a variety of undesirable consequences that may be caused by its breach, either personally (e.g. distress) or socially (e.g. unfairness). Informational privacy is a utility, also in the sense of providing an essential condition of possibility of good human interactions, e.g. by preserving human dignity or by providing political checks and balances.

The ownership-based interpretation argues that informational privacy needs to be respected because of each person's rights to bodily security and property, where 'property of x' is classically understood as the right to exclusive use of x. A person is said to *own* his or her information (information about him- or herself)—recall Virginia Woolf's 'infinitely the dearest of our *possessions*'—and therefore to be entitled to control its whole life-cycle, from generation to erasure through usage.[10]

The two approaches are not incompatible, but they stress different aspects of informational privacy. The first is more oriented towards a consequentialist assessment of privacy protection or violation. The second is more oriented towards a 'natural rights' understanding of the concept of privacy itself, in terms of private or intellectual property. Here is a typical example:

Perhaps the final issue is that concerning information ownership: should information about me be owned by me? Or should I, as a database operator, own any information that I have paid to have gathered and stored? (Forester and Morrison, 1994, p. 102)

Unsurprisingly, they both compare privacy breach to trespass[11] or unauthorized invasion of, or intrusion in, a space or sphere of personal information, whose accessibility and usage ought to be fully controlled by its owner and hence kept private. A typical example is provided by the border-crossing model of informational privacy developed by Gary T. Marx since the late nineties (2005).

The reductionist interpretation is not entirely satisfactory. Defending the need for respect for informational privacy in view of the potential misuse of the information acquired is certainly reasonable, especially from a consequentialist perspective, but it may be inconsistent with pursuing and furthering social interests and welfare. Although it is obvious that some public personal information may need to be protected—for example, against profiling or unrestrained electronic surveillance—it remains unclear, on a purely reductionist basis, whether a society devoid of any informational privacy may not be a better society, with a higher, common welfare.[12] It has been argued, for example, that the defence of informational privacy in the home may actually be used as a subterfuge to hide the dark side of privacy: domestic abuse, neglect, or mistreatment.

---

[10] The debate on the ownership-based interpretation developed in the seventies, see Scanlon (1975) and Rachels (1975), who criticize Thomson (1975), who supported an interpretation of the right to privacy as being based on property rights.

[11] See Spinello (2005) for an assessment of the use of the trespassing analogy in computer-ethical and legal contexts. Charles Ess has pointed out to me that comparative studies have shown such spatial metaphors to be popular only in Western contexts.

[12] Moor (1997) infers from this that informational privacy is not a core value, i.e. a value that 'all normal humans and cultures need for survival', but then other values he lists as 'core' are not really so in his sense, e.g. happiness and freedom. According to Moor, privacy is also intrinsically valuable, while being the expression of the core value represented by security.

Precisely because of reductionist-only considerations, even in democratic societies such as the UK and the USA, it tends to be acknowledged that the right to informational privacy can be overridden when other concerns and priorities, including business needs, public safety, and national security, become more pressing. All this by putting some significant hermeneutic pressure on the 'arbitrary' clause that qualifies article 12 of The Universal Declaration of Human Rights which states that:

> No one shall be subjected to *arbitrary* interference with his privacy, family, home or correspondence, nor to attacks upon his honour and reputation. Everyone has the right to the protection of the law against such interference or attacks. (emphasis added)

The ownership-based interpretation also falls short of being entirely satisfactory. Three problems, already encountered in Section 3.4.5, are worth highlighting here:

(1) the issue of informational contamination undermining passive informational privacy. This is the unwilling acquisition of information or data (e.g. mere noise) imposed on someone by some external source. Brainwashing may not occur often, but junk mail, or the case of a person chatting loudly on a mobile nearby, are unfortunately very common experiences of passive privacy breach, yet no informational ownership seems to be violated;

(2) the issue of informational privacy in public contexts. As we saw in Section 3.4.5, informational privacy is often exercised in public spaces, that is, in spaces which are socially, physically and informationally public: anyone can see what one is doing downtown (Patton, 2000). How could a CCTV system be a breach of someone's privacy if the agent is accessing a space that is public in all possible senses anyway? and

(3) the metaphorical and imprecise use of the concept of 'information ownership', which cannot quite explain the lossless acquisition (or usage) of information: contrary to other things that one owns, one's personal information is not lost when acquired by someone else. Analyses of privacy based on 'ownership' of an 'informational space' are metaphorical twice over.

12.5.4 The ontological interpretation of informational privacy and its value

Both the reductionist and the ownership-based interpretation fail to acknowledge the radical change brought about by digital ICTs. They belong to an industrial culture of material goods and of manufacturing/trading relations. So they are overstretched when trying to cope with the new challenges offered by an informational culture of services and usability. Warren and Brandeis (1890) had already realized this limit with classic insightfulness:

> [W]here the value of the production [of some information] is found not in the right to take the profits arising from publication, but in the peace of mind or the relief afforded by the ability to prevent any publication at all, *it is difficult to regard the right as one of property, in the common acceptation of the term.* (p. 25, emphasis added)

More than a century later, in the same way as the information revolution is best understood as a fourth revolution that fundamentally re-ontologizes the infosphere and its inhabitants, informational privacy requires an equally radical re-interpretation, one that takes into account the essentially informational nature of human beings and of their operations as informational social agents.

Such a re-interpretation is achieved by considering each person as constituted by his or her information—as I argued in the previous chapter—and hence by understanding a breach of one's informational privacy as a form of aggression towards one's personal identity. Such an ontological interpretation of informational privacy is consistent with the fact that ICTs can both erode and reinforce informational privacy, and hence that a positive effort needs to be made in order to support not only Privacy Enhancing Technologies (PET) but also 'poietic' (i.e. constructive) applications, which may allow users to design, shape, and maintain their identities as information agents. The information flow needs some friction in order to keep firm the distinction between the macro multi-agent system (the society) and the identity of the micro multi-agent systems (the individuals) constituting it. Any society (even a utopian one) in which no informational privacy is possible is one in which no personal identity can be maintained and hence no welfare can be achieved, social welfare being only the sum of the welfare of the individuals involved. The total 'transparency' of the infosphere that may be advocated by some reductionists—recall the example of the glass house and of our mentally super-enhanced students—achieves the protection of society only by erasing all personal identity and individuality, a 'final solution' for sure, but hardly one that the individuals themselves, constituting the society so protected, would be happy to embrace freely. As Cohen (2000) has rightly remarked,

> the condition of no-privacy threatens not only to chill the expression of eccentric individuality, but also, gradually, to dampen the force of our aspirations to it. (p. 1426)

The advantage of the ontological interpretation over the reductionist one is then that consequentialist concerns may override respect for informational privacy, whereas the ontological interpretation, by equating its protection to the protection of personal identity, considers it a fundamental and inalienable right,[13] so that, by default, the presumption should always be in favour of its respect. As we shall see, this is not to say that informational privacy is never negotiable in any degree.

Looking at the nature of a person as being constituted by that person's information enables one to understand the right to informational privacy as a right to personal immunity from unknown, undesired, or unintentional changes in one's own identity as an informational entity both *actively* and *passively*. Actively, because collecting, storing, reproducing, manipulating, etc. one's information amounts now to stages in stealing, cloning or breeding someone else's personal identity. Passively, because

---

[13] For a different view see Volkman (2003).

breaching one's informational privacy may now consist in forcing someone to acquire unwanted data, thus altering her or his nature as an informational entity without consent.[14] Brainwashing is as much a privacy breach as mind-reading.

The first difficulty facing the ownership-based interpretation is thus avoided: in either case, the ontological interpretation suggests that one's informational sphere and one's personal identity are co-referential, or two sides of the same coin. There is no difference because 'you are your information', so anything done to your information is done to you, not to your belongings. It follows that the right to informational privacy (both in the active and in the passive sense just seen) shields one's personal identity. This is why informational privacy is extremely valuable and ought to be respected.

Heuristically, violations of informational privacy are now more fruitfully compared to a digital kidnapping rather than trespassing, as argued in Section 3.4.5. A further advantage, in this change of perspective, is that it becomes possible to dispose of the false dichotomy qualifying informational privacy in public or in private contexts. To repeat here what has been already argued in Section 3.4.5, a piece of information constitutes an agent context-independently, so Alice is perfectly justified in wishing to preserve her integrity and uniqueness as an informational entity even in entirely public places. Trespassing makes no sense in a public space, but kidnapping is a crime independently of where it is committed. The second problem affecting the ownership-based interpretation is also solved.

As for the third problem, one may still argue that an agent 'owns' his or her information, yet no longer in the metaphorical sense seen above, but in the precise sense in which an agent *is* her or his information. 'My' in 'my information' is not the same 'my' as in 'my car' but rather the same 'my' as in 'my body' or 'my feelings': it expresses a sense of constitutive *belonging*, not of external *ownership*, a sense in which my body, my feelings, and my information are part of me but are not my (legal) possessions. It is worth quoting Warren and Brandeis (1890) once again, this time at length:

[T]he protection afforded to thoughts, sentiments, and emotions ... is merely an instance of the enforcement of the more general right of the individual to be let alone. It is like the right not to be assaulted or beaten, the right not to be imprisoned, the right not to be maliciously persecuted, the right not to be defamed [or, the right not to be kidnapped, my addition]. In each of these rights ... there inheres the quality of being owned or possessed and ... there may be some propriety in speaking of those rights as property. But, obviously, they bear little resemblance to what is ordinarily comprehended under that term. *The principle ... is in reality not the principle of private propriety but that of inviolate personality ... [T]he right to privacy, as part of the more general right to the immunity of the person, [is] the right to one's personality.* (pp. 31, 33, emphases added)

This ontological conception has started being appreciated by more advanced information societies where identity theft is the fastest-growing white-collar offence

---

[14] This view is close to the interpretation of privacy in terms of protection of human dignity defended by Bloustein (1964).

(see Section 1.2). Informational privacy is the other side of identity theft, to the point that, ironically, for every person whose identity has been stolen (around 10 million Americans are victims annually) there is another person (the thief) whose identity has been 'enhanced'.

Problems affecting companies such as Google or Facebook and their privacy policy convey a similar picture. As Kevin Bankston, staff attorney at the Electronic Frontier Foundation, once remarked:

> Your search history shows your associations, beliefs, perhaps your medical problems. *The things you Google for define you*.... data that's practically a printout of what's going on in your brain: What you are thinking of buying, who you talk to, what you talk about. (quoted in Mills (2005), emphasis added)

As anticipated, the ontological interpretation reshapes some of the assumptions behind our still 'industrial' conception of informational privacy. Three examples are indicative of this transition.

If personal information is finally acknowledged to be a constitutive part of someone's personal identity and individuality, then one day it may become strictly illegal to trade in some kinds of personal information, exactly as it is illegal to trade in human organs (including one's own) or slaves. The problem of child pornography may also be revisited in light of an ontological interpretation of informational privacy. At the same time, one might relax one's attitude towards some kinds of 'dead personal information' that, like 'dead pieces of oneself', are not really or no longer constitutive of oneself. One should not sell one's kidney, but can certainly sell one's hair or be rewarded for giving blood. Recall the experiment of the journalist at *The Economist*. Very little of what Sam had discovered could be considered ontologically constitutive of the person in question. We are constantly leaving behind a trail of personal data, pretty much in the same sense in which we are shedding a huge trail of dead cells. The fact that nowadays digital ICTs allow our data trails to be recorded, monitored, processed, and used for social, political, or commercial purposes is a strong reminder of our informational nature as individuals and might be seen as a new level of ecologism, as an increase in what is recycled and a decrease in what is wasted.

At the moment, all this is just speculation and in the future it will probably be a matter of fine adjustments of ethical sensibilities, but the third Geneva Convention (1949) already provides a clear test of what might be considered 'dead personal information': a prisoner of war need only give his or her name, rank, date of birth, and serial number and no form of coercion may be inflicted on him or her to secure any further information, of any kind. If we were all considered 'prisoners of the information society', our informational privacy would be well protected and yet there would still be some personal data that would be perfectly fine to share with any other agent, even hostile ones.

A further issue that might be illuminated by the ontological interpretation is that of confidentiality. The sharing of private information with someone, implicitly or explicitly, is based on a relation of profound trust that binds the agents involved. This coupling is achieved by allowing the agents to be partly constituted, ontologically, by the same information. Visually, the informational identities of the agents involved now overlap, at least partially, as in a Venn diagram. The union of the agents forms a single unity, a supra-agent, or a new multi-agent individual. Precisely because entering into a new supra-agent is a delicate and risky operation, care should be exercised before 'melding' oneself with other individuals by sharing personal information or its source, i.e. common experiences. Confidentiality is a bond that is hard and slow to forge properly, yet resilient to many *external* forces when finally in place, as the supra-agent is stronger than the constitutive agents themselves. Relatives, friends, classmates, fellows, colleagues, comrades, companions, partners, team-mates, spouses, and so forth may all have experienced the nature of such a bond, the stronger taste of a 'we'. But it is also a bond very brittle and difficult to restore when it comes to *internal* betrayal, since the disclosure, deliberate or unintentional, of some personal information in violation of confidence can entirely and irrecoverably destroy the privacy of the new, supra-agent born out of the joining agents, by painfully tearing them apart. I shall return to the topic of trust and confidentiality at the end of this first part of this chapter.

The third and final issue can be touched upon rather briefly, as it has already been mentioned above: the ontological interpretation stresses that informational privacy is also a matter of construction of one's own informational identity. The right to be left alone is also the right to be allowed to experiment with one's own life, to start again, without having records that mummify one's personal identity forever, taking away from the individual the power to mould it. Every day, a person may wish to build a different, possibly better, 'I'. We never stop becoming ourselves, so protecting a person's informational privacy also means allowing that person the freedom to construct and change herself, ontologically.[15]

12.6 Informational privacy, personal identity, and biometrics

On 12 September 1560, the young Montaigne attended the public trial of Arnaud du Tilh, an impostor who was sentenced to death for having faked his identity. Many acquaintances and family members, including his wife Bertrande, had been convinced for a long while that he was Martin Guerre, returned home after many years of absence. Only when the real Martin Guerre came home was Arnaud's actual identity finally ascertained.

Had Martin Guerre always been able to protect his personal information, Arnaud du Tilh would have been unable to steal his identity. Clearly, the more one's

---

[15] In this sense, Johnson (2001) seems to be right in considering informational privacy an essential element in an individual's autonomy. Moor (1997), referring to a previous edition of Johnson (2001), disagrees.

informational privacy is protected the more one's personal identity can be safeguarded. This new qualitative equation is a direct consequence of the ontological interpretation. Personal identity also depends on informational privacy. The difficulty facing our contemporary society is how to combine the new equation with the other equation, introduced in Section 14.3, according to which informational privacy is a function of the ontological friction in the infosphere. Ideally, one would like to reap all the benefits from

(1) the highest level of information flow; and hence from
(2) the lowest level of ontological friction;

while enjoying

(3) the highest level of informational privacy protection; and hence
(4) the highest level of personal identity protection.

The problem is that (1) and (4) seem incompatible: facilitating and increasing the information flow through digital ICTs and the protection of one's personal identity is bound to come under increasing pressure. You cannot have an identity without having an identikit. Or so it seems, until one realizes that the information flowing in (1) consists of all sorts of data, including *arbitrary* data *about* oneself (e.g. a name and surname) that are actually shareable, whereas the information required to protect (4) can be *ontic* data—that is, data *constituting* someone (e.g. someone's DNA), or constituting the *interpretation* of someone as an informational entity—that are hardly shareable by nature.[16] Enter biometrics.

Personal identity is the weakest link and the most delicate element in our problem. Even nowadays, personal identity is regularly protected and authenticated by means of some *arbitrary* data, *randomly* or *conventionally* attached to the bearer/user, like a mere label: a name, an address, a Social Security number, a bank account, a credit card number, a driving licence number, a PIN, and so forth. Each label in the list has no ontologically constitutive link with its bearer; it is merely associated with someone's identity and can easily be detached from it without affecting the individual. The rest is a mere consequence of this 'detachability'. The more the ontological friction in the infosphere decreases, the swifter these detached labels can flow around, and the easier it becomes to grab and steal them and use them for illegal purposes. Arnaud du Tilh had stolen a name and a profile, and succeeded in impersonating Martin Guerre for many years in a rather small village, within a community that knew him well, fooling even Martin's wife, apparently. Eliminate all personal interactions and identity theft becomes the easiest thing in the world.

A quick and dirty way to fix the problem would be to clog the infosphere by slowing down the information flow; building some traffic-calming device, as it were. It seems the sort of policy popular among some IT officers and middle-ranking bank

---

[16] On the tripartite distinction between information *as*, *about*, or *for* reality see Floridi (2004).

managers, keen on not allowing this or that operation for security reasons, for example. However, as with all counter-revolutionary or anti-historical approaches, 'resistance is futile': trying to withstand the evolution of the infosphere only harms current users and, in the long run, fails to deliver an effective solution.

A much better approach is to ensure that the ontological friction continues to decrease, thus benefiting all the inhabitants of the infosphere, while safeguarding personal identity by data that are not arbitrary labels for, but rather constitutive traits of, the person in question. Arnaud du Tilh and Martin Guerre looked very similar, yet this was as far as biometrics went in the sixteenth century. Today, biometric digital ICTs are increasingly used to authenticate a person's identity by measuring informationally the person's physiological traits—such as fingerprints, eye retinas and irises, voice patterns, facial patterns, hand measurements, or DNA sampling—or behavioural features, such as typing or gait patterns. Since they also require the person to be identified to be physically present at the point of identification, biometric systems provide a very reliable way of ensuring that the person is who the person claims to be; of course not always, and not infallibly—after all Montaigne used the extraordinary case of Martin Guerre to challenge human attempts ever to reach total certainty—but far more successfully than any arbitrary label can. It is a matter of degree.

All this is not to say that we should embrace biometrics as an unproblematic panacea. There are many risks and limits in the use of such technologies as well (see e.g. Alterman (2003)). But it is significant that digital ICTs, in their transformation of the information society into a digital community, are partly restoring, partly improving (see the case of Martin Guerre) that reliance on personal acquaintance that characterized relations of trust in any small community. By giving away some information, one can safeguard one's identity and hence one's informational privacy, while taking advantage of interactions that are personalized (through preferences derived from one's habits and behaviours) and customized (through preferences derived from one's expressed choices). In the digital community, you are a recognized individual, whose tastes, inclinations, habits, preferences, etc. are known to the other agents, who can adapt their behaviour accordingly.

As for protecting the privacy of biometric data, again, no rosy picture should be painted, but if one applies the 'Geneva Convention' test, it seems that even the worst enemy could be allowed to authenticate someone's identity by measuring her fingerprints or his eye retinas. These seem to be personal data that are worth sacrificing in favour of the extra protection they can offer for one's personal identity and private life.

Once a cost–benefit analysis is taken into account, it makes sense to rely on authentication systems that do not lend themselves so easily to misuse. In the digital community, one is one's own information and can be (biometrically) recognized as oneself as one was in the small village. The case of Martin Guerre is there to remind us that mistakes are still possible. But their likelihood decreases dramatically the more biometric data one is willing to check. On this, Penelope can teach us a final lesson.

When Odysseus returns to Ithaca, he is identified four times. Argos, his old dog, is not fooled and recognizes him despite his disguise as a beggar. Then Eurycleia, his wet-nurse, while bathing him, recognizes him by a scar on his leg, inflicted by a boar when hunting. He then proves to be the only man capable of stringing Odysseus' bow. All these are biometric tests no Arnaud du Tilh would have passed. But then, Penelope is no Bertrande either. She does not rely on any 'unique identifier' but finally tests Odysseus by asking Eurycleia to move the bed in their wedding-chamber. Odysseus protests that this is impossible: he himself had built the bed around a living olive tree, which is now one of its legs. This is a crucial piece of information that only Penelope and Odysseus ever shared. By naturally relying on it, Odysseus restores Penelope's full trust. She recognizes him as the real Odysseus not because of who he is or how he looks, but, ontologically, because of the information that they have in common and that constitutes both of them as a couple. Through the sharing of this piece of information, identity is restored and the supra-agent is reunited. There is a line of continuity between the roots of the olive tree and the married couple. For Homer, their bond was ὁμοφροσύνη (like-mindedness); to Shakespeare, it was the 'marriage of true minds'. To us, it is informational privacy that admits no ontological friction.

12.7 Four challenges for a theory of informational privacy

As anticipated, in this second part of the chapter I wish to consider a number of challenges that seem to confront any theory of informational privacy and how an ontological approach might deal with them. The perspective is meta-theoretical: problems concerning informational privacy itself are not under discussion here. The account will not be exhaustive, not merely because this would be impossible, but mainly because it would be useless. For the challenges to be taken into account are only those substantial enough to run the risk of undermining a theory of informational privacy, or sufficiently interesting to cast a better light on why a theory is particularly valuable. Since there are several that satisfy these criteria, I shall proceed rather schematically. Finally, no degree of importance should be inferred from the order of presentation, although I shall make an effort to proceed from more general to more specific challenges, and try to link them in a unifying narrative.

12.7.1 Non-Western approaches to informational privacy

One is often reminded that different cultures and languages may not share similar conceptions of privacy in general, and of informational privacy in particular. Indeed, it has become fashionable to state that privacy is a Western invention of the eighteenth-century. Thompson, for example, recalls that:

In 'The Structure of Everyday Life', Fernand Braudel states that 'privacy was an eighteenth-century innovation'; [and that] in 'The Structural Transformation of the Public Sphere',

Habermas asserts that the public sphere was an eighteenth-century invention. (Thompson, 1996, p. 29)

Yet this is only partly true, for the history of privacy is far more complex and nuanced, as the monumental work by Ariáes and Duby (1987) testifies.

In connection with the suggestion that 'privacy' might be a matter (and obsession) limited to Western cultures, global differences may also be unduly emphasized, even when they represent a healthy reminder that no assumption should be too readily made when it comes to such a basic issue (Ess, 2005). For example, the word 'privacy' is certainly imported in Thai (Kitiyadisai, 2005) and in Japanese (Nakada and Tamura, 2005), but so it is in other European languages such as Italian or Spanish. And one may easily build a case for a general difference between a Mediterranean and a more northern-European sense of privacy. Such generalizations are often amusing but rarely informative. The truth is that no one would find it reasonable to compare, for example, Eastern and French cuisine. Similar comparisons between over-generic (e.g. Western, Eastern) and more focused (e.g. French, Buddhist, Thai) categories are better left behind, if one wishes to understand what really is at stake conceptually.

The difficult solution here seems to navigate between self-deprecation and chauvinism, while avoiding the adoption of some form of more or less hidden relativism, which would merely be synonymous for a substantial failure in achieving a real dialogue. Perhaps the key is a constructive commitment towards the identification and uncovering of those common and invariant traits that unify humanity universally, at all times and in all places. Like 'friendship', for example, 'privacy' is a slippery concept, which seems to qualify a variety of phenomena that may change from place to place and through time; and yet, this is no argument against its presence in virtually any given culture. In this respect, the ontological approach, developed in the first half of this chapter, offers two advantages.

First, instead of trying to achieve an impossible 'view from nowhere', the approach seeks to avoid assuming some merely 'local' conception of what Western philosophical traditions dictate as 'normality'—no matter whether this is understood as post-eighteenth century or not—in favour of a more neutral ontology of entities modelled informationally. By referring to such a 'lite' ontological grounding of informational privacy, the theory allows the adaptation of the former to various conceptions of the latter, working as a potential cross-cultural platform. This can help to uncover different conceptions and implementations of informational privacy around the world in a more neutral language, without committing the researcher to a culturally charged position.

Second, since the ontological theory of privacy relies on an informational ontology, it may more easily resound with a humanity that is increasingly used to the re-ontologizing impact of global ICTs. Teenagers from all over the world are nowadays more likely to communicate by relying on their shared experiences with online entertainment, for example, than by referring to their parents' conceptions of reality based on dolls and plastic figures of WWII soldiers. In a few generations, an informational ontology will seem obvious to the point of being trivial. This is not to say that a global and uniform sort of digitally pasteurized culture will be dawning on us any time soon. As Saussure clearly demonstrated with respect to languages, diachronic forces of appropriation and re-appropriation inevitably articulate, particularize and localize any apparently global trend. No universal language or culture should be expected to arise across all the various information societies around the world. However, in the same way that people increasingly often speak not only their own idioms and native dialects but also some form of basic English good enough to communicate with each other, likewise, an informational ontology will probably represent the shared *koiné* among future netizens.

12.7.2 Individualism and the anthropology of informational privacy

Western alleged 'individualism' may be seen as a specific form of parochialism, determined by a deeply ingrained and yet utterly contingent anthropology, obsessed with individuals, their needs and desires, their egotisms, and their market-driven, cost–benefit-oriented, logocentric behaviours. The latter is a caricature and a rather unsophisticated one at that, I concede, but it is not too far from a decent sketch of some culturally shortsighted and mono-ethnic work that circulates even in some applied ethics studies. The broad challenge here is whether there can be any sense in talking of a theory of informational privacy without the private subject, to paraphrase the title of a famous article by Popper on epistemology without the knowing subject. My short answer is negative: informational privacy requires a privacy holder, but with a crucial qualification.

What most critics of 'individualism' seem to overlook, perhaps blinded by an understandable eagerness to redress the situation, is that the concept of 'individual' is not the same as the concepts of 'person', 'subject', 'agent', 'mind', 'soul', or 'self'. All these can be used interchangeably, of course, and not necessarily mistakenly so. But when some generic allusion is made to the alleged absence of any concept of any sort of individuality in non-Western cultures or philosophies, or when theories of privacy (including the informational variety) are criticized for being oblivious to the patent lack of any privacy holders in some non-Western countries, then the ethicist needs to reach for his finest pencil and re-draw some distinctions, even at the risk of being pedantic.

First, facts are not norms: if things are such that a culture, a piece of legislation, or a philosophy lacks any conception of a privacy holder, this is no reason to argue that it should not acquire one. A specific example may help. We saw that article 12 of the Universal Declaration of Human Rights states that:

> [N]o one shall be subjected to arbitrary interference with his privacy, family, home or correspondence, nor to attacks upon his honour and reputation. Everyone has the right to the protection of the law against such interference or attacks.

Now the Declaration was adopted in 1948 by the General Assembly of the United Nations, and that date might be taken as the beginning of a universal theory of privacy, not limited to Western countries and cultures. However, the African (Banjul) Charter on Human and Peoples' Rights, adopted in 1981 by the Organization of African Unity (OAU), which is quite clearly modelled on the Universal Declaration, contains no reference to privacy or cognate concepts. From a normative point of view, it seems that this is a shortcoming, that the shortcoming is suspicious, and that it would be good if the Charter could be amended. The document does not prove that it is ethically acceptable that privacy rights in Africa should not be recognized.

Second, there are mainstream and influential traditions, within Western cultures and philosophies, that value (if not privilege) the community over the individual. Space here allows only for a few quick reminders. Greek and Roman philosophies are primarily social, to the extent that they defended the role of the *polis* and of the *res publica* as the real contexts where someone becomes oneself. Christianity is intrinsically ecclesiastical[17] and Judaism congregational (God relates to the whole people of Israel). The very concept of democracy takes something away from the individual to emphasize the centrality of the multi-agent system. It would be easy to add other examples.

What goes under the label of 'Western individualism' is to be understood not so much in terms of the centrality of the single self, but rather in terms of the raising of a sense of personal responsibility, which co-develops with political activities (Greece), legal systems (Rome), religious beliefs (Judaism/Christianity), and epistemic practices (Scientific Revolution) and is often supposed to be monitored by an omniscient God, who can see everything you do better than any omniscient Big Brother ever imagined, for 'His eyes are on the ways of men; he sees their every step' (*Job* 34:21) and he 'knows what you need before you ask him' (*Matthew* 6:8,32).

This leads to a third point: personal responsibility is not unknown to other cultures, far from it. If I may be allowed to draw some more caricatures: in many non-Western cultures or religions it is up to the *individual* to see that he or she reincarnates into, or transmigrates to, higher forms of life. And responsibility is not 'dispersed' in a vaporous sense of fuzzy subjectivity if you feel the pressure of committing suicide for having failed, again, as an *individual*, to uphold some specific standards or fulfil some expectations, or if you are invited, as an embodied and embedded agent, to annihilate your subjectivity, which therefore must be there in the first place (Hongladarom, 2006). Not every philosophy of the subject is subjectivist, nor is every philosophy of the 'I' also a philosophy of the 'me', and not every philosophy that talks of agents is necessarily committed to the existence of substantial selves. Yet a lot of bad press concerning poor Monsieur Descartes, for example, takes advantage of such confusions. Where there is personal responsibility there is also an individual capable of shouldering it, but then there is some conception of a single human being, different from society, capable of desiring some form of privacy for his or her own life.

---

[17] 'Ecclesia' simply meant 'assembly' in Greek, etymologically 'the body of the select counsellors'. Solon originally coined it as the name given to the public formal assembly of the Athenian people.

Superficial contrasts between Western and non-Western cultures both trivialize ostensible differences and obscure important commonalities, distorting central notions of the individual and of individual responsibility. It seems it is high time to re-shelve supermarket spiritualism where it belongs, i.e. the department of astrology, comfort food, and Western parochialism.

The ontological theory of informational privacy can help in this process in that it does not presuppose either a personalist or a substantialist conception of the agents involved in moral actions. We saw in Chapter 7 that agents need not be persons, they can be organizations, for example, or artificial constructs, or hybrid syntheses. And they do not need to consist of some self-like sort of entity, as they may be constituted by bundles of properties and processes, as I argued in Chapter 11. Once again, this 'lite' ontology can be adapted to further interpretations and cultural needs. It helps to frame the discussion in a minimalist way that does not exclude *a priori* some interlocutors.

12.7.3 *The scope and limits of informational privacy*

Under this heading it is useful to list a family of problems that highlight how some theories end up either shrinking or inflating the concept informational privacy.

First, there are some insightful and conclusive criticisms to Rachels (1975) and Fried (1970), moved by Reiman (1976) in the context of his broader criticism of Thomson and her 'ownership-based' theory of informational privacy (Thomson, 1975).[18] According to what Reiman labels the Rachels–Fried theory,

> Only because we are able to withhold personal information about—and forbid intimate observation of—ourselves from the rest of the world, can we give out the personal information—and allow the intimate observations—to friends and/or lovers, that constitute intimate relationships. On this view, intimacy is both signalled and constituted by the sharing of information and allowing of observation *not shared with or allowed to the rest of the world*. If there were nothing about myself that the rest of the world did not have access to, I simply would not have anything to give that would mark off our relationship as intimate. (Reiman, 1976, pp. 31–2)

---

[18] According to Thomson (1975), the right to privacy is a derivative right that follows from one's other rights, and especially one's rights to one's property. As one of OUP's anonymous referees remarked:

> It is crucial to understand that this [Thomson's] account is in no way to be taken as a theory of privacy; it is a theory of privacy *rights* that seeks to account for privacy rights in terms of the antecedent foundations of natural rights theory. In that sense, it is not necessarily even in competition with Floridi's view, which attempts to account for the condition of privacy rather than providing a political or legal philosophy of rights.

Thomson is also criticized by Scanlon (1975), while Rachels (1975) criticizes both. Reiman, coming last in the debate, is able to show the shortcomings of all three. Introna (1997) seems to agree with, and update, Reiman's position, if from a more Foucaultian perspective, while Johnson (1992) seeks to reconcile Benn's Kantian approach (Benn, 1975) to privacy in terms of protection of selfhood with Reiman's care-oriented approach. A very valuable contribution is provided by Cohen (2000), who develops a clear and sharp criticism of theories of informational privacy based on the concepts of ownership, control/choice and freedom of speech. The article is particularly interesting as it shows how such interpretations of informational privacy may 'back-fire' and allow, if adopted, solid reasons in favour of a more relaxed and market-friendly attitude towards personal data processing, especially in the USA.

Intimacy is certainly an important aspect of informational privacy (Inness, 1996). Yet, Reiman rightly argues that a 'market-oriented' analysis of privacy as a sort of intimacy-purchasing currency ('moral capital', in Fried's terminology) is both contingent on what has been defined above as a form of parochialism (the market orientation of values, in this case) and undermined by a logical fallacy.

If things were as the Rachels–Fried theory suggests, then people would be most intimate with, for example, doctors, lawyers, psychoanalysts, or priests, with whom they share all sorts of personal information they would not dare to share with anyone else, including those with whom they are actually most intimate. This is a *reductio ad absurdum*. For I agree with Reiman that the real difference is made by the relation of *caring*, not by the mere amount or type of information exchanged. And it is precisely the relation of caring that regulates what and how much information one is willing to share with someone with whom one enjoys an intimate relationship. It is well known that sometimes one can speak more freely with a stranger precisely because there is very little intimacy and not in view of establishing any.

Furthermore, anyone intrinsically unable to enter into any social relationship—like a comatose or seriously mentally ill person (recall the example of Mary in Chapter 6)—would be *de facto* deprived of any informational privacy, since the latter depends on the former (Reiman, 1976, p. 36), in the same sense in which some old banknotes, that cease to be legal tender, can no longer be used to purchase any goods. Allegedly, 'privacy creates the moral capital which we spend in friendship and love' (Fried, 1970, p. 25). But if you can no longer be a customer, you do not need it.

Rachels and Fried fail to take into account forms of informational privacy that we would like to consider both genuine and important. Yet others may end up inflating the concept of informational privacy in ways that turn out to be unrealistic (things stand differently) and then vacuous (nothing counts as privacy-unrelated). This is the case when *any* informational process concerning a person becomes a breach of that person's informational privacy. Again, Reiman provides an early and very valuable analysis of this sort of problem in his lucid criticism of Benn (1975). Let me illustrate it by using an everyday example.

Imagine that Alice and Bob are neighbours. If Alice sees Bob's car parked outside the house, a theory of informational privacy needs to be able to avoid counting this as necessarily a case of privacy breach. The same holds true for the case in which Bob drives away at a specific time in the afternoon and, without him knowing it, he is inadvertently seen by Alice, who is doing some gardening. If all cases of access to information about someone become cases of infringement of the informational privacy of that someone, we merely erase the conceptual distinction between being informed about someone's business and infringing someone's informational privacy, and thus deprive ourselves of the possibility of explaining when the former does not amount to the latter and what ought to be done when it does. A theory of informational privacy needs a criterion of discrimination to be able to explain why some information processes do *not* count as violations of informational privacy.

A third difficulty of 'scope', affecting several theories of informational privacy based on some version of personal information ownership/control, concerns inferential processes. Consider our simple example. Suppose Alice is informed that, if Bob leaves the house, Bob's wife, Carol remains alone in the house. Imagine next that Alice sees Bob driving away and Carol going back into the house. She is therefore informed that Carol is alone in the house. Information is closed under entailment, as logicians like to say. So seeing Bob driving away triggers a process that ends by breaching Carol's privacy. Now, what interests us here is the opposite process. Precisely because one may infer from Bob's absence Carol's state as the only person in the house, where does Carol's ownership of, or right to control, 'information about herself' end? It seems it should include Bob's localization as well. This generates a cascade of further difficulties, two of which are worth stressing.

On the one hand, there is a collapse of the naïve idea that information I about a group of people S might be easily partitioned into a finite set of disjoint pieces of information $\{I_1, \ldots, I_n\}$, whose union is I, about the individuals $\{i_1, \ldots, i_n\}$ constituting S. In other words, a great deal of personal information overlaps and covers many people at once: information about Bob's absence is information about Carol's solitude in the house, and *vice versa*, so these pieces of information cannot be merely owned or controlled by either Bob or Carol disjointly. Facebook's difficulties in managing individuals' privacy when group pictures are uploaded and tagged is a clear illustration of this problem. This calls for a refined theory of control closure among distributed systems (Turilli, 2007).

On the other hand, speaking of co-ownership or shared control of personal information becomes meaningless once it is clear that—even if semantic information is defined as embedding truth and 'false information' merely means 'not information', as in Floridi (2011a)—there is still an endless amount of information that can be inferred (and hence retro-engineered) starting from some initial information. Inferential closure plus co-ownership or shared control make the concept of 'personal information' too foggy to be of much use and applicability.

How the ontological theory of informational privacy avoids these difficulties may be explained in the following terms. Anyone defending the following two theses:

(a) that false information is genuine information;

and

(b) that informational privacy is based on ownership/control of information about oneself;

is also forced to conclude that, since

(c) 'being informed' is closed under implication,

then

(d) any informational process whatsoever is an infringement of one's informational privacy.

Yet, this is a *reductio ad absurdum*. And if one seeks to avoid it by weakening condition (a) into:

(a★) only 'true' information is genuine information,

and condition (c) into:

(c★) inferential closure may fail sometimes,

this is still insufficient to make (d) reasonably constrained. There still remains a huge amount of information that seems to belong to individuals exclusively and should fall under their personal control. The only way out is to drop (b), but this is exactly what the ontological theory of informational privacy does. Agents do not own their information but are constituted by it.

12.7.4 Public, passive, and active informational privacy

As we saw in Chapter 3, it may seem an oxymoron but a theory of informational privacy should be able to explain and support 'public informational privacy', i.e. privacy in public, as Nissenbaum (1998, 2010) and Margulis (2003) have convincingly argued. The difficulty here is represented by the need to abandon some naïve conceptions of privacy in terms of metaphorical private vs. public 'spheres'. Contrary to what intuition may initially dictate, by moving in and out of the 'public sphere' (e.g. by going to the pub or staying home) Alice is not *ipso facto* re-adjusting, each time, the degree of informational privacy to which she has a justified claim, but only the degree of informational privacy for which she can have a reasonable expectation. Many people, who would be embarrassed to appear naked in front of strangers, find showering at the gym with other unknown users unproblematic. The degree of informational privacy one may enjoy is patently determined also by the social context, as we have seen above, but it should not be confused with it. Likewise, there is of course a difference between private (non-public) personal information, which might be highly sensitive, such as one's own medical records, and public personal information, which is not necessarily confidential or intimate, such as one's own gender, race and ethnic group. And in public, one's informational privacy is more easily at stake than in private, obviously. But the fragility of one's informational privacy in public and of one's public personal information—both so readily subject to computerized processing (gathering, exchanging, mining, matching, merging, etc.)—is a fundamental reminder that we should be more, and not less, concerned about the phenomenon of 'public privacy'. After all, recent American and European history is full of tragic abuses of 'public information' (Seltzer and Anderson, 2001).

The reader may recall that the ontological theory of personal identity and informational privacy developed in this and in the previous chapter tackles this difficulty by comparing privacy to other rights such as personal safety. One has a right to personal safety both in private and in public, although, in public contexts, expectations that this right will be respected might be much lower than in private contexts.

We have already encountered what I have called 'passive informational privacy', when discussing the need for a theory to account for, and safeguard, one's identity as an informational entity not only from operations of cloning in public but also from attempts at corruption, again, especially in the public sphere. Providing someone with some information may easily mean violating that person's informational privacy, in two senses.

On the one hand, each of us has a fundamental right not to know: that is why violent scenes, disturbing news, pornography, advertising, unwanted reports, or spoilers (the final of the World Cup is over but one does not wish to know the result in order to enjoy it later on TV) and, I may add, mere idiocy, of which there is an overabundance throughout all media, may be suffered as contaminations of one's own self, as breaches of one's own informational privacy, brainwashing of the worst kind. Silence is hugely undervalued in our world; witness the difficulty of finding a restaurant, a pub, or a bar without some kind of background music.

On the other hand, each of us has a fundamental duty to ignore (or pretend not to know): in human societies privacy is also fostered through tacit agreements. TMI, as the younger generation says: *too much information*, more than one wants to have. We 'politely' ignore—e.g. do not bring up in conversation—moments we all witness and know about, ranging from keeping our eyes straight ahead at the urinal to never speaking of, say, marital acts that we know (and sometimes have evidence to confirm) must take place, etc. Again, no theory of informational privacy is complete that cannot account for such phenomena.

Finally, by 'active informational privacy' in the public sphere I mean to refer to those practices that facilitate and foster the development of individuals, by guaranteeing relevant conditions of informational privacy construction. What the latter may be varies from culture to culture and through time, but it seems quite clear that the right to informational privacy is not merely a negative right not to be *x*-ed, but also a positive right to *x*-ing. Parents know this only too well when they decide that their children's rooms, or that space in the tree house, are off-limits. It is respect for such conditions of possibility of other's informational privacies that marks the presence of that *caring* attitude already highlighted in the previous chapters.

12.8 Non-informational privacies

Let me now close this second part with two last comments. One concerns non-informational kinds of privacy. In Section 12.2, I outlined three other kinds of privacy, physical, mental, and decisional. An overlapping taxonomy distinguishes between *accessibility privacy*, understood as the freedom from intrusion and/or the right to be left alone in one's own physical space, and *decisional privacy*, understood as the freedom from interference in one's own choices and decisions, or the right to determine one's own course of actions, especially in relation to sexual options and reproductive alternatives (Schachter, 2003). Now, it seems natural to expect that theories of informational privacy, once mature, will make a sincere and robust effort to coordinate their findings and conclusions with those of other theories of other forms of privacy, in order to gain a comprehensive and coherent view of privacy in all its major aspects. And yet this seems an area largely unexplored. As usual, talking of Wittgensteinian family resemblances (Solove, 2002) only helps to postpone the problem: for those who stress the differences will then concentrate on the mere 'resemblance', whereas those who stress the similarities will keep looking for the common traits.

The second observation concerns a lower level of analysis. In this chapter, I have been concerned with challenges concerning a theory of informational privacy. Moving from this meta-level to the object level of problems regarding informational privacy itself, I would like to suggest that, depending on one's theory, some practical difficulties may be turned into hermeneutic opportunities, providing a metaphorical keyhole through which one may look at other phenomena otherwise difficult to investigate. By this I mean that a careful study of privacy infringements may provide an indirect method to probe whatever lies beneath, if anything, much like the study of unhealthy brains helps to understand the proper functioning of healthy ones. This is generally true of any theory that reduces or (more moderately) relates informational privacy to some other phenomena. For example, a theory that interprets informational privacy in terms of ownership/control will also be able to understand the latter more accurately by studying the pathology of the former. In our case, if informational privacy is indeed strictly connected to personal identity—as I have argued—then the study of its pathology, i.e. of informational privacy breaches, will offer valuable insights into the nature and dynamics of personal identity itself. In both cases, as far as the ontological theory is concerned, there is still much work to be done.

CONCLUSION

Privacy does not play a significant role in standard macroethics because it is the property of a class of entities as patients, not of actions. It becomes a central issue only within a culture that begins to recognize that entities are clusters of information and that privacy is a fundamental concept referring to the integrity and well-being of an informational entity as a patient. Privacy is not only an individual's problem, but may be a group's problem, a company's, or a corporation's problem, or a whole nation's problem, since all these entities have their nature fully determined and constituted by the information they are. How does the problem of privacy arise then? Within the infosphere, entities form a web of dependencies and symbiotic relations. The data output by data collection and analysis processes can easily become the input of other, or even the same, information processes (no hierarchy is implied). Complex relations among data-producers, data-collectors, data-processors, and data-consumers constitute an ecosystem in which data may be recycled, collated, matched, restructured and hence used to make strategic decisions about individuals. In this scenario, questions of informational privacy become increasingly urgent the easier it becomes to collect, assemble, transmit, and manipulate huge quantities of data. Note that cases in which privacy and confidentiality are broken because the information in question is legally or ethically significant are cases which society may agree to tolerate: for instance, we may all agree that, in special circumstances, bank accounts may be checked, computer files searched, or telephones bugged. The interesting point, for a theoretical foundation of information ethics, is not that information may have some legal consequences. Typically, privacy and confidentiality are treated as problems concerning S's ownership of some information, the information being somehow embarrassing, shameful, ominous, threatening, unpopular, or harmful for S's life and well-being. Yet this is very misleading, for the nature of the information in question is quite irrelevant. It is when the information is as innocuous as one may wish it to be that the question of privacy acquires its clearest value. A husband, who reads his wife's diary without her permission and finds in it only memories of their love, has still acted wrongly. The source of the wrongness is not the consequences, nor any general maxim concerning personal privacy, but a lack of care and respect for the individual as an informational entity. Yet this is not the familiar position we find defended in literature. Rather, a person's claim to privacy is usually justified on the basis of a logic of ownership and employment: a person possesses her own information (her intimately related facts) and has a right to exercise full control over it, for example to sell it, disclose it, conceal it, and so forth. It follows that the moral problem is normally thought to consist both in the improper acquisition and use of someone else's property, and in the instrumental treatment of a human being, who is reduced to numbers and lifeless collections of information. Sometimes, it is also argued that privacy has an instrumental value, as a necessary condition for special kinds of social relationships or behaviours, such as intimacy, trust, friendship, sexual preferences, religious or political affiliations, or intellectual choices. The suggestion is finally advanced that a person has a right to both exclusive ownership and unique control/use of her private information and that she must be treated differently from a mere packet of information. According to IE, however, this view is at least partly mistaken and fails to explain the problem in full. Instead of trying to stop agents treating human beings as informational entities, we should rather ask them to realize that, when they deal with personal and private information, they are dealing with human beings themselves, and should therefore exercise the same care and show the same ethical respect they would exercise and show when dealing with other people, living bodies or environmental elements. We have seen that a person, a free and responsible agent, is after all a packet of information. She is equivalent to an information microenvironment, a constantly elastic and permeable entity with centres and peripheries but with boundaries that are neither sharply drawn nor rigidly fixed in time. What kind of microinfosphere am I? Who am I? I am my, not anyone's, self. I am 'me', but who or what is this constantly evolving object that constitutes 'me', this selfhood of mine? A bundle of information. Me-hood, as opposed to type-self-hood and to the subject-oriented I-hood (the Ego), is the token-person identified as an individual patient from within, is an individual self as viewed by the receiver of the action. We are our information and when an informational entity is a human being at the receiving end of an action, we can speak of a me-hood. What kind of moral rights does a me-hood enjoy? Privacy is certainly one of them, for personal information is a constitutive part of a me-hood. Accessing information is not like accessing physical objects. Physical objects may not be affected by their manipulation, but any cognitive manipulation of information is also performative: it modifies the nature of information by automatically cloning it. Intrusion in the me-hood is therefore equivalent to a process of personal alienation: the piece of information that was meant to be and remain private and unique is multiplied and becomes public, it is transformed into a dead piece of my self that has been given to the world, acquires an independent status and is no longer under my control. Privacy is nothing less than the defence of the personal integrity of a packet of information, the individual, and the invasion of an individual's informational privacy, the unauthorized access, dispersion, and misuse of her information is an infringement of her me-hood and a disruption of the information environment that it constitutes. The violation is not a violation of ownership, of personal rights, of instrumental values, or of consequentialist rules, but a violation of the nature of the informational self, an offence against the integrity of the me-hood, and the efforts made by the individual to construct it as a whole, accurate, autonomous entity independent from, and yet present within, the world. The intrusion is disruptive not just because it breaks the atmosphere of the environment, but because our information is an integral part of ourselves, and whoever owns it possesses a piece of ourselves, and thus undermines our uniqueness and our autonomy from the world. There is information that everyone has about us, but this is only our public side, the worn side of our self, and the price we need to pay to society to be recognized as its members.

In the same way as the fourth revolution is best understood as a fundamental re-ontologization of the infosphere and its inhabitants, informational privacy requires an equally radical re-interpretation, one that takes into account the essentially informational nature of human beings and of their operations as social agents. Such a re-interpretation is achieved by considering each individual as constituted by his or her information, and hence by understanding a breach of one's informational privacy as a form of aggression against one's personal identity.

In the next chapter, we shall see how inforgs can interact in the infosphere to give rise to forms of distributed morality.

13

Distributed morality

> And I let myself go in a dream of lands where every force should be so regulated, every expenditure so compensated, all exchanges so strict, that the slightest waste would be appreciable; then I applied my dream to life and imagined a code of ethics which should institute the scientific and perfect utilisation of man's self by a controlling intelligence.
>
> André Gide, *The Immoralist* (1960), pp. 71–2.

SUMMARY

Previously, in Chapter 7, I argued that standard perspectives on 'mindless morality'—ethical issues involving artificial, synthetic, hybrid multi-agents, from companies to webbots—run two risks.

The first is that they might be unduly constrained by an anthropocentric conception of agency, thus overlooking the increasing importance of artificial agents as legitimate sources of morally loaded actions. I dealt with this problem in that chapter, arguing that our information ethics should include the analysis of the design and behaviour of artificial agents, as part of a larger strategy to understand a range of new ethical issues not only in technological contexts but also in ethics in general.

The second risk, also due to an excessive focus on the moral life of a stand-alone, individual, human agent, is that standard macroethics might fail to develop a satisfactory investigation of *distributed morality* (DM) in multi-agent systems (MAS). As I anticipated, this is the specific topic investigated in this chapter.

In Section 13.1, I shall introduce the basic idea of DM, by relying on a comparison with the well-known phenomenon of distributed knowledge in epistemic logic. I shall then explain the difference made by the occurrence of DM by discussing the moral scenario before and after its introduction (Sections 13.2 and 13.3 respectively). Next (Section 13.4), I shall provide some elementary examples of DM that should help to illustrate the phenomena in question more vividly and intuitively. In Section 13.5, I shall argue that the biggest challenge posed by DM concerns the possibility of harnessing its power in the right way. In Section 13.6, I shall outline a theory of morally enabling environment (*infraethics*) that can facilitate the occurrence and dynamics of DM. In the concluding section, I shall stress that the scope and magnitude of the ethical issues that we are, and will be, facing is such that it requires equally powerful

MAS—capable of dealing with them through the impact of their proper DM-based actions—as well as morally enabling environments that are friendly towards, and can facilitate, MAS' distributed morality.

13.1 Introduction: the basic idea of distributed morality

There is a sense in which cases of distributed morality have always been with us. *Collective responsibility*, for example, according to which a whole group of people is held responsible for some of its members' actions, even when the rest of the group has had no involvement at all (not even passively) in such actions, is a rather familiar concept in the *Old Testament*. The same applies to *social* or *group actions* and to (the theory of) *unintended consequences*. However, if these and similar phenomena are understood as being entirely reducible to the sum of (some) human, individual, and already morally loaded actions—and I agree with Narveson (2002) that sometime they might be—then this is not what I will be concerned with in this chapter. As explained in the introduction, in the following pages I intend to use 'distributed morality' (DM) to refer only to cases of moral actions that are the result of otherwise morally neutral or at least morally negligible (on this distinction, see below) interactions among agents constituting a multi-agent system, which might be human, artificial or hybrid. A comparison to a very elementary, classic case of distributed knowledge in epistemic logic (Halpern and Moses, 1990; Fagin et al., 1995) may help to clarify the basic idea.

Consider the case in which Alice knows only that [P ∨ Q], for example that 'the car is in the garage or Carol has it', whereas Bob only knows that ¬ P, i.e. that 'the car is not in the garage'. Neither Alice nor Bob knows that Q, only the supra-agent (with 'supra' as in 'supranational') Alice + Bob or Alob, knows that Q. It is the aggregation of Alice's and Bob's epistemic states that leads to Alob knowing that Q. Or, more precisely, Alob is the agent that is perceived to know that Q at the level of abstraction at which we observe Alice and Bob as a single agent. This is what happens regularly when one states, for example, that a company knew that such and such was the case. Now, suppose Alice causes a set of actions $\{a_1, \ldots, a_n\}$, and Bob causes another set of actions $\{b_1, \ldots, b_m\}$ to the effect that the supra-agent Alob causes a set of actions $\{c_1, \ldots, c_o\}$. The question about 'distributed morality' is this: can 'big' morally loaded actions (in our example, Alob's actions) be caused by many, 'small', morally neutral or morally negligible interactions (in our example, Alice's and Bob's actions)? I hold the answer to be yes, and the rest of the chapter is dedicated to supporting and explaining it. A good step forward is to start from a scenario in which there is no DM and then see what difference its introduction makes.

13.2 The old ethical scenario without distributed morality

Let us follow common practice and assume, for the sake of simplicity, that for every action a, a can be either morally Evil (E(a)), Good (G(a)), or Neutral (N(a)). A moment of reflection shows that, for the *deontologist*, it is quite easy to fill up the grey oval (Figure 16), representing the set of all actions that are morally neutral. This is because, as is well known, morally good actions done out of a sense of convenience, or interest, or inclination or any other heteronomous reason, to use Kant's terminology, are stripped of their positive moral value. Slightly more formally,[1] let us represent the deontologist's evaluative tendency to demote actions from G(a) to N(a) with the symbol $\{\dashrightarrow\}$, thus:

$$G(a) \dashrightarrow N(a) \qquad (i)$$

Graphically, (i) is represented by the *D-tendency* in Figure 16.

Following a similar reasoning, it is easier for the *intentionalist* to demote good to neutral ('great, but was not meant'), as in (i), but also evil to neutral ('sad, but was not meant'), so we have:

$$E(a) \dashrightarrow N(a) \qquad (ii)$$

Graphically, both (i) and (ii) are represented by the *I-tendency* in Figure 16.

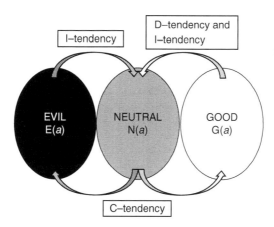

Figure 16. The old ethical scenario without distributed morality

---

[1] For the logically minded reader, these are not formulae but mere abbreviations. They could be transformed into formulae by adopting a quantification ranging over the domain of all actions occurring in the system under observation, but this would be cumbersome and provide no further insights. The same holds true for an analysis in terms of deontic logic.

As for the *consequentialist*, it is quite difficult to ensure that ultimately there is any *a* that is neither E nor G, but N. This is so because all actions have consequences and the latter inevitably have some moral value (at least in the weak but not yet negligible sense that they lead to other actions the consequences of which have moral value), so we have two tendencies to promote actions:

$$N(a) \dashrightarrow G(a) \quad \text{(iii)}$$

$$N(a) \dashrightarrow E(a) \quad \text{(iv)}$$

Graphically, both (iii) and (iv) are represented by the *C-tendency* in Figure 16.

Now, trend (i) is one of the traditionally counterintuitive aspects of Kantian ethics, which requires a theory of praise in order to make (i) more palatable. Trend (ii) might be welcome in many contexts of 'mindful morality', where it grounds the concepts of exculpation and forgiveness. Trends (iii) and (iv), in their full strength and if left unmodified, lead to the unacceptable conclusion that there are really no neutral actions at all, but only actions that are morally loaded, either positively or negatively. This is too implausible to be acceptable as it is, for it would force us to consider as morally significant a boundless number of *prima facie* non-moral actions, from the way Alice scratches her head to how she opens the door of a car. In order to rescue the consequentialist, we need to ring-fence both (iii) and (iv).

An obvious safety measure is provided by the concept of the *morally negligible* (the drop in the ocean effect, to oversimplify): many, if not most, actions are *actually* neither morally good nor morally evil (they are not subject to either trend described in (iii) and (iv)) because their actual effects are too small to be morally significant or (inclusive or) because they mutually cancel each other. A spy scratching her head might be intentionally decreeing the death of an individual, but that is an extraordinary case. Likewise, Alice might open the door of a car in such a way, or in such circumstances, that her action might count as morally approvable (perhaps she helped someone in real difficulty), yet this too seems to be the exception rather than the rule. Finally, moral agents often do and undo things in such ways that the end result is still negligible.

In order to implement the idea of morally negligible consequences, and thus ground the possibility of morally neutral actions, let us introduce (see Figure 17) two *moral thresholds* in our model: one makes it more difficult to apply $N(a) \{\dashrightarrow\} G(a)$, while the other has the same function with respect to $N(a) \{\dashrightarrow\} E(a)$. Now actions need to be morally significant in order to move from being neutral to being morally loaded. More formally, the two arrows that graphically describe the *C-tendency* become *vectors*: they have not only a direction but also a strength, which needs to be sufficiently high in order to overcome the relevant threshold.

Specifying how actions can be, or become, morally significant—conversely, establishing the right level at which the thresholds can be overcome—is a serious difficulty, comparable to the problem of identifying individual utilities when single rational agents need to make personal choices. It is certainly a major issue for the consequentialist, who probably needs to deal with it more contextually than she might be happy to

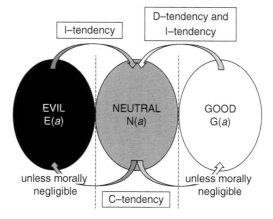

Figure 17. The old ethical scenario with moral thresholds and morally negligible actions

admit initially, as the unresolved debate on rule consequentialism indicates. Luckily, all this need not concern us here, since our goal is to gain a better understanding of DM. For this purpose, it is interesting to note that, once we model the applications of (iii) and (iv) as being constrained by some thresholds, we obtain one more new concept, that of *moral inertia*: most actions are morally neutral and tend to stay that way either because of the two thresholds, if one is a consequentialist; or because of the *I-tendency*, if one is an intentionalist; or because of the *D-tendency*, if one is a deontologist. These are all the details we need from the old scenario. We shall now use the concepts of *morally negligible*, *moral threshold*, and *moral inertia* in order to introduce a new variable in our model, that of distributed morality, a task for the next section.

13.3 The new ethical scenario with distributed morality

Recall that we wish to consider actions that might be performed by human, artificial, or hybrid multi-agent systems, so that you and I, as well as artificial agents (e.g. some kinds of webbots online), a corporation, or an individual driving a car with the help of a GPS, may all count as potential sources of possibly morally loaded actions. Because we are adopting such a MAS-oriented approach, we cannot rely on a system of moral evaluations based on intentionality or motive-related analysis. After all, the MAS in question might be totally mindless, so that any talk of beliefs, desires, intentions, and motivations would be merely metaphoric. Indeed, we are interested in adopting a uniform, minimalistic level of abstraction such that even human individuals might be treatable as mindless agents. Minimalism should not be confused with reductionism. Not every pizza can be reduced to a pizza margarita, but once you know how to cook at least a pizza margarita you can always add all the extra toppings you wish. The consequence of a mindless approach is that we need to evaluate actions not from a sender but from a receiver perspective: actions (including MAS', artificial and supra-agents') are assessed on the basis of their impact on the environment and its inhabitants. With these adjustments in place, let us return to the three concepts introduced in the previous section.

Because most actions are morally negligible, that is, because they remain below a given moral threshold, it follows that possibly evil actions (the subset of neutral actions labelled ◊Evil in Figure 18, where the diamond ◊ is the symbol borrowed from modal logic to mean 'possibly') may in the end be ineffective, that is, fail to bring about the evil that they could potentially generate. From a receiver's perspective, this is another way of saying that environments can be *morally resilient*—as we saw in Section 4.6—or, to paraphrase Paul of Tarsus (1 *Corinthians* 13), that goodness (understood as the absence of evil, hence including also neutrality) is *fault-tolerant*. An elementary example is provided by speeding on the motorway: a possibly evil action fails to become actually evil thanks to the resilience of the overall environment. The driver is morally irresponsible not because of the effects of his action—we assume that his reckless driving turned out to have no nasty consequences—but because of his unwarranted reliance on the fault-tolerance of the rest of the system. This is why his behaviour cannot be universalized:[2] the system can bear only so much pressure before collapsing.

At the same time, precisely because most actions are morally negligible and remain below a given moral threshold, possibly good actions (the subset of neutral actions labelled ◊Good in Figure 18) can equally fail to be effective. Environments can be *morally inert*: below a given threshold, possibly good actions never actually make a (significant, lasting or indeed any) difference but remain neutral. In other words, potential goodness can be too weak to become actual goodness, as we saw in Chapter 10. In this case, some forms of charity provide a good example of ineffectiveness.

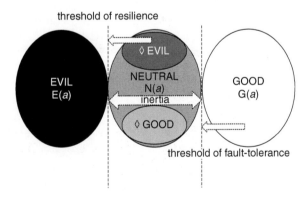

Figure 18. The new ethical scenario of distributed morality

---

[2] Universalization is an obvious factor that can help in such a strategy. By universalization I refer here to the normative coordination of the possibly good, distributed actions of a multi-agent system: agents constituting a MAS ought to implement, optimize, and coordinate their actions in such a way as to make them converge on the achievement of a morally good output. There are of course several other ways of understanding ethical universalization, see my reply to Stahl in Floridi (2008f, ch. 16).

To summarize (Figure 18), on the one hand, environments are morally inert or morally fault-tolerant. On the other hand, many MAS' actions often turn out to be morally neutral, in the sense of having insufficient strength to overcome the two thresholds introduced in the previous section. This might be because:

a. they are morally unloaded (value-free, in a different vocabulary); or
b. they are insufficiently morally loaded (have some moral value, but still fail to overcome the threshold); or
c. they mutually off-set each other.

We have seen in Section 14.1 that, unless Alice and Bob interact properly, their distributed knowledge cannot emerge, for it is held neither by Alice nor by Bob alone. Likewise, unless Alice and Bob interact properly, their distributed action may remain below the threshold of the morally negligible. The overall result is that, in this new scenario, neutrality works as a powerful attractor, and many actions are simply unable to escape their neutral status. In many cases, it is only by aggregating and merging individual courses of action that a moral difference is made. Note that, at this stage, such difference could be for the best (moral goodness) or for the worst (moral evil). Note[3] also that the aggregation in question is not one-way. Some evils emerging from DM might be further aggregated to such actions as to generate morally good outcomes. Likewise, some morally good actions reached through DM might be further aggregated in such a way as to cause evil. Clearly, in all these cases, the correct management of DM is both a challenge and an opportunity. Before discussing it, let me complete the description of DM by briefly presenting a few concrete illustrations.

13.4 Some examples of distributed morality

A classic and well-known example of negative DM is represented by the tragedy of the commons (Hardin, 1968, 1998). I have already analysed its digital version insofar as it applies to the infosphere in Greco and Floridi (2004), and I shall not discuss it here, where I wish to focus instead on some examples of positive DM. Just for the sake of illustrative simplicity, they are all based on quantitative analyses, in terms of moral benefits that can easily be quantified economically. In each of the following cases, MAS' actions, which are morally negligible in themselves, give rise to aggregated morally good actions:

(1) the shopping Samaritan: (RED);
(2) plastic fidelity: the Co-operative Bank;
(3) the power of giving: JustGiving;
(4) socially oriented capitalism: P2P lending.

Let's have a look.

---

[3] I am grateful to Massimo Durante for having called my attention to this important point.

13.4.1 The shopping Samaritan: (RED)

Perhaps the best way to present (RED) is by quoting the website of the project:

(RED) is a simple idea that transforms our incredible collective power as consumers into a financial force to help others in need.

Each time you buy a (RED) product or service, at no extra cost to you, the company who makes that product will give up to fifty per cent of its profit to buy and distribute antiretroviral medicine to our brothers and sisters dying of AIDS in Africa. Every dollar goes straight to Africa. Straight to the people who need it. Straight to keeping them alive so that they can go on taking care of their families and contribute socially and economically to their communities.... Since (RED)'s launch in 2006, over 5 million people have been impacted by HIV and AIDS programs supported by your (RED) purchases.[4]

Partners in the programme include American Express, Apple, Armani, Converse, Dell, GAP, Motorola, Nike, Starbucks, and many others.

13.4.2 Plastic fidelity: the Co-operative Bank

The next example of positive DM is represented by a fidelization programme, promoted by the Co-operative Bank in the UK. The bank offers a number of credit cards, linked to specific charities, including Amnesty International, Oxfam, and Greenpeace. By using the card, the customer ensures that:

- the chosen charity receives £15 for every account opened;
- a further £2.50 is received if the account is used within six months; plus
- 25p for every £100 spent using the card and 25p for every £100 transferred to the card.

These might seem drops in the desert, but, for example, between 1994 (the year the scheme was launched) and 2007 the Co-operative Bank's Oxfam-affiliated credit cards contributed £3 million towards Oxfam's work around the world.[5]

13.4.3 The power of giving: JustGiving

It can be expensive to run charities. In the UK, their management and administration typically represents between 5 per cent and 13 per cent of their total expenditure.[6] So a company that provides online fundraising tools to enable the electronic management of charitable donations, like JustGiving in the UK and its twin organization FirstGiving in the USA, can make a huge difference. Not only can it facilitate the process of fundraising and lower its costs, it can also provide visibility and support, as well as suggestions and solutions for extra funding opportunities. Here is some evidence.

---

[4] (Red), 'The (Red) Manifesto' <http://www.joinred.com/aboutred>.
[5] Oxfam, Ways to Donate <http://www.oxfam.org.uk/donate/other-ways-to-donate/oxfam-credit-cards>.
[6] CharityFacts, http://www.charityfacts.org/charity_facts/charity_costs/index.html.

Since 2000, JustGiving has provided its service for more than 9,000 UK registered charities, raising over £770 million. The administrative function includes the automatic reclaiming of Gift Aid on all donations from UK taxpayers. JustGiving's stated goal is to

> allow ordinary people to raise extraordinary amounts of money.... Charity Times claimed the company had 'transformed the face of donating in the UK'.[7]

In the 'business of beneficence' (Rockefeller) agents need to be frugal with their wasteful consumption but generous with their fruitful interactions.

13.4.4 Socially oriented capitalism: peer-to-peer lending

The last example concerns P2P lending, also known as social lending, person-to-person lending or community lending. This is lending online occurring between individuals directly, without the intermediation of an institute (usually a bank). P2P lending as a macroscopic phenomenon is really made possible only by the Internet, the enabling technology. There are two models, each illustrating a case of DM. In the marketplace model, online intermediaries, such as Prosper in the USA or Zopa in the UK, put lenders and borrowers in touch, who go through an auction-like process to negotiate a loan. In the community model, lenders and borrowers are already acquainted with each other, and online intermediaries such as Virgin Money US (formerly CircleLending) only help them to formalize a personal loan. In both models, we see distributed actions being aggregated to make a difference in the lives of the receivers.

13.5 The big challenge: harnessing the power of DM

The previous examples show how actions that are morally negligible in themselves may become morally significant, if properly aggregated. I have already mentioned that harnessing the power of DM is a challenge but also an important opportunity. This is represented by the possibility of strengthening environmental resilience and fault-tolerance, while weakening inertia, so that possibly evil but still neutral actions are blocked below the moral threshold, while possibly good but still neutral actions are enhanced above the moral threshold. Such a twofold manoeuvre requires ethical policies of

(a) *aggregation* of possibly good actions, so that the latter might reach the critical mass necessary to make a positive difference to the targeted environment and its inhabitants; as well as

(b) *fragmentation*, so that possibly evil actions might be isolated, parcelled, and neutralized.

Such policies are socially furthered by

---

[7] Wikipedia, 'JustGiving' <http://en.wikipedia.org/wiki/Justgiving>.

(c) incentives and disincentives, which represent the political and legislative side of the ethical discourse, and

(d) technological mechanisms that work as 'moral enablers'.

Regarding (c), since the moral behaviour of large number of agents has always been a concern, there is a long tradition of trial and error, social and political thinking (under the label social or public choice theory), legislation, ethical norms, and mass behaviour (think of the phenomenon of 'social pressure' or 'peer pressure') that can help significantly in shaping and orienting DM in the right direction. I shall not expand on this point in this chapter, but it is obviously of crucial importance.

Regarding (d), however, much work still needs to be done, for the following reason. DM is made *increasingly* possible by multi-agent systems, which in turn are made *increasingly* possible by extended, pervasive and intensive interactions. These interactions are *increasingly* enhanced, facilitated, and multiplied by ICTs. And all these 'increasingly's' explain why it is really only in advanced information societies that we can more readily and frequently witness the occurrence of DM phenomena. The sceptical reader only need recall how often the news reports about forces such as 'the markets', 'investors', 'public opinion', and so forth, and how tightly interwoven interactions have become in our globalized information society (more on this in the next chapter). ICTs are a most influential empowering factor behind the emergence of DM, working as powerful moral enablers, as I shall explain in some detail in the next section. Individuals are more and more connected and interactive in onlife environments, so that DM phenomena become progressively more frequent and important. For example, in 2011, 20.7 per cent of the European Union population accessed the Internet, by a laptop, while being away from both home and the office[8] (see Figure 19), and that is because our world is becoming our infosphere. We no longer login or logout, we are always *onlife*. Nevertheless, ICTs as moral enablers are not (at least not yet) designed in such a way as to meet the serious challenge posed by the correct management of DM. At the risk of trivializing a much more complex issue by using an elementary illustration, P2P technology, for example, can be used in order to aggregate neutral actions and make them overcome either threshold in both directions, towards evil or towards goodness. More controversially, the debate on network neutrality seems to be a case in which old prejudices against diversification are going to hinder the development of morally good, distributed dynamics (Turilli et al., 2012).

It might be that some specific technologies will always maintain their dual nature. Web 2.0 applications may be used to aggregate all sorts of interests and interactions, even the darkest ones. Very plausibly, at least part of the solution rests in the intelligent synthesis between three factors:

---

[8] Eurostat—Community survey on ICT usage in Households and by Individuals, <http://scoreboard.lod2.eu/index.php?scenario=2&indicators%5B%5D=i_iuport+IND_TOTAL+%25_ind&countries%5B%5D=EU27#chart>.

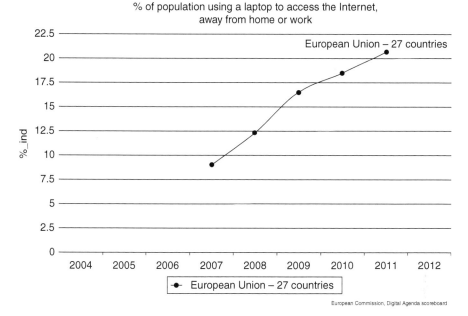

Figure 19. EU population accessing the Internet, away from home or work

Source: Eurostat—Community survey on ICT usage in Households and by Individuals. <http://scoreboard.lod2.eu/index.php?scenario=2&indicators%5B%5D=i_iuport+IND_TOTAL+%25_ind&countries%5B%5D=EU27#chart>

(1) a more profound and detailed understanding of the logical dynamics of DM and hence new forms of civil education;[9]
(2) better design of our technological moral aggregators, as argued for example by Adam (2005) in her discussion of privacy in relation to DM, by Turilli (2007) in terms of ethical protocols design, and by Cavoukian (2009), Pagallo and Bassi (2010), and Pagallo (2012) insofar as privacy might be approached from a design perspective; and
(3) improved ethical policies of incentives and disincentives.

Equally plausibly, it seems that part of the solution will also depend on the development of social and technological infrastructures (also known as meta-technologies, more on this in the next section) that will foster the right sort of distributed morality. This is the last point I wish to analyse in this chapter.

---

[9] See e.g. Erasmus International Institute MSH Nord-Pas-de-Calais, LCD'07—Workshop on Logics and Collective Decision Making, 13–14 March 2007, Lille, France.

13.6 Distributed morality and the enabling infraethics

There is a long and well-established tradition in ethics that seeks to identify, explain, and defend moral values, in order to develop and justify, on their basis, universal, normative analyses of morally loaded actions, and hence support reasonable, if sometimes competing, interpretations of the morally good life and its achievability. One crucial aspect, which seems to have been underemphasized by this tradition, is the analysis, implementation, and furthering of the non-moral factors that can facilitate morality and hinder immorality.

The idea may be quickly introduced by comparing it to a phenomenon well known to economists and political scientists. When one speaks of a 'failed state', one refers not only to the failure of a *state-as-a-structure* to fulfil its basic roles, such as exercising control over its borders, collecting taxes, administering justice, providing schooling, and so forth. One also refers to the collapse of a *state-as-an-infrastructure* or environment, which makes possible and fosters the right sort of social interactions; that is, one may be referring to the collapse of (certainties about) the rule of law, of acceptable ways of dealing with economic transactions, of default expectations about the protection of human rights, of a sense of political community, of civilized dialogue among differently minded people, of modes of communication to reach peaceful resolutions of ethnic, religious, linguistic, or cultural tensions, and so forth. All these expectations, attitudes, practices, in short such an implicit 'socio-behavioural infrastructure', which one may take for granted, provide a vital ingredient for the success of any complex society. It plays a crucial role in socio-political contexts, comparable to the one that we are now accustomed to attributing to physical infrastructures in economics. By analogy, it seems time to acknowledge that the morally good behaviour of a whole population of agents is *also* a matter of 'ethical infrastructure' or *infraethics*.[10] This is to be understood not as a kind of second-order ethical discourse or metaethics, but as a first-order framework of implicit expectations, attitudes, and practices that *can* facilitate and promote morally good decisions and actions. Examples include trust, respect, reliability, privacy, transparency, freedom of expression, openness, fair competition, and so forth. I highlighted 'also' and 'can' above because it is important to understand that such an infraethics is *not* necessarily morally good in itself. Any successful complex society, be it the City of Man or the City of God, has an implicit infraethics. Even a society in which the entire population consisted of angels, that is, perfect moral agents, needs norms for collaboration, coordination, and cooperation. Theoretically, that is, when one assumes that morally good values and the infraethics that promotes them may be kept separate (an abstraction that never occurs in reality but that facilitates our analysis here), a society in which the entire population consisted of Nazi fanatics could rely on high

---

[10] This is related to, but not to be confused with, what Jonsen and Butler (1975) meant by 'infraethics', which they understood as a particular level of ethical enquiry concerning public ethics, see Daniels (1996), p. 341.

levels of trust, respect, reliability, privacy, transparency, and even freedom of expression, openness, and fair competition. Clearly, what we want is not just the successful mechanism provided by the right infraethics, but also the coherent combination between it and morally good values, such as civil rights. To rely on an analogy: the best pipes may improve the flow but they do not improve the quality of the water, and water of the highest quality is wasted if the pipes are rusty or leaky.

In sociology, economics, politics, and law studies increasing attention has been paid in the last few decades to so-called enablers such as education, health, safety and security, property rights and credit opportunities, clear legislation, and reliable implementation of the law, especially in developing countries. The lack of similar studies about the need for, and the nature of, an infraethics is understandable, given the troublesome history of human priorities, but it also seems to be time to redress it.

Within this general context, the specific point I wish to address, in relation to the phenomenon of DM, may be clarified rather simply by means of two questions. Suppose we have a general view of what morally good is and of the sort of distributed morality that might bring it about, then what exactly may facilitate the implementation of the latter? And what exactly may hinder the sort of DM that could bring about the morally evil? Of course, the two questions are as strictly related as the two sides of the same coin. Indeed, they may be further simplified by labelling the referent of the 'what' in each of them as an infraethics (understood as the ensemble of moral enablers) and then rephrase them thus: given a dynamic, moral system in which DM plays a significant role, what is the right sort of enabling infraethics that can foster it?

An enquiry into the nature and logic of the right sort of infraethics, its interactions and operations within a dynamic system, and its positive effects on DM does not seek to uncover the morally good and evil, but rather presupposes a satisfactory understanding of both. It addresses a different problem, namely what sort of facilitating framework makes the morally good more likely to occur, and then become more stable and permanent, i.e. to take root; and what sort of hindering framework makes the morally evil more unlikely to occur or, when it has occurred, to remain unstable and more transient, so to wash away more quickly and easily? Now, investigations into ICTs, their personal, social, and ethical impact, and hence into ICE issues, have helped us, both historically and theoretically, to unveil cases of moral facilitation and thus identify moral enablers to an unprecedented extent and with a much higher level of clarity. Examples include information availability and accessibility, trust online (Taddeo, 2009, 2010), information transparency (Turilli and Floridi, 2009) and openness (as in open source software (Chopra and Dexter, 2008)), information privacy (see Chapter 12), and the relation between forgetfulness and forgiveness.

Unsurprisingly, issues of moral facilitation that seem too complex to tackle if we use a first-order logic become rather unproblematic once we adopt a second-order logic. Trust, for example, becomes very easily treatable if understood as a second-order relation (and hence an enabler), rather than a first-order one (Taddeo, 2010). However, the temptation of interpreting specific moral enablers, e.g. trust or transparency, in terms of *meta-values*, that is, as if they were values qualifying other values, should be resisted. I argued above that an infraethics is not a metaethics. Logically speaking, a more fruitful way to represent specific moral enablers is by relying on the apparatus of modal semantics, and to treat them as agents in themselves, which operate between possible worlds (PWs). Such enabling agents, when properly designed and regulated, can act as promoters and facilitators of the morally good. At worst, they can prevent, neutralize, or at least limit the paths to evil, that is, undesirable transitions from some PWs to other, morally worse PWs. Or (in the logical, inclusive sense of the disjunction), at best, they can foster, enhance, and consolidate desirable transitions from some PWs to other, morally better PWs, the paths to goodness. The other temptation, to understand specific moral enablers as *infra-values*, i.e. values that underpin other values and make them possible, should also be resisted. On the contrary, moral enablers are better understood as intra-components of the moral system, metaphorically comparable to the lubricant of the moral machinery. They work at the same level as moral values, neither below nor above them, as integral parts of the dynamic moral system, even if they themselves are not moral values.

ICE has cast a powerful light on a less visible side of the ethical discourse, the rich and fertile humus that provides nourishment and strength to the roots of moral interactions. It follows that we have now the opportunity to understand that, in ethics, moral facilitation is a much more influential, macroscopic and perhaps necessary phenomenon—not merely limited to ICE contexts—than we suspected in the past, a phenomenon that lies hidden behind the more visible scenes of many moral interactions. No determinism is involved, but freedom may be exercised more fully and in better directions if the right moral enablers are in place and work properly. Recall the analogy with physical infrastructures: they can help the economy of a country enormously, even if they do not determine the nature of the businesses in question. Once again, this is not an entirely new phenomenon. Within our information societies, moral enablers may often have an ICT nature, hence their study and implementation may be best carried out by an Information Ethics, but they do not need to be only ICT-based. To use the previous example, trust has always been a moral enabler. The fact that only in recent years have we focused so much on its ethical role is largely due to the impact of ICTs, which have worked as a magnifier.

CONCLUSION

Many more examples could be provided of cases of infraethical phenomena that facilitate the emergence of DM and positive moral behaviours. Consider fourth-generation bikesharing, for instance.

The advances and shortcomings of previous and existing bikesharing models have contributed to a growing body of knowledge about this shared public transportation mode. Such experiences are making way for an emerging fourth-generation bikesharing model or demand-responsive, multimodal systems. These systems build on the third generation and emphasize (a) flexible, clean docking stations; (b) bicycle redistribution innovations; (c) smartcard integration with other transportation modes, such as public transit and carsharing; and (d) technological advances including Global Positioning System (GPS) tracking, touch screen kiosks, and electric bikes. (Shaheen et al., 2010, pp. 165–66)

Clearly, it is a whole ensemble of facilitators that need to be designed, coordinated, and implemented for an infraethics to become possible, and such infraethics can make a difference in terms of DM only if a sufficient number of agents become involved. Bikesharing is a healthy and environmentally good thing and a morally positive trend, but it requires advanced ICT applications, no component of the system in itself would make any difference, and if only a few users were to take advantage of it, the environmental benefits would be virtually nil. As stressed above, the risk of misuse and moral hazard are also never entirely absent. To simplify, such bicycles, for example, could be used to rob a bank or may initially lead to more traffic-related accidents. Yet it seems obvious that the advantages vastly outweigh the risks.

The proper shaping and steering of DM through the design of the right sort of infraethics appear to be an important challenge that will deserve much more intellectual work, education, and political attention. In its scope and influence, DM is a largely unprecedented phenomenon, which characterizes advanced information societies, not because it never did or could occur in the past—this would be of course both factually and theoretically wrong—but because ICTs have just begun to make DM a much more common, pragmatically influential, and epistemologically salient phenomenon. Instances of DM that were 'too weak' and sporadic in the past to be worth much attention or ethical analysis are now playing an increasingly important role in our lives, and will be more and more influential in the future.

The conclusion is that an information society is a better society if it can implement an array of moral enablers, an infraethics that is, that can support and facilitate the right sort of DM, while preventing the occurrence and strengthening of moral hinderers. Agents (including, most importantly, the State) are better agents insofar as they not only take advantage of, but also foster the right kind of moral facilitation properly geared to the right kind of distributed morality. It is a complicated scenario, but refusing to acknowledge it will not make it go away.

There are both practical and theoretical problems affecting the development of a theory of distributed morality, of its moral enablers, and of their correct implementation. One may need to consider, for example, the global nature of information societies and the necessity to negotiate interactions with alternative, pre-existing moral traditions. Likewise, consistency and partial-ordering in terms of priority among different instances of DM and several moral enablers are certainly issues of crucial importance, as the debate between defenders of information privacy and defenders of information transparency highlights. Despite these difficulties, however, the study and actual development of DM and the corresponding infraethics are challenges worth tackling. The nature of the ethical issues facing humanity is increasing in scope, magnitude, and seriousness. Big issues call for big agents. We need powerful, multi-agent systems that, by aggregating and controlling their distributed actions, can cope ethically well with macroscopic, global moral issues. DM is a new phenomenon whose importance will only grow steadily. The sooner we learn how to harness its power explicitly the better. Infraethical environments in which moral enablers can flourish that support the right sort of MAS and DM will be better equipped to deal with our uncertain future. They may actually play a big role in how we solve some of the most pressing and intractable, ethical problems at a global level. One only needs to be reminded of the international crises involving financial institutions to concede that we are dealing with extremely powerful agents, whose actions and reactions affect the whole world for good and evil. When the ethical behaviour of such agents is in question, it is normal to turn to business ethics (BE) as the applied field that deals with the relevant sorts of moral investigations. This is the topic of the next chapter.

14

Information business ethics

> The fate of information in the typically American world is to become something which can be bought or sold.
>
> Norbert Wiener, *The Human Use of Human Beings* (1954), p. 113.

SUMMARY
Previously, in Chapter 13, I discussed the logic, the genesis, and the implications of distributed morality. I analysed the nature of distributed morality, as a feature of moral agency, and explained how it can arise in multi-agent systems comprising also non-human agents. I concluded the chapter by exploring some of the implications of the occurrence of distributed morality in the infosphere in general and in the information society in particular. Now, it seems obvious that business organizations are among the most influential multi-agent systems affecting the well-being of the infosphere. Quite naturally, both computer ethics and business ethics (BE) deal with the ethical impact of ICTs. However, so far, they have remained largely independent. In this chapter, I shall argue that information ethics can provide a good, foundational approach to both. I shall articulate and defend an informational approach to the conceptual foundation of business ethics, by using ideas and methods developed in the previous chapters, in view of the convergence of the two fields in an increasingly hyperconnected society. This brings to completion the line of reasoning begun in Chapter 7, where I defended the importance of extending our conception of moral agents to artificial entities as well (and therefore to business agents), and further expanded in Chapter 13, where we saw the systemic, distributed nature of aggregate actions emerging from the interactions between different moral agents of various kinds.

The task of providing business ethics with an informational foundation is made pressing by the realization that we live in a global, networked, information-based society in need of a distributed, information-based business ethics. This is not to say that information or computer ethics and business ethics have not been conversing for some time (see e.g. Coates (1982); Langford (1999)). It goes without saying that the emergence of a global information society, with its ICT-based ethical challenges, and the growing importance of ICT-intensive and networked business interactions, have made academic and practical barriers between the two ethical disciplines increasingly porous (De George, 2003, 2006). Nor does it mean that ethicists, policy-makers, lawyers, and business people more generally have failed to recognize the intrinsically hybrid nature of many of the key ethical issues with which they deal. It is widely acknowledged that privacy, copyright, informed consent, transparency and disclosure, P2P, digital divide, and so forth (see e.g. Ennals (1994); Mason et al. (1995); Vaccaro (2006); Vaccaro and Madsen (2009)) can be fully understood only if they are placed at the intersection between information, computer, and business ethics (Hodel et al., 1998). Rather, the exact point in question is that, despite their obvious commonalities, overlapping interests, and joint concerns (Wong, 2000), information ethics and business ethics have not yet converged on a shared, conceptual foundation of their investigations. Such common roots are what I hope to disclose in this chapter, which is structured as follows. In Section 14.1, I shall develop an informational analysis of business agents and processes. This will lead, in Section 14.2, to the identification of the three main ethical questions to be addressed by an information-based business ethics, namely:

(1) what goods or services are provided?
(2) how are they provided? and
(3) what impact do (1) and (2) have?

Answers to these questions offer evidence about the moral performance of the system. However, in order to motivate, prompt or cause the system to improve its behaviour, one has to identify the main points where normative pressure can be exercised. Such points are analysed in Section 14.3. In Section 14.4, I shall argue that *profit* is neither part of the *definition*, nor the *function*, of business, but a *goal* that provides no ethical guidance by itself. By this I do not mean to deny an obvious truth, namely that profit and trade play a crucial role in any business, ordinarily speaking (see e.g. Floridi (2012a)). What I wish to highlight, instead, is that, in order to understand what it means for a business to be morally good, we need to understand its deepest philosophical roots (ontology), and realize that business is an ethically good force insofar as it embodies the human, poietic drive in favour of systemic growth and well-being, and against wastefulness (of opportunities, of resources, of demands, of supplies, and so forth), that is, in favour of the *flourishing* of, and against the *destruction, corruption, pollution,* and *depletion* of, (parts of) the natural and man-made environment in which a business operates. In other words, business ethics should be seriously concerned about the poietic and anti-entropic vocation of human agents. I shall argue that a business agent is increasingly morally good the more successful it is in implementing four pro-flourishing and anti-wastefulness principles, mediated from Chapter 4. In the conclusion, I shall offer a positive note about the respectful, caring, but also fostering and constructionist role that business agents may play as stewards for the realities that they can positively affect.

14.1 Introduction: from information ethics to business ethics

We saw in Chapter 1 that the informationalization of our environment, of human society, and of ordinary life has created entirely new realities, made possible unprecedented phenomena and experiences, provided a wealth of extremely powerful tools and methodologies, raised a wide range of unique problems and conceptual issues, and opened up endless possibilities hitherto unimaginable. As a result, it has also deeply affected our moral choices and actions, affected the way in which we understand and evaluate moral issues and pose fundamental ethical problems, whose complexity and dimensions are rapidly growing and evolving. It would not be an exaggeration to say that many of the new ethical problems we are facing today are related to the information revolution. In general, I agree with Martin and Freeman (2004) when they argue that 'business ethicists are uniquely positioned to analyse the relationship between business, technology, and society'. However, I would also contend that information ethics, as developed in the previous chapters, offers an innovative and flexible methodology, which turns out to be particularly well suited to model some foundational aspects of business ethics in the new context of a highly hyperconnected society. Such a methodological approach is based on four main features we have encountered in the previous chapters. I shall briefly review them here for the convenience of the reader.

(1) *The nature of moral agents.* IE, like BE, defends a much less anthropocentric concept of agents, which also include non-human (artificial) and non-individual (distributed) entities, as well as networked, multi-agent systems and hybrid agents (e.g. companies, institutions). This goes hand in hand with BE's stress on business organizations as ethical agents in and of themselves.

(2) *The nature of moral receivers.* IE argues in favour of a more inclusive and less biologically biased concept of potential receivers of moral actions as 'centres of ethical worth', which now encompasses not only humans or living entities, but also engineered entities and their networks. Again, this is in line with BE's interest in the fate of business organizations and their environments, as well as with BE's expansion of the concept of receivers to include not only shareholders but also stakeholders of various kinds (Freeman, 1984). This enables now the expansion of classic stakeholder theory (Phillips, 2003) to informational entities, the fabric of their networks, and, ultimately, to the whole environment or infosphere.

(3) *The nature of the environment.* IE offers an informational, network-based conception of the environment, which now includes both natural and artificial (synthetic, man-made) ecosystems. It reconceptualizes reality in terms of the *infosphere*. From a BE perspective, it is useful to consider the whole environment understood informationally, that is, as constituted by all informational entities (thus also

including information agents like us or like companies, governments, etc.), their properties, and the network of their interactions, processes, and mutual relations. In this book, we have seen that the infosphere is also a concept that is rapidly evolving. The alert reader will have noticed a drift from a semantic (the infosphere understood as a space of contents) to an ontological conception (the infosphere understood as a hyperconnected environment populated by informational entities).

(4) *The nature of the moral relations and interactions.* IE supports an environmental, receiver-oriented approach. It is the well-being of the receiver of the moral action that, in principle, ought to contribute to the possible guidance of the agent's ethical decisions, and potentially constrain and orient the agent's moral behaviour. The receiver of the action is placed at the core of the ethical discourse, at the centre of the ethical network, while the 'transmitter' of any moral action (the agent) is moved to its periphery. This approach resonates with a variety of BE's more advanced views, which can be employee-, customer-, shareholder-, and stakeholder-centred (Wood et al., 2008).

These four features make IE and BE highly compatible and invite the application of IE to the informational analysis of BE, as we shall see in the next section.

14.2 The informational analysis of business

The first step consists in revisiting the definition of 'business' from a network-based, informational perspective. There are, of course, two main senses to be taken into account: business as an *agent*, that is, as a node in the network, and business as an *activity*, that is, as a relational process in the network. The standard definition of 'business' as an agent states that:

[1] Business (agent) = $_{def.}$ the provider of goods or services to customers.

When 'business' is to be understood as a process, activity, or interaction, rather than as the agent that is its source, the following definition is equally unproblematic:

[2] Business (activity) = $_{def.}$ the provision of goods or services to customers.

Although the two definitions [1] and [2] are uncontroversial, it is worth highlighting the fact that neither contains any reference to *profit*, which therefore turns out to be a feature that is neither necessary nor sufficient to qualify something as a business agent or process. The importance of this remarkable absence is not often appreciated in full, and I will return to it later in the chapter. At this stage, [1] and [2] may seem obvious, but that is the nature of all starting points of an adequate logical analysis: they should be uncontroversial to the point of being trivial. What follows immediately from [1] and [2] is that one can define business-agents as the source of business-activities. Wherever some provision of goods or services to customers occurs, there we find a business-agent, whether this is an individual selling coconut water on a beach in Rio de Janeiro, a school charging tuition fees in Oxfordshire, or a multinational corporation refining crude oil in Alaska.

Once [1] is understood on the basis of [2], it becomes possible to analyse [2] in relational terms: the agent is defined in terms of the activity that characterizes it, and the activity is defined in terms of the ternary[1] relation that constitutes it. To put it simply, we want to be able to state that x counts as a business if and only if, if x is an agent and y is a good or service and z is a customer, then x provides y to z. Using classic, first-order predicate logic, the set of entities in the infosphere on which x, y, and z range, and the following abbreviations:

- $A\ (x) = x$ is an agent
- $B\ (x) = x$ is a business
- $C\ (z) = z$ is a customer
- $D\ (y) = y$ is a (deliverable) good or service
- $P\ (x, y, z) = x$ provides y to z

we obtain:

[3] $\forall x\ (B\ (x) \leftrightarrow (A\ (x) \wedge \exists y\ \exists z\ (D\ (y) \wedge C\ (z) \wedge P\ (x, y, z))))$

The formula in [3] expresses more precisely what is stated above more informally. The advantage is that it makes it easier to appreciate four major features that we shall need in the rest of the chapter.

First, [3] should be understood as allowing, as perfectly possible, cases in which $x = y = z$. In other words, this means that the three variables could be replaced by the same constant, as in the extreme (and rather unlikely) case in which a corporate business sells (parts of) itself to (some other parts of) itself. Of course, normally x, y, and z will be interpreted as different constants. This is as it should be, since our model would be extremely inadequate if it could not accommodate the rather common case in which a business sells a product, which could also be a business, to a customer, which could also be another business. In short, the formula allows for the highest degree of interpretative flexibility.

Second, [3] is process- or relation-centred, as required above: first comes the concept of business as a transaction, which then defines the related elements as business agent and customer, not *vice versa*.

Third, from [3], it is simple to obtain a customer-centred (or receiver-centred) model, as illustrated by Figure 20 below (the Figure is a simplification and is not meant to suggest that the number of businesses and customers is the same).

Finally, one can apply to [3] a standard move in predicate logic, whereby ternary relations are reduced to combinations of binary relations, e.g. by transforming '5 is

---

[1] This is a relation that needs three elements to be satisfied, such as Germany is between France and Poland.

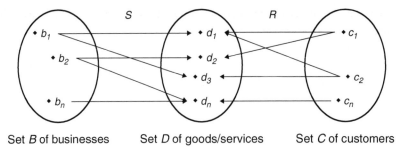

Figure 20. The relational analysis of business

between 4 and 6' into '"5 is bigger than 4" and "5 is smaller than 6"'. It follows that, although it might seem that business as an activity might necessarily require a ternary relation P in order to connect three, non-empty sets of related elements—namely B, constituted by the business agents, C, constituted by the customers, and D, constituted by the deliverable goods and services—one can actually transform P into a conjunction of two relations S and R without any loss of logical adequacy. The economic interpretation of S and R would normally be in terms of 'supplying' and 'demanding'. Given the informational approach adopted in this article, we shall use a different semantics and read S as 'sending' and R as 'receiving'. This enables one to interpret unsold goods, for example, as messages sent but not received (communication loss), or just-in-time production of only the requested goods as a reduction in the redundancy in the message sent, and so forth. The result is shown in Figure 20.

We are now ready to simplify the analysis further and obtain the initial model which, once transformed from static to dynamic, will serve us throughout the remainder of this chapter. I shall refer to it as the *concentric ring model* (see Figure 21). Note that this is not a Venn diagram, where the smaller unit is completely contained within the larger one, nor a 'layer' diagram, like a wedding cake. Rather the three sets C, D, B could be seen as flat washers surrounding a disk or three doughnuts, if a 3D model can help to

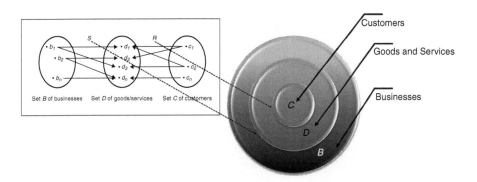

Figure 21. The concentric ring model of business

visualize it more easily. *C* is like a filled doughnut. It is surrounded by *D*, which can be imagined to be similar to a ring doughnut (technically, a torus). In its turn, *D* is surrounded by *B*, also shaped like a ring doughnut. The threshold between *C* and *D* is the relation *R*, and the threshold between *D* and *B* is the relation *S*. We have finally obtained an accurate but also intuitive representation, as a single object, of our logical model.

The concentric ring model, illustrated in Figure 21, places customers at the centre of all business activities. This is a valuable feature. Another valuable feature is that it shows constraints placed on *C* by both the elements of *D* that happen to be available (one cannot get an iPhone if there are none available) and by *B*. It suggests a power relationship that *B* holds both over *D* and *C*. Finally, the model further highlights the absence not only of profit as part of the definition of business (whether as an agent or as an activity it does not matter), as expected, but also of two other aspects that have been extensively discussed in the literature on business ethics: the problem of (fair) prices, which now appear to be a property of elements of *D*, and the nature/identity of business agents, that is, the elements of the set *B*, now defined as the sources of good/services. The model, however, still has one major limit: it is merely static. So it fails to take into account the interactions between business agents and customers over time and within a shared, hyperconnected environment. This is the last refinement that needs to be provided.

The parameters in our dynamic model are obviously time (the *x* axis) and the number of interactions between the various elements (the *y* axis). By placing the concentric ring models or *whirlpools* (their influence on the surrounding environment proceeds like decreasing waves) obtained above, in such a 2-dimensional space, we finally reach a more accurate description of the development of business interactions in real life, one

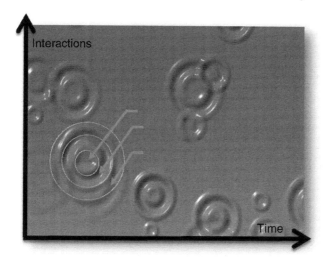

Figure 22. The whirlpool model of business

284 THE ETHICS OF INFORMATION

which will suffice to explain and facilitate critical analysis as it is developed in the rest of this chapter. I shall refer to this model as the *whirlpool model*. Figure 22 provides an illustration. Note that it is like a snapshot of a dynamic system: the whirlpools should be imagined as constantly increasing and then decreasing through time.

We now have a sufficiently detailed analysis of the object whose properties need to be investigated. Given the complex, informational, hyperconnected scenario represented by the whirlpool model, what are the main kinds of ethical questions faced by business agents? This is the key issue addressed in the next section.

14.3 The WHI ethical questions: what, how, and impact

Consider Figure 22. It seems evident that the fundamental questions to be addressed by a business ethics that is informationally modelled concern:

(1) What goods/services are provided?
(2) How are they provided?
(3) What impact do (1) and (2) have on both natural and artificial environments?

Figure 23 illustrates how the questions may be located in our model. Let me briefly comment on each of them.

By asking (1), an ethical theory concentrates on the product and hence shows that questions regarding the moral nature of the sender ('is the so and so business agent/source morally good or evil?') are still important but can be dealt with as secondary. As

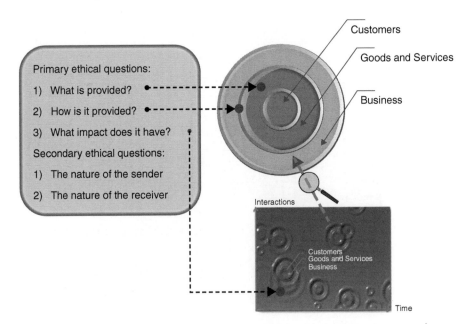

Figure 23. The WHI ethical questions: what, how, and impact

we saw in Chapter 9, this is perfectly in line with mainstream ethical theories, which consider 'good' and 'evil' as properties qualifying primarily actions and their outcomes, and only secondarily their sources. We qualify someone as mainly good or evil depending on whether his or her actions or their effects are mainly good or evil. Accordingly, in our model, business agents, their states and plans, strategies and policies, are identified as morally good or evil not in themselves, but only in a derivative sense, that is, following the assessment of their actions and the corresponding outcomes.

By asking (2), the same ethical theory addresses the moral nature of the process that leads to a particular output. Whereas question (1) concerns the nature of the elements of the set D, question (2) concerns the nature of the relation S between B and D.

Finally, no ethical analysis would be complete without a careful investigation of question (3), that is, the impact that the supply of specific goods/services, and hence the presence of goods/services themselves, have on the hyperconnected environment within which the business agents operate in interplay with their customers.

Depending on how one answers questions (1)–(3), there follows a different ethical evaluation of the business agent under investigation. However, if one wishes to modify that agent's behaviour, the WHI questions are of little help. They might signal that something needs to be done, but they cannot help in achieving the required modifications. For such a pragmatic goal, the whirlpool model needs to identify what one may call the right *points of normative pressure*, the topic of the next section (but see also Sections 4.5.4 and 6.2).

14.4 Normative pressure points

We have just seen what the most fundamental ethical questions that might be asked in BE are, when BE is approached from an IE perspective. I also anticipated that it would be a mistake to think that they are also the points where normative pressure can be applied to the system. Answers to the WHI questions may indicate how well (or indeed how badly) the overall system is performing, morally speaking, but they only contribute informatively to the process of guiding and shaping the system. Pragmatically, insofar as the processes of motivating, fostering, causing, or preventing new conducts are concerned, we need to identify ways in which the performance of the system may be successfully affected. In other words, we need to identify the main points where normative pressure can be exercised with some hope for success. There are three such points (or a combination of them) since, ideally, normative pressure should be exercisable on each of the three sets constituting the model.

(1) *Educational pressure on C*. One might exercise pressure on the system by educating or informing customers about the answers to the WHI questions. The availability, accessibility, and transparency of more and better information about:
- what goods/services are provided,

- how they are provided, and
- the impact that their provision has, or might have, on the overall environment or infosphere,

helps customers to make and shape their choices, and hence provides a significant way of influencing the moral behaviour of business agents. This is what drives not only standard competition, but also phenomena such as *ethical consumerism* and *fair trade certification* (Crane and Matten, 2007).

(2) *Prescriptive pressure on D*. The system may also be influenced by indicating *what goods/services ought to be provided* (of course both positively and negatively). This often means identifying requirements, specifications or standards that ought to be satisfied by the provided goods/services. We shall see that quality certification and control, as well as the indication of what features goods/services should or should not have, can be not only an ethical but also a legal issue.

(3) *Proscriptive pressure on B*. Finally, business agents may be influenced directly, through moral (or legal, more on this later) proscriptions about *what ought not to be done*.

As anticipated, in each of the previous three cases pressure can be exercised not only morally, but also legally. The point is worth clarifying. So far, I have analysed the ethical business system as if only two types of agents were involved in the network, businesses, interpreted as senders, and customers, interpreted as receivers. The State (and by this I refer to any entity with the legal power and legitimacy to impose its decisions on the whole whirlpool system, at least in principle) represents a third set of agents. Ideally, by issuing laws, imposing sanctions or disincentives, and offering rewards or incentives, the State is expected to play the role of facilitator, regulator, and referee of the communication system. Thus, the role of a liberal State in a modern democracy is *proscriptive* and *proactive* when it comes to the behaviour of business agents: it focuses on what business agents should not do (the legal don'ts) and on incentives to facilitate specific behaviours which are morally good or beneficial to the whole system or society. However, the role becomes *prescriptive* and *reactive* when the goods or services provided are concerned: the liberal State legislates on what they ought to be (the legal dos) and on the sanctions that might be imposed, if such legislation is not respected. The State's intervention in the network is usually hugely influential, as it adds a third dimension: the system acquires not only ethical norms but also legal rules or laws, and hence corresponding incentives, disincentives, prohibitions, and sanctions (Nelson, 2006). Of course, ethical norms aim at inviting endorsement, whereas legal rules seek to enforce compliance. The former are supposed to foster moral behaviour, the latter might be entirely neutral about it. How does this third dimension affect our model? An example will help to clarify the issue.

Consider the tobacco industry and more specifically the cigarette business. B is represented by tobacco companies, which send (i.e. produce and sell) goods in D, let us say cigarettes, which are received (i.e. bought) by C, individual customers. Answers to the three WHI ethical questions are well known and do not have to be rehearsed here. What about the normative pressure points and the State's intervention? Through taxation, the setting of age limits, the indication of non-smoking areas or the constraint on advertising forms and targets, the State exercises an external control on the system which is not, in and of itself, of a moral nature, but that has the function of facilitating moral behaviour. There is nothing intrinsically right or wrong, morally speaking, in smoking a cigarette in a cinema. By making it illegal, however, the State makes it easier to give up a bad habit, to prevent fire hazards, to diminish passive smoking, and so forth, and these effects have a morally positive value, insofar as they are good for the system in general and its individual receivers in particular. Smoking is unhealthy, can easily become an addiction and hence a moral vice, and the State seeks to make the choice of smoking more difficult and responsible, without infringing individual liberties. The same applies to wine, beer, and spirits. It should really apply to other recreational drugs as well, such as cannabis. Next, by making it compulsory to provide health information regarding smoking—for example health education in schools, documentaries on TV, or health warnings on cigarette packets—the State exercises direct moral pressure on the system through the '*education of C*', one of the normative pressure points discussed above. In terms of *prescriptions*, the second pressure point, consider the ethical choice of producing only self-extinguishing cigarettes, available since the 1930s. This could be a moral choice of the producers, instead of becoming a proactive legal requirement brought about by some legislation, as is already the case in several American states, such as New York, Massachusetts, and California, and it might soon happen in the EU. After all, burning cigarettes are among the main causes of fires.

Finally, in terms of *proscriptions*, 'Joe Camel' provides an illustrative example. In 1987, RJR (R.J. Reynolds Tobacco Company) created Joe Camel as the mascot for the brand of Camel cigarettes. In 1991, the American Medical Association reported that the campaign had been particularly popular among five- and six-year-olds, who apparently could recognize Joe Camel more easily than Mickey Mouse, Fred Flintstone, Bugs Bunny, or even Barbie (Fischer et al., 1991). This led the association to invite RJR to terminate the Joe Camel campaign. Although RJR initially declined, in 1997, after further appeals, the Joe Camel campaign was replaced by a somewhat more adult campaign. The point is that RJR never lost a legal battle on this issue, but acted in a way that, whatever the ultimate motives might have been, had a positive moral outcome.

14.5 The ethical business

The previous analysis has provided the formal framework within which BE can be interpreted informationally. Some of the features of the model already cast a different light on the ethical agenda of BE in an information society. Three issues—(1) the nature of business agents, (2) the fairness of prices of goods or services, and (3) the obtainment of profit (Friedman, 1970)—have been shown to be of much less pressing concern than (4) the nature of the delivered goods or services, (5) the ways in which they are produced, and (6) what impact their provision might have on the overall environment. This is interesting, since it seems that much of contemporary BE has been focused on (1)–(3) rather than on (4)–(6). What the previous formal analysis cannot provide yet is actual *content* (as opposed to a *formal analysis*), that is, a coherent vision, based on explicit ethical principles, of what an ethical business in the information society should be like. This is the last but perhaps most important contribution offered by IE.

It seems uncontroversial that, in order to be just and fair, laws and regulations should be at least compatible with, if not directly based on, morally right norms (ethical prescriptions and proscriptions), which in turn may be expected to depend on morally good principles. But what sort of morally good principles should guide business agents? How are we going to know when the answers to the WHI questions are morally unsatisfactory? And how can we judge whether, and what sort and degree of, normative pressure should be exercised, and in which direction, if the behaviour of the agents in the hyperconnected system or infosphere is deemed unsatisfactory and needs to be rectified? Much seems to depend on how we understand the ultimate nature of business itself, its function, goal, and role in society, in other words, its deep ontology.

Recall the definition of business given in [1]. It applies both to for-profit and to non-profit business organizations. Indeed, following [1], even business organizations with negative net profit do not, for this reason, stop qualifying as business agents. So profit is clearly not part of the essence of a business, not in the sense in which having three sides is part of the essence of a triangle. Unfortunately, this point is often overlooked, by objecting that being profitable might not be part of the *nature* of what may count as business, but it does capture its basic or primary *function*. At least in the case of for-profit enterprises, that very qualification refers exactly to what the task or function of a business is, although not its necessary and sufficient conditions. If this is granted, then one might further argue that the function, i.e. profit, determines the moral quality of the function-bearer, namely the business. With an analogy, our opponent might argue that the definition of a knife does not include 'being sharp', but since a knife's function is to cut, then the sharper the knife, the better it cuts, and so the better that knife should be judged to be.

The reasoning is muddled, for it fails to distinguish between a functional analysis ('for cutting') and a teleological analysis (cutting in order to achieve which goal?). Suppose, just for a moment, that profitability could qualify as the primary function of business. A sharp knife is a very good knife functionally speaking, but a morally bad tool in the hands of a serial killer. A very profitable business would be a very good business functionally speaking, but still a moral disaster if it sells slaves or child pornography. Clearly, if profit is understood as the function of business, this leaves unanswered any moral question worth asking. We have begun to rectify the confusion. From a functional perspective, profit may be (and often is) the much desired *effect* of a well-run business, whose primary function, nevertheless, remains that of providing goods and services to customers. If the distinction is unclear, the following analogy should help. People who argue that profit is the primary function of business are as mistaken as those who argue that the primary function of sexual intercourse is pleasure. Naturally, pleasure plays a very important and positive role, and of course animals may pursue sexual pleasure only for its own sake or, in the case of humans, for mental reasons as well. All this, however, should not blind us to the fact that sexual intercourse has a reproductive function: pleasure, like profit, is only the effect.

So far we have established that profit is part of neither the essence nor the function of business. Recall that we are trying to understand on what moral principles a business agent could be ethically evaluated. Now, our opponent has a further reply. Let us admit that the distinction between essence, function, and goal of a business is sound. Profit might be transformed into the teleological goal of a business, that is, into its purpose or mission. A morally good business would then be one that takes due care of its goal—being profitable—in view of the advantage that this brings for its shareholders. Here, we find the most dangerous mistake, because it is the least visible. First, let me clarify a final point about the function of business. Above, we assumed that profit could play such a role. We conceded this only for the sake of argument. Our opponent has now re-interpreted profit as playing the role of a business' goal. This has left empty the role of function. We know from [1] and [2] that the function of business is to provide goods and services to customers. Like the knife before, the more successful a business is, in providing goods or services to its customers, the better it is *functionally* speaking. Of course, our opponent is still waiting for an answer to his objection that profit might be the purpose orientating the function. This requires one further distinction.

It is a truism that for-profit businesses have profit as their main, if not only, ostensible goal. This truism, however, should not mislead us into thinking that we have finally hit upon the foundation for an ethical evaluation. Consider the difference between:

(a) what is good for x; and
(b) what x is good for.

We are interested in the case in which x = business. Following (a), profit is certainly what is good for business. Without it, business can much less easily grow in size, develop in quality, and flourish as a rewarding and successful enterprise for the people involved. Yet this is a factual remark, similar to what we said above about the sharp knife. Profit, in this sense, still does not enable one to draw any distinction between morally good or evil businesses. For profit is ethically blind: it rewards any business that pursues it successfully. Profit would still be a good thing for business even if the business were that of trading slaves. The ethical question is addressed once we move to (b). What is a business, for which profit is indeed a good thing, good for? The answer must arguably come from a consideration of the contribution made by the business in question to the network in which it is embedded, its overall environment, by which I mean not just its physical or natural habitat, but ultimately the whole ecosystem affected by that business, by its practices and its products or, in the informational vocabulary of IE, the region of the infosphere that is affected by that business. Of course, (b) can be answered purely factually, by arguing that profit is what business is good for. In this case, there is no circularity but rather mere consistency. The problem is not logical, but conceptual: we are still failing to touch upon the moral question. Profit is good for business and business is good for profit. The ethical blindness is still there, as we are still unable to distinguish between a morally good and a morally evil business. In order to provide such a normative evaluation, we need to address (b) in such a way as to step into the realm of ethical principles. We need to decide what a business is *morally* good for. This is the deeper ontological question that really matters, ethically speaking, and it is one that, following IE, can be addressed from a receiver-oriented perspective.

From the receiver-oriented perspective supported in this book, one can see that business is the art of matching supply and demand and, in so doing, fostering human flourishing and avoiding wastefulness. By wastefulness, I mean any kind of *destruction, corruption, pollution*, and *depletion* of (parts of) reality; that is, any form of impoverishment on the side of the receivers of the business activities. We have already encountered several times such wastefulness in terms of metaphysical entropy. It follows that a business agent is increasingly morally good the more successful it is in implementing the following environmental principles, mediated by a more abstract and inclusive analysis provided by IE:

1. wastefulness ought not to be caused in the world (the infosphere)
2. wastefulness ought to be prevented in the world (the infosphere)
3. wastefulness ought to be removed from the world (the infosphere)
4. the flourishing of entities as well as of the whole infosphere ought to be promoted by preserving, cultivating, and enriching their properties.

The reader will recognize them as the business-related translations of the more general e-nvironmental principles analysed in Section 4.7. These four principles play a twofold role. On the one hand, they can provide clarification when answers to the WHI questions are morally unsatisfactory. On the other hand, they can indicate how the behaviour of the overall system could be improved. To put it more simply, they can be used as a yardstick by which to measure when business matters are not going morally well, and how they can be rectified to go morally better.

CONCLUSION

As I remarked in Section 1.4, we are living in a hyperconnected environment (infosphere) that is becoming increasingly *synchronized* (time), *delocalized* (space), and *correlated* (interactions). It is an environment in need of an information ethics that might qualify as global, as I shall argue in the next chapter. Previous revolutions (especially the agricultural and the industrial ones) created macroscopic transformation in our social structures and physical environments, often without much foresight. The information revolution is no less dramatic. We shall be in deep trouble if we do not take seriously the fact that we are constructing the new environment that will be inhabited by future generations. We should be working on an ecology of the infosphere. Unfortunately, I suspect it will take some time and a whole new kind of education and sensitivity to realize that the infosphere is a common space, which needs to be preserved and improved to the advantage of all (for advancements in this direction see e.g. Wood and Logsdon (2008)). One thing seems unquestionable though: business is part of the human exception, like the open-ended use of language or tools. We are the only animals that do business. Other animals trade favours at most, they do not engage in financial transactions. So civilizations and societies are often evaluated on the basis of how friendly they have been towards this special feature of human life. It is to be hoped that the information society will be judged, by future generations, as business-friendly, and that such friendliness will be repaid by the respect and care exercised towards the infosphere by the business agents inhabiting it (Crane and Matten, 2004).

15
Global information ethics

> Society can only be understood through a study of the messages and the communication facilities which belong to it; ... in the future development of these messages and communication facilities, messages between man and machines, between machines and man, and between machine and machine, are destined to play an everlasting part.
>
> Norbert Wiener, *The Human Use of Human Beings* (1954), p. 16.

SUMMARY

Previously, in Chapter 7, I defended the importance of extending our conception of moral agents in order to include artificial entities as well, from software agents to hybrid systems like corporations. In Chapter 13, I discussed the nature of distributed morality, as an emerging feature of a more extended conception of moral agency. In Chapter 14, I considered business agents and their moral behaviour in the infosphere from an informational perspective. The last task is to consider what happens to artificial agents (including business agents) and distributed morality when the infosphere becomes a globalized environment. In this chapter, I shall argue that IE can provide a successful approach for coping with the challenges posed by our increasingly globalized and information-based reality. After a brief review of some of the most fundamental transformations brought about by the phenomenon of globalization (Section 15.1), I shall distinguish, in Sections 15.2 and 15.3, between two ways of understanding global information ethics: as an *ethics of global communication* or as a *global-information ethics*. I shall then argue, in Sections 15.4 and 15.5, that cross-cultural, successful interactions among micro- and macro-agents call for a high level of successful communication; that the latter requires a shared *ontology* friendly towards the implementation of moral actions; and that this is provided by IE. There follows in Section 15.6 an account of *ontic trust*, the hypothetical pact among all agents and patients presupposed by IE. In the conclusion, I shall highlight the importance of e-nvironmentalism or the marriage of *physis* and *techne*.

15.1 Introduction: from globalization to information ethics

Globalization is a phenomenon too complex even to sketch in this brief introduction. For a very synthetic but well-balanced and informed overview, I would recommend Held and McGrew (2001). In their terminology (see Held et al., 1999), I am a subscriber to the transformationalist approach, according to which

> globalization does not simply denote a shift in the extensity or scale of social relations and activity. Much more significantly, argue the transformationalists, it also involves the spatial re-organization and re-articulation of economic, political, military and cultural power. (Held and McGrew, 2001, p. 324)

So I hope that I shall be forgiven if I am rather casual about many features that would deserve full attention in another context. Here, I wish to highlight just six key transformations characterizing the processes of globalization. I shall label them *contraction, expansion, porosity, hybridization, synchronization,* and *correlation*. They provide the essential background for making sense of the thesis developed in the rest of the chapter, which is that IE can provide a successful approach for coping with the challenges posed by our increasingly globalized infosphere.

15.1.1 Contraction

The world has gone through alternating stages of globalization, growing, and shrinking, for as long as humanity can remember. Here is a reminder:

[I]n some respects the world economy was more integrated in the late 19th century than it is today... Capital markets, too, were well integrated. Only in the past few years, indeed, have international capital flows, relative to the size of the world economy, recovered to the levels of the few decades before the first world war.[1]

The truth is that, after each 'globalization backlash' (think of the end of the Roman or British Empires), the world never really went back to its previous state. Rather, by moving two steps forward and one step back, sometime towards the end of the last century the process of globalization reached a point of no return. Today, revolutions or the collapse of military or financial empires can never shrink the world again, short of a complete unravelling of human life as we know it. Globalization is here to stay.

Globalization has become irreversible mainly thanks to radical changes in world-wide transport and communications (Brandt and Henning, 2002). Atoms and bytes have been moving increasingly rapidly, frequently, cheaply, reliably, and widely for the past fifty years or so. This dramatic acceleration has shortened the time required for many interactions: economic exchanges, financial transactions, social relations,

---

[1] '1897 and 1997—the Century the Earth Stood Still', *The Economist*, 18 December 1997 <http://www.economist.com/node/455942>.

information flows, movements of people, and so forth (Hodel et al., 1998). And this acceleration has meant a more compressed life and a contracted physical space. Ours is a smaller world, in which one may multi-task fast enough to give and have the impression of leading parallel lives. We may regain a nineteenth-century sense of time and space only if one day we travel to Mars.

15.1.2 Expansion

Human space in the twenty-first century has not merely shrunk, though. ICTs have also created a new digital environment, which is constantly expanding and becoming progressively more diverse. Again, the origins of this global, transnational common space are old. As we saw in Chapter 1, they are to be found in the invention of recording and communication technologies that range from the alphabet to printing, from photography to television. Yet it is only in the last few decades that we have witnessed a vast and steady migration of human life to the other side of the screen. Some time ago, when you asked 'where were you?' it became normal and common to receive the answer 'on line'. Nowadays, many people live *onlife* and that question has further evolved to the point of not making much sense any more. Meanwhile, it used to be pointless to ask someone about her location when calling her over the telephone, yet today it is common to begin a conversation with such a request, for a telephone number says nothing about the physical whereabouts of the receiver. Globalization also means the emergence of this sort of single virtual space, shareable in principle by anyone, anytime, anywhere. The infosphere has begun to pervade any space.

15.1.3 Porosity

An important relation between our contracting physical space and our expanding, virtual environment is that of *porosity*. Imagine living as a flat figure on the surface of an endless cylinder. You could travel on the surface of the cylinder as a two-dimensional space, but not through it. So in order to reach any other point on the cylinder, the best you could do would be to follow the shortest path (geodesic) on the cylindrical surface. The empty space inside the cylinder would be inconceivable, as would a third dimension. Imagine now that the surface became porous, and hence that a third dimension were added. The geodesics would be revolutionized, for you could travel through the vacuum encircled by the cylinder and reach the other side, in a straight line, thus significantly shortening your journeys. To use the rather apt vocabulary of surfing, you would be *tubing*: space would be curling over you, forming a 'tube', with you inside the cylindrical space. From a 2D perspective, you would literally move in and out of space. This sort of porosity characterizes the relation now between physical and virtual space. It is difficult to say where one is when one is 'tubing', but we know that we can travel through cyberspace to interact with other physical places in a way that would have been inconceivable only a few decades ago. The more porous our environments become, the clearer it is that we are living in a global infosphere.

15.1.4 Telepresence

In our porous environments, being elsewhere in a variety of ways, as a mere passive observer or more and more interactively, is an ordinary experience and this is also what globalization means. I discussed telepresence at length in Chapter 3, so I won't comment on this topic any further.

15.1.5 Synchronization

In a world in which information and material flows are becoming so tightly integrated and enmeshed, it is not surprising to see global patterns emerging not only from well-orchestrated operations—consider the tedious experience of the launch of a major blockbuster, with interviews in magazines, discussions on TV programmes, advertisements of merchandise, and by-products throughout the world, special food products in supermarkets and fast-foods, etc.—but also inadvertently, as the result of the accidental synchronization of otherwise chaotic trends. All of a sudden, the world reads the same novel, or wears the same kind of trousers, or listens to the same music, or eats the same sort of food, or is concerned about the same problems, or cherishes the same news, or is convinced that it has the same disease. Some of this need not be the effect of any plan by some Big Brother, secret agency, powerful multinational, or any other *deus ex machina* that is scheming behind the scenes. After all, worldwide attention span is very limited and flimsy, and it is very hard to compete for it. The truth is that some global trends merely arise from the constructive interference of waves of information that accidentally come into phase, and hence reinforce each other to the point of becoming global, through the casual and entirely contingent interaction of chaotic forces. It may happen with the stock markets or the fashion industry, or dietary trends. Once a silly video reaches a critical mass of viewers on YouTube, it is almost impossible to remove it from its top position, for example. The recurrent emergence of temporarily synchronized patterns of human behaviour, both transculturally and transnationally, is a clear sign of globalization, but not necessarily of masterminded organization. There is no intelligent plan, evil intention, autonomy, or purposeful organization hiding in the billion snowflakes that become an avalanche. Social group behaviour is acquiring a global meaning. The distributed power that generates Wikipedia is the other side of the dark, mindless stupidity of millions of slaves of fashions and trends. Viruses have no intelligence and the 'viral' popularity of something says a great deal about those who contracted the 'virus'.

15.1.6 Correlation

We know we live in a hyperconnected reality. Imagine a safety net, like the one used in a circus. If it is sufficiently tight and robust, the heavier the object that falls into it, the larger the area of the net that is stretched, sending waves of vibration throughout the net. Globalization also refers to the emergence of a comparable net of correlations among agents all over the world, which is becoming so tight and sensitive that the time lag in the transmission of the effects of an event 'dropping' onto it is fast shortening, to the point that sometimes there is almost no distinction between what counts as local or remote. Global often means not everywhere but actually delocalized, and in a delocalized environment social friction is inevitable, as there is no more room or allowance for absorbing the effects of agents' decisions and actions. If anyone moves, the global boat rocks, from Alexandria to Tripoli to Damascus.

15.2 Globalizing ethics

If we consider now the profound transformations just sketched, it would be rather surprising if they did not have serious implications for our moral lives (see Weckert (2001) and Ess (2002)). In a reality that is more and more physically contracted, virtually expanded, porous, hybridized, synchronized and correlated, the very nature of moral interactions, and hence of their ethical analysis, is significantly altered. Innovative forms of agency are becoming possible; new values are developing and old ones are being reshaped; cultural and moral assumptions are ever more likely to come into contact if not into conflict; the very concepts of what constitutes our 'natural' environment and our enhanced features as a biological species are changing; and unprecedented ethical challenges have arisen, just to mention some macroscopic transformations in which globalization factors, as sketched above, play an important role.

What sort of ethical reflection can help us to cope successfully with a world that is undergoing such dramatic changes? Local approaches are as satisfactory as burying one's head in home values and traditions. The ethical discourse appears to be in need of an upgrade to cope with a globalized, hyperconnected world. Each ethical theory is called upon to justify its worldwide and cross-cultural suitability. This seems even more so if the theory in question seeks to address explicitly the new moral issues that arise from the digital revolution, as is the case with IE. The question is whether, in a world that is fast becoming more and more a globalized infosphere, information ethics can provide a successful approach for dealing with its new challenges. I shall argue in favour of a positive answer. But to make my case, let me first clarify what *Global Information Ethics* may mean.

15.3 Global-communication ethics vs. global-information ethics

There are at least two ways of understanding global information ethics: as an *ethics of global communication* (Smith, 2002) or as a *global-information ethics* (Bynum and Rogerson, 1996). Since I shall concentrate only on the latter, let me briefly comment on the former first.

Global Information Ethics, understood as a Global-Communication Ethics, that is, as an ethics of worldwide communication, may be seen as a commendable effort to foster all those informational conditions that facilitate participation, dialogue, negotiation, and consensus-building practices among people, across cultures and through generations. It is an approach concerned with new and old problems, caused or exacerbated by global communications or affecting the flow of information. Global-Communication Ethics is therefore a continuation of policy by other means. It is Habermasian in nature. And it does not have to be reduced to a mere gesture towards the importance of mutual respect and understanding (meeting people and talking to each other can hardly do any harm and often helps). It is, however, faced by the serious problem of providing its own justification. What sort of ethical principles of communication and information are to be privileged, and why? Is there any macroethics (e.g. some form of consequentialism or deontologism or contractualism) that can rationally buttress a Global-Communication Ethics? And isn't any attempt at providing such a macroethics just another instance of 'globalization' of some values and principles (usually based on some version of the Enlightenment) to the disadvantage of others? Without decent theorization, the risk is that we will reduce goodness to goodiness and transform the ethical discourse into some generic, well-meant sermon. At the same time, a robust foundation for a Global-Communication Ethics may easily incur the problem of failing to respect and appreciate a plurality of diverse positions. The dilemma often seems to be left untouched, even when it is not overlooked. The good news is that it may be possible to overcome it by grounding a Global-Communication Ethics on a Global-Information Ethics.

15.4 Global-information ethics and the problem of the lion

If we look at the roots of the problem, it seems that,

(1) in an increasingly globalized world, successful interactions among micro- and macro-agents belonging to different cultures call for a high level of successful communication;
(2) but successful, cross-cultural communications among agents require, in their turn, not only the classic three 'e's—*embodiment, embeddedness*, and hence *experience* (a sense of 'us-here-now')—but also a shared *ontology* (more on this presently); and yet
(3) imposing a uniform ontology on all agents only seems to exacerbate the problem, globalization becoming synonymous with ontological imperialism.

By 'ontology' I do not mean to refer here to any metaphysical theory of Being, of what there is or there isn't, of why there is what there is, or of the ultimate nature of reality in itself. All this would require a form of epistemological realism (some confidence in some privileged access to the essential nature of things) that I do not hold (Floridi, 2011a), and that, fortunately, is not necessary to make my case. Rather, I am using 'ontology' to cover the outcome of a variety of processes that allow an agent to appropriate (be successfully embedded in), semanticize (give meaning to and make sense of), and conceptualize (order, understand and explain) her environment, through a wealth of levels of abstraction. In simplified terms, one's ontology is one's world: that is, the world as it appears to, is experienced by and interacted with, the agent in question.

How an ontology is achieved specifically and what sort of philosophical analysis is required to make sense of its formation is not a relevant matter in this context. What is relevant is that agents can talk to each other only if they can partake to some degree in a shared ontology anchored to a common reality to which they can all refer. More technically, this means that two agents can communicate only if they share at least some possible level of abstraction. Imagine two solipsistic minds, α and β, disembodied, unembedded, and devoid of any experience. Suppose them living in two entirely different universes. Even if α and β could telepathically exchange their data, they could still not *communicate* with each other, for there would be absolutely nothing that would allow the receiver to interpret the sender. In fact, it would not even be clear whether any message was being exchanged at all.

The impossibility of communication between α and β is what Wittgenstein (2001) had in mind, I take it, when he wrote that 'if a lion could talk, we could not understand him'. The statement is obviously false—because we share with lions a similar form of embeddedness and embodiment, and hence experiences like hunger, fear, pain, or pleasure—if one fails to realize that the lion is only a place-holder to indicate an agent utterly and radically different from us, like our α and β. The lion is a Martian, someone you simply cannot talk to because he is 'from another ontology'.[2]

From this perspective, the famous phrase *hic sunt leones* (here there are lions) acquires a new meaning. It occurred on Roman maps to indicate unknown and unexplored regions beyond the southern, African borders of the empire.[3] In a Wittgensteinian sense, the Romans were mapping the threshold beyond which no further communication was possible at all. They were drawing the limits of their ontology. What was beyond the border, the *locus* inhabited by the lions, was nothing, a non-place. Globalization has often meant that what is not inglobate simply isn't, i.e. fails to exist.

We can now formulate the difficulty confronting a Global-Information Ethics as *the problem of the lion*: cross-cultural communication, which is the necessary condition for any further moral interaction, is possible only if the interlocutors partake in a common ontology. When Crusoe and Friday meet, after twenty-five years of Crusoe's solitude, they can begin to communicate with each other only because they share the most basic ontology of life and death, food and shelter, fear and safety. Agents may be strangers to

---

[2] If it took endless time and effort to decipher the hieroglyphics, imagine what sense an extraterrestrial being could make of a message in a bottle like the plaque carried by the Pioneer spacecraft: <http://www.en.wikipedia.org/wiki/Pioneer_plaque>.

[3] Unfortunately, we do not have African maps drawn from the 'lions' perspective'. The Da Ming Hun Yi Tu, or Amalgamated Map of the Great Ming Empire, the oldest map of Africa known so far, dates back 'only' to 1389.

each other ('stranger' being an indexical qualification). They do not have to speak the same language, empathize or sympathize. But they do need to share at least some basic appropriation, semanticization and conceptualization of their common environment, as a minimal condition for the possibility of any further, moral interaction. So can information ethics provide a solution to the problem of the lion? The short answer is yes; the long one is more complicated and requires more space.

15.5 Global information-ethics and its advantages

Information ethics, as described in the previous chapters, has many advantages to offer when it comes to the new challenges posed by globalization. Let me outline four of them relevant to solving the 'problem of the lion'.

1) Embracing the new informational ontology Not only do we live in a world that is moving towards a common informational ontology, we also experience our environment and talk and make sense of our experiences in increasingly informational ways. *Information is the medium.* This calls for an ethics, like IE, that, by prioritizing an informational ontology, may provide a valuable approach to decoding current moral phenomena and orienting our choices.

2) Sharing a minimal, horizontal, lite ontology There is a risk, by adopting an ontocentric perspective, as IE suggests, that one may be merely exchanging one form of 'centrism' (American, Athenian, Bio, British, Enlightenment, European, Greek, Male, Roman, Western, you-name-it) with just another, perhaps inadvertently, thus failing to acknowledge the ultimate complexity, diversity, and fragility of the multicultural, ethical landscape with which one is interacting. We have seen how the problem of the lion may become a dilemma. This justified concern, however, does not apply here because IE advocates a *minimal* informational ontology, which is not only timely, as we have just seen, but also tolerant of other local ontologies, and interfaceable with them. Thick cultures with *robust, vertical ontologies*—i.e. deep-seated, often irreconcilable, fundamental conceptions about human nature, the value and meaning of life, the nature of the universe and our place in it, society and its fair organization, religious beliefs, and so forth, in short a *human project*—can more easily interact with each other if they can share a *lite* and *horizontal ontology* as little committed to any particular *human project* as possible. The identification of an absolute, ultimate, monistic ontology, capable of making all other ontologies merge, is just a myth, and a violent one at that. There is no such thing as a commitment-free position with respect to the way in which a variety of continuously changing agents appropriate, conceptualize, and semanticize their environment. Yet the alternative cannot be some form of relativism. This is no longer sustainable in a globalized world in which choices, actions, and events are delocalized. There simply is not enough room for 'minding one's own business' in a network in which the behaviour of each node may affect the behaviour of all nodes. The approach to be pursued seems rather to be along the lines of respect for, and tolerance towards, diversity and pluralism and identification of a minimal common ontology, which does not try to be platform-independent (i.e. absolute), but cross-platform (i.e. portable).

As in Queneau's *Exercises in Style*, we need to be able to appreciate both the ninety-nine variations of the same story[4] and the fact that it is after all the same story that is being recounted again and again. This plurality of narratives need not turn into a Babel of fragmented voices. It may well be a source of pluralism that enriches one's ontology. More eyes simply see better and appreciate more angles, and a thousand languages can express semantic nuances that no global Esperanto may ever hope to grasp.

3) Informational environmentalism The ontocentrism supported by IE means that at least some of the weight of the ethical interpretations may be carried by (outsourced to) the informational ontology shared by the agents, not only by the different cultural or intellectual traditions (vertical ontologies) to which they may belong. Two further advantages are that all agents, whether human, artificial, social, or hybrid, may be able to share the same minimal ontology and conceptual vocabulary; and then that any agent may take into account ecological concerns that are not limited to the biosphere.

4) Identifying the sources and targets of moral interactions One of the serious obstacles in sharing an ontology is often how the sources and targets of moral interactions (including communication) are identified. The concept of person or human individual, and the corresponding features that are considered essential to his or her definition, might be central in some ontologies, marginal in others, and different in most. IE may help foster communication and fruitful interactions among different, thick, vertical ontologies by approaching the problem with conceptual tools that are less pre-committed. For when IE speaks of agents and patients, these are neutral elements in the ethical analysis that different cultures or macroethics may be able to appropriate, enrich, and make more complex, depending on their conceptual requirements and orientations. It is like having an ontology of agency that is open source, and that anyone (indeed any inforg) can adapt to its own proprietary *human project*.

15.6 The cost of a global-information ethics: postulating the ontic trust

It would be silly to conclude at this point that a Global IE may provide an answer to any challenge posed by the various phenomena of globalization. This would be impossible. Of course, there will be many issues and difficulties that will require

---

[4] See Queneau (2008). On a crowded bus, a narrator observes a young man with a long neck in a strange hat yell at another man whom he claims is deliberately jostling him whenever anyone gets on or off the bus. The young man then sits down in a vacant seat. Two hours later the same narrator sees that same young man with another friend, who is suggesting that the young man have another button put on his overcoat.

substantial extensions and adaptations of IE, of its methodology, and of its principles. All I have tried to do is to convince the reader that such a great effort to apply IE as a global ethics would be fruitful and hence worth making.

It would be equally wrong to assume that the adoption of IE as a fruitful approach to global challenges comes at no conceptual cost. Every ethical approach requires some concession on the part of those who decide to share it and IE is no exception.

The cost imposed by IE is summarizable in terms of the postulation of what I shall define as the *ontic trust* binding agents and patients. A straightforward way of clarifying the concept of ontic trust is by drawing an analogy with the concept of 'social contract'.

Various forms of contractualism (in ethics) and contractarianism (in political philosophy) argue that moral obligation, the duty of political obedience, or the justice of social institutions have their roots in, and gain their support from, a so-called 'social contract'. This may be a real, implicit or *merely hypothetical* agreement between the parties constituting a society (e.g. the people and the sovereign, the members of a community, or the individual and the state). The parties accept to agree to the terms of the contract and thus obtain some rights in exchange for some freedoms that, allegedly, they would enjoy in a hypothetical state of nature. The rights and responsibilities of the parties subscribing to the agreement are the terms of the social contract, whereas the society, state, group, etc. is the artificial agent created for the purpose of enforcing the agreement. Both rights and freedoms are not fixed and may vary, depending on the interpretation of the social contract.

Interpretations of the theory of the social contract tend to be highly (and often unknowingly) anthropocentric (the focus is only on human, rational, individual, informed agents) and stress the coercive nature of the agreement. These two aspects are not characteristic of the concept of ontic trust, but the basic idea of a fundamental agreement between parties as a foundation of moral interactions is sensible. In the case of the ontic trust, it is transformed into a primeval, entirely hypothetical *pact*, logically predating the social contract, that all human (I shall drop henceforth this specification, unless this generates confusion) agents cannot but sign when they come into existence, and that is constantly renewed in successive generations.

There are important and profound ways of understanding this *Ur-pact* religiously, especially but not only in the Judeo-Christian tradition, where the parties involved are God and Israel or humanity, and their old or new *covenant* ($\delta\iota\alpha\theta\acute{\eta}\kappa\eta$) makes it easier to include and accommodate environmental concerns and values otherwise overlooked from the strongly anthropocentric perspective *prima facie* endorsed by contemporary contractualism. However, it is not my intention to endorse or even draw on such sources. I am mentioning the point here in order to shed some light both on the origins of contractualism and on a possible way of understanding the ontocentric approach advocated by IE. The sort of pact in question can be understood more precisely in terms of an actual trust.

Generally speaking, a trust in the English legal system is an entity in which someone (the trustee) holds and manages the former assets of a person (the trustor, or donor) for the benefit of some specific persons or entities (the beneficiaries). Strictly speaking, nobody owns the assets, since the trustor has donated them, the trustee has only legal ownership, and the beneficiary has only equitable ownership. Now, the logical form of this sort of agreement can be used to model the ontic trust, in the following way:

- the assets or 'corpus' is represented by the world, including all existing agents and patients (the infosphere);
- the donors are all past and current *generations* of agents;
- the trustees are all current *individual* agents;
- the beneficiaries are all current and future *individual* agents and patients.

By coming into being, an agent is made possible thanks to the existence of other entities. It *is* therefore bound to all that already is, both *unwillingly* and *inescapably*. It *should be so* also *caringly*. Unwillingly, because no agent wills itself into existence, though every agent can, in theory, will itself out of it. *Inescapably*, because the ontic bond may be broken by an agent only at the cost of ceasing to exist as an agent. Moral life does not begin with an act of freedom but it may end with one. *Caringly* because participation in reality by any entity, including an agent—that is, the fact that any entity is an expression of what exists—provides a right to existence and an invitation to respect and take care of other entities. The pact then involves no coercion, but a mutual relation of appreciation, gratitude and care, which is fostered by the recognition of the dependence of all entities on each other. A simple example may help to clarify further the meaning of the ontic trust.

Existence begins with a gift, even if possibly an unwanted one. A foetus will be initially only a beneficiary of the world. Once she is born and has become a full moral agent, Alice will be, as an individual, both a beneficiary and a trustee of the world. She will be in charge of taking care of the world, and, insofar as she is a member of the generation of living agents, she will also be a donor of the world. Once dead, she will leave the world to other agents after her and thus become a member of the generation of donors. In short, the life of an agent becomes a journey from being only a beneficiary to being only a donor, passing through the stage of being a responsible trustee of the world. We begin our career of moral agents as strangers to the world; we should end it as friends of the world.

The obligations and responsibilities imposed by the ontic trust will vary depending on circumstances but, fundamentally, the expectation is that actions will be taken or avoided in view of the well-being of the whole infosphere understood as a conceptualization of Being.

The ontic trust is what is postulated by the approach supported by IE. We saw that in IE, the ethical discourse concerns any entity, understood informationally, that is, not only persons, their cultivation, well-being, and social interactions, not only any form of life and its habitat, but also any expression of Being. The ontic trust (and the corresponding ontological equality principle among entities, see Section 4.5) means that any form of reality (any instance of information/*Being*), simply by the fact of *being* what it is, enjoys a minimal, initial, overridable, equal right to exist and develop in a way that is appropriate to its nature. The acceptance of the ontic trust requires a disinterested judgement of the moral situation from a patient-oriented and non-anthropocentric perspective (see Section 4.5). The ontic trust is respected whenever actions are impartial, universal and 'caring' towards the world. In the history of philosophy, this position can already been found advocated by Plato, Stoic and Neoplatonic philosophers, as well as Spinoza. I am told it is a view that resonates with the spiritual and religious positions of some non-Western cultures. I shall return to this point in Section 16.19.

CONCLUSION

An interesting way of looking at the history of cultures is in terms of increasing distance of human life from the natural course of events, thanks to an ever-thickening layer of technological mediations. A culture (not necessarily a *good* culture, let alone a civilization) emerges when a society is able to detach itself from the physical world (*physis*), and generate sufficient resources to express itself with some stability. From the division of labour to sheer oppression, from the invention of tools to the creation of weapons, there must be at least a fissure between surviving and living, where the seeds of a culture can take roots non-ephemerally. A culture therefore can be pre-historical (no recordings, see Chapter 1) but hardly pre-technological. 'Hardly' because, exceptionally, such breaking away from *physis* may be achievable by barehanded individuals in unaided contexts. In theory, nothing prevents extraordinary people from planting some cultural seeds even when life is flattened into survival two-dimensionally, here and now. In practice, however, cultures tend to emerge and flourish only behind the dam provided by some *techne*. Even embittered stylites need pillars on which to stand, and peasants to bring food.

Once cultures are sufficiently advanced and reflect critically on their technological conditions of possibility, they seem to encounter two traps.

One is the trap of *nostalgia* for some primordial authenticity. It leads a culture to believe that the future could improve if only there were no dam. Pristine nature, virgin phenomena, the humanly untouched are seen, from the safe distance afforded by *techne* itself and the accumulation of its benefits, as the regulative ideal one should at least aspire to implement. *Physis* appears as the Promised Land, lost because of *techne*, and hence progressively reclaimable only through the increasing removal of the latter. With a caricature: if only technology could be eradicated, our problems would be solved. Aristotle has a wonderful analogy that can be borrowed to chastise such an illusion: it is as foolish as the dove's, which believed it could fly faster if there were no air.

The other trap is the *hubris* of ultimate power. In this case, a better future allegedly lies in ever-smarter and more efficient uses of *physis*. The space on the other side of the dam is now seen as a reservoir of resources, which *techne* can appropriate and process more and more effectively. The other caricature in the diptych suggests that our problems would be solved if only we had more and better technologies. In this case, Hegel provides the insightful analogy which can be recycled to expose the mistake: the mastery culture foolishly fails to grasp that its increased technological reliance on the resources of the enslaved *physis* inevitably leads to the empowerment of the latter and its ultimate rebellion. Predictably, *physis* will take revenge, if the culture that exploits it is too careless.

In both cases, nostalgia or hubris, a lack of reflective discernment causes a seriously dangerous misunderstanding about the ecological relationship between *physis* and *techne*. Only together can they create the environment in which humanity may flourish. Ultimately, the equation is simple: culture requires resources, and resources are acquired and managed through technology. No *techne*, no resources, no culture. The question is not which side of the relation to drop or disregard, *physis* or *techne*, but how to negotiate the fine balance that can harmonize both.

Unfortunately, locating the traps and describing their nature is much easier than identifying the right course of action for the development of a successful marriage between *physis* and *techne*. Fortunately, it is exactly here that philosophy can help by exercising its mediating role, in the following sense.

The three-player game between *physis*, *techne*, and philosophy can be complex, in the same way as chess is complex: not because of the basic rules, but because of the sophisticated ways in which they can be applied in a variety of circumstances, full of implicit constraints, consequences, and potential pitfalls. Indeed, the rules are trivial: avoid the two traps, nurture the right sort of *physis* (after all, we do wish to eradicate malaria once and for all, if possible), develop the right sort of *techne* (e.g. renewable, non-polluting energy), and foster the right sort of culture. And all this thanks to the right sort of philosophy, which can enable us to make the right sorts of move that guarantee a harmonious and healthy decoupling of human history from natural events. The superficial simplicity of all these 'the right sort's is misleading, as many examples of mistaken decisions and misdirected policies too often testify. The game is incredibly hard to master, not least politically and economically. And as if all this were not already bad enough, it is a game against the clock, in which time is running out, and one on which the future well-being of humanity is increasingly dependent. Nobody sufficiently informed can fail to feel a deep sense of urgency. This is why more philosophy is required, not less. We need to know the state of the game better, we need more discernment about the next promising moves, and we need more originality in envisioning the feasible strategies that may be successful. Analyses, syntheses, insights, and intellectual creativity: this is how philosophy as conceptual design can help us to semanticize and build a world in which, to use the previous metaphor, the marriage of *physis* and *techne* may be successful and bear fruit. We need to be stubbornly intellectual.

Whether *physis* (nature, the world) and *techne* (applied knowledge, technology) may be reconcilable is not a question that has a predetermined answer, waiting to be divined. It is more like a practical problem, whose feasible solution needs to be devised. With an analogy, we are not asking whether two chemicals could mix but rather whether a marriage may be successful. There is plenty of room for a positive answer, provided the right sort of commitment is made.

It seems beyond doubt that a successful marriage between *physis* and *techne* is vital and hence worth our effort. Information societies increasingly depend upon technology to thrive, but they equally need a healthy, natural environment to flourish. Try to imagine the world not tomorrow or next year, but next century, or next millennium: a divorce between *physis* and *techne* would be utterly disastrous both for our welfare and for the well-being of our habitat. This is something that technophiles and green fundamentalists must come to understand. Failing to negotiate a global, fruitful, symbiotic relationship between technology and nature is not an option. Fortunately, a successful marriage between *physis* and *techne* is achievable. In this and in the previous chapters, I have sought to articulate and defend an *e-nvironmental* or *synthetic ethics* according to which the fight against metaphysical entropy, be this the destruction, impoverishment, vandalism, or waste of both natural and human (including historical and cultural) resources, is our main priority. Now, some macroethics, especially in the Christian tradition, seem to assume that the moral game, played by agents in their environments, may be won absolutely, i.e. not in terms of higher scores, but by scoring perhaps very little as long as no moral loss or error occurs, a bit like winning a football match by scoring only one goal as long as none is conceded. A moral analogue of the Italian *catenaccio*. It seems that this absolute view has led different parties to underestimate the importance of successful compromises (imagine an environmentalist unable to accept any technology responsible for some level of carbon-dioxide emission, no matter how it may be counterbalanced otherwise). The more realistic and challenging view from IE is that moral evil is unavoidable, so that the real effort lies in counterbalancing through more moral goodness. This invites us play the moral analogue of Brazilian *futebol*.

We should resist any Greek epistemological tendency to treat *techne* as the Cinderella of knowledge; any absolutist inclination to accept no moral balancing between some unavoidable evil and far more goodness; or any modern, reactionary, metaphysical temptation to drive a wedge between naturalism and constructionism by privileging the former as the only authentic dimension of human life. We are shaping the infosphere from within, as the only ethical inforgs fully responsible for its future. The challenge is to reconcile our roles as agents within nature and as stewards of nature. The good news is that it is a challenge we can meet. The odd thing is that it has required a fourth revolution (see Chapter 1) to realize that we have such a hybrid nature.

The constructive part of this book is now complete. However, there is still one final step to be taken. It is a defence of information ethics from some misunderstandings and objections that have appeared since the late nineties,[5] when I first introduced IE as a foundationalist approach to computer ethics and an e-nvironmental macroethics. This is the task of the next and last chapter, where I shall address some main criticisms.

---

[5] Fourth International Conference on Ethical Issues of Information Technology (Department of Philosophy, Erasmus University, The Netherlands, 25–27 March 1998). This was published as Floridi (1999a).

16

In defence of information ethics

> Only insofar as human beings live under the guidance of reason, do they always necessarily agree in nature.
>
> Spinoza, *Ethica Ordine Geometrico Demonstrata* (2000, Pt. IV, Proposition 35, translation modified).

SUMMARY

Previously, in Chapters 1 to 15, I outlined the nature and scope of information ethics. My goal in this chapter is not to convince the reader that no reasonable disagreement is possible about the value of IE as a specific approach to computer ethics or, more generally, as a macroethics. On the contrary, several of the theses defended in this book might be interesting precisely because they are also open to discussion. Rather, my goal is to remove some ambiguities, possible confusions, and mistaken objections that might prevent the correct evaluation of IE in its various interpretations, so that disagreement can become more constructive. There is not much point in mapping here the structure of the chapter in detail since, after a brief introduction in Section 16.1, I shall deal with nineteen objections in as many sections. Every section opens with a formulation of the objection, followed by a reply. The objections are not listed in order of importance, but rather follow as far as possible a logical order of narrative. It is clear that much more and better work needs to be done in IE before one may justifiably claim to have a full theory. My hope is that I have been able to show that such work is worth our intellectual efforts.

16.1 Introduction: addressing the sceptic

During the past two decades or so, information ethics has become a lively area of philosophical research, attracting an increasing amount of interesting work. Twenty years of sustained, international research of high standards is a long time in any academic field. In information ethics, this is even more so, given the fast-paced and radical transformations involving ICTs and their ethical implications. As a result, IE has certainly widened its philosophical scope. It now interacts with many other ethical fields, from business ethics to environmental ethics, from medical ethics to the ethics of nanotechnologies, from the ethics of cyberwar to the ethics of e-research. IE has also deepened its conceptual insights. These now involve dialogues with other philosophical and ethical traditions, such as Platonism, Neo-Platonism, Stoicism, Spinozism, deontologism, consequentialism, contractualism, virtue ethics, other intellectual cultures such as Buddhism, as well as analyses and discussions of metaphysical, epistemological and logical topics, from digital physics to the method of levels of abstraction, from structural realism to the philosophy of information. The list could easily be expanded. In this rich and varied context, informed and reasonable debates and disagreements are not only to be expected but are also welcome. For they are clear evidence of a healthy market of ideas, open to different and sometimes contrasting views, and they can foster our understanding and help to guide sound judgements. The interested reader will find plenty of such discussions in this chapter.

In addressing potential misunderstandings about, and objections against, IE as I have formulated it in this book, I shall not address actual authors but only conceptual issues. So I shall adopt the rhetorical device of talking about a hypothetical sceptic, to whom I shall attribute misunderstandings and objections alike. None of the objections discussed in this chapter are of my own invention, and they have all appeared in print, with only one exception, number eight, which was moved by one of OUP's anonymous referees. In some cases, the text is actually quite close to the original, with some minor rewording to fit the context. The reader interested in tracking some of the actual sources of the contents of this chapter may wish to consult: Mathiesen (2004), Siponen (2004), Himma (2004a), and the following collections, Boltuc (2008), Ess (2008), Allo (2010, 2011), and Demir (2010, 2012).

16.2 IE is an ethics of news

Objection

IE defends the intrinsic moral value of semantic objects, like news, texts, or computer files, and this is untenable.

Reply

IE is sometimes understood as if it defended the view that the ethical discourse should also take into account artificial, digital, or informational realities, besides human and biological agents and patients. This is a reasonable but slightly mistaken view. The thesis actually advocated by IE is significantly different, and perhaps even more radical (so, to a sceptic, it will look even less credible). It consists in arguing for a change in the level of abstraction at which the ethical discourse may *also* be fruitfully developed. IE supports a development from biocentrism to ontocentrism, where the latter is expressed in terms of an informational metaphysics. To put it simply, according to IE, the effort to be made consists not merely in adding new agents and patients to the list of already ethically qualifiable entities, but in changing perspective altogether, and interpreting all agents and patients informationally, thus including humans, animals, social agents, and engineered entities as well. It is an extension achieved not in terms of addition, but in terms of modification of the interface through which we analyse moral interactions. It follows that the three fundamental issues of *applicability*, *inclusivity*, and *extensibility* of IE[1] should really be answered (see Section 16.11) after the following question: does it help to adopt an information-based (metaphysical) approach to ethics? As I anticipated, IE defends a firm answer in the positive, and the latter points in the direction in which further research should be developed. What does follow from such shift from a biocentric to an ontocentric perspective and from an informational interpretation of the latter is that, by defending the intrinsic moral worth of *informational entities*, IE does not refer to the moral value of well-formed and meaningful data such as an email, the *Britannica*, or Newton's *Principia*, or some science-fiction robot such as *Star Wars*' C3PO and R2D2. What IE suggests is that we adopt an informational approach (technically, a level of abstraction) to the analysis of *Being* in terms of a minimal common ontology, whereby human beings as well as animals, plants, artefacts (and hence emails, the *Britannica* or Newton's *Principia*), and so forth are all interpreted, *insofar as they are entities*, as informational entities. IE is not an ethics of the BBC news or some artificial agent *à la* Asimov. Of course, it remains open to debate whether the informational level of abstraction adopted is correct. For example, the choice and hence its implications have been criticized by Johnson (2006); whereas Capurro (2006) has argued against the ontological stance adopted by IE. Yet, the clarification should help in eliminating a potential source of misunderstanding.

If IE is correct in treating humans as informational entities, isn't this overly reductivist? Surely we are not just clusters of information or biological databases. This is the next objection.

16.3 IE is too reductivist

Objection

IE reduces people to mere numbers.

Reply

I still recall one conference in the nineties when a famous computer ethicist compared me to a sort of Nazi, who wished to reduce humans to numbers, pointing out that the Nazis used to tattoo six-digit identity tags onto the left arms of the prisoners in their Lager. This is rhetorical nonsense. Likewise, suggesting that 'surely we are not just...' is an old line that was already flawed when used against Darwin. Of course, we are not just animals, but treating human beings as animals or, if you find the expression

---

[1] Applicability: whether IE can help in how we deal with everyday moral issues. Inclusivity: whether IE's treatment of patients and agents as informational entities can improve our analysis and understanding of the full spectrum of applied ethical problems. Extensibility: whether IE's ethical principles may be applied to material (as opposed to digital) contexts. See Vaccaro (2008).

infelicitous, adopting the LoA at which one analyses humans as bipedal primates belonging to the mammalian species *Homo sapiens*, does not detract one iota from our nature. On the contrary, such *naturalization* makes it more likely that we might understand and appreciate a side of ourselves otherwise easily neglected. Now, in the same way as we are biological organisms, which share with their natural environment a long history of co-evolution, likewise, we can look at ourselves (change the LoA) as informational organisms, or *inforgs*. And this *informatization* also helps us to appreciate our nature and our relation to reality, now understood as the infosphere. The minimalism advocated by IE is methodological. It means to support the view that entities can be analysed by focusing on their lowest common denominator, represented by an informational ontology. Other perspectives can then be evoked in order to deal with other, more human-centred moral values and responsibilities. In the case of human entities, the LoA adopted returns a special kind of informational organism. Once again, such informational analysis of human agents should not be confused with a Shannon-like sense of information, jumping out of the frying pan into the fire. When Wiener famously described human beings as 'patterns that perpetuate themselves' (Wiener, 1954, p. 96), the patterns in question may be analogue and continuous 'persistent information patterns' that have little to do with Shannon's and other similar mathematical approaches to quantitative data, their probability distributions, and so forth. This is good news, because Shannon's information entails a view of the ultimate nature of reality as necessarily discrete and possibly deterministic (this is also known as digital physics). This neo-Pythagorean ontology is hardly tenable nowadays (Floridi, 2009a). Instead, and more constructively, it is possible to show that an ontological conception of information as relational patterns is much more satisfactory, and provides IE with a minimalist, structural ontology that is more successful philosophically and more easily reconcilable with our current scientific knowledge (Floridi, 2008g).

If individuals are their information, we still do not know exactly which information may or may not count as constituting a person, and IE fails to solve this difficulty. This is the next objection.

16.4 IE fails to indicate what information constitutes an individual

Objection

IE leaves open this central and exceptionally difficult question as to what information is or ought to be considered constitutive of the individual.

Reply

We have already met this kind of objection in Section 6.4.1. This new objection is largely correct, but the good news is that the question it highlights might be left at least partly open (and hence up to us to decide, case by case) for a good reason.

What the sceptic is demanding is a full ontology that will tell us, with certainty and precision, in a variety of disparate and possibly complex cases, when some specific data are (or fail to be) part of what constitute an individual (and mind that the individual in question need not be a *single* individual, it could easily be a married couple, a company, a team, a social group, and so forth). Now, in formal or engineered cases (e.g. in set theory or in the car industry), this is achievable, as one can be fairly sure about what does and what does not constitute the class of rational numbers, or the specific VW Polo parked in the garage, for example. But even in everyday life, when we think we should know better because we are dealing with concrete and very well-known entities, the fuzziness and slippery nature of the boundary between what counts in or out bubbles up everywhere, and it reminds us of our epistemic limits, if not of the ontic vagueness of reality. On 6 September 2001, for example, the European Parliament adopted an initiative called '25 years' Application of Community Legislation for Hill and Mountain Farming' in which it urged

> the Commission to lay down an exact definition [of hill and mountain farming] based on the criteria of height (in metres), slope, shortened growing seasons, and appropriate combinations of those criteria....[2]

More than a decade later, that definition is still to be found. But then, if it is so hard to agree on an exact and uncontroversial understanding of what counts, for legislative and economic purposes, as hill and mountain farming, how much harder can it be to define all and only the information that constitutes a person? The request for necessary and sufficient conditions is often natural but, equally often, needs to be resisted, if one wishes to be reasonable and accurate rather than precise and inflexible. We saw in the previous section that Wiener famously described human beings as 'patterns that perpetuate themselves'. I have argued for a very similar view in IE: we are homeostatic information patterns, bent on resisting all forms of entropy, thermodynamic and metaphysical. Yet, patterns tend to lack sharp or clear-cut edges and, being in constant dynamic evolution, they can easily be polymorphic. The waves on the beach are quite clearly individual waves, but any attempt to fix the precise drops of water that constitute each wave would be a pointless exercise. So, does all this mean that, although the sceptic is right, his demand is bound to remain unsatisfied? This would be overly pessimistic, for two reasons.

On the one hand, we might not have a definition, but we can rely on our intelligent understanding and experience, as in the case of the waves. Just ask a surfer whether there are waves, and what sorts of wave there are. Certainly, borderline, complex, or extreme cases may test our capacity of discernment, but then, IE should be praised for opening our eyes to these new perspectives and for providing the right approach to

---

[2] European Parliament, European Parliament Resolution on 25 years' Application of Community Legislation for Hill and Mountain Farming (2000/2222(INI)), 6 September 2001 <http://www.eur-lex.europa.eu/LexUriServ/LexUriServ.do?uri=OJ:C:2002:072E:0354:0359:EN:PDF>.

tackle such difficulties (essentially: humans are, or at least might also be considered and treated as, informational objects), rather than blamed for failing to deliver a secure route to a final answer, a request which to some may appear supererogatory. For example, an influential verdict by Germany's highest court in February 2008 on whether the government might have the right to check remotely a citizen's computer has, according to many commentators, established a new 'fundamental right' for the 21st century, according to which a person's 'private sphere' includes her computer, even when that person is online.[3] This is going in the direction indicated by IE.

On the other hand, it might be possible to enrich IE with some guidelines, which could bring some coherence to the law relevant to data representations. This is not the place to provide them, but I suppose an example might be useful to illustrate the sort of research that one may wish to see developed in the near future. In brief, the task is to identify some criteria (recall the example of the definition of hill and mountain farming based on height, slope, and shortened growing seasons) that could help us in determining when some information is (or fails to be) constitutive of an individual. Here is an initial proposal.

Let me first remind the reader about the concept of inverse function. To do so, a simple example will suffice. If the function is $f(x) = x^2$, and x ranges over the domain of non-negative natural numbers (that is, $x \geq 0$), then the inverse function is $f^{-1}(y) = \sqrt{y}$. For example, if we have $f(3) = 3^2 = 9$, then the inverse function is $\sqrt{9} = 3$. More precisely, if f is a function whose domain is the set X and whose range is the set Y, then the inverse of f is the function f^{-1} with domain Y and range X, defined by the following rule: if $f(x) = y$ then $f^{-1}(y) = x$. The obvious but powerful property that an inverse function enjoys is that of uniquely identifying the input x of another function based only on its output y, for all $y \in Y$. In plain English, a function leads you from x to y and an inverse function leads you back, from y to x. Not all functions have an inverse function. As we all know, things in life may or may not be reversible: you cannot un-break the eggs you mistakenly broke (no reverse function available here), but you can empty the bottle you mistakenly filled. The precise concept of reversibility, as a reverse relation that leads us back to where we were, is what we need. If some information is personal, it is linked to the person to whom it belongs in a way similar to that in which y is related to x through a function f. But then, at least in theory, one might be able to move not only from the person's properties (say, Alice's) to the information (say, some credit card information), but also back, and uniquely identify the properties from the information. This backward route is often blocked on purpose. Alice might be a conservative who voted for a particular presidential candidate, but the electoral system

---

[3] See e.g. the report and comments on the Spiegel International Online website: 'The World from Berlin: Germany's New Right to Online Privacy', *Spiegel Online*, 28 February 2008 <http://www.spiegel.de/international/germany/0,1518,538378,00.html>. Ralf Bendrath's blog entry is enlightening: 'Germany: New Basic Right to Privacy of Computer Systems', 28 February 2008 <http://www.bendrath.blogspot.com/2008/02/germany-new-basic-right-to-privacy-of.html>.

makes sure it is not possible to work out who the person behind that conservative vote is. Abstract or obliterate sufficient details (some information), and the output becomes irreversible. This is privacy through irreversibility. It follows that a possible way of answering the sceptic's request might be to develop criteria of personal information identification on the basis of their reversibility. Fingerprints certainly seem to qualify rather well, as do DNA and retina patterns. What about some statistical data, such as the average number of times I check my email daily? They might as well; it all depends on whether they can be reverse-engineered, as output, in order to obtain at least some information about the original input. As I mentioned at the beginning of this section, the use of *phronesis* is not optional.

If IE can find a way of dealing with the identification of the information that is constitutive of an individual, doesn't the previous reply show that ethical evaluation is a matter of human convention? This is the next objection.

16.5 IE's de-anthropocentrization of the ethical discourse is mistaken

Objection

IE fails to see that whether processes count as moral action is a matter of human convention. In terms of computer ethics, it fails to recognize that the meanings that humans give to particular operations of a computer system are contingent.

Reply

This is a dangerous form of relativism. I do not believe that it is a matter of human convention whether any process (such as raping a child) counts as a moral action. Ethical discourse normally starts from *the fact* that it is. I like that. On 12 October 2007, during a shooting exercise at the South African Army's Combat Training Centre, at Lohatlha, in the Northern Cape, a computerized Oerlikon 35mm MK5 anti-aircraft gun went out of control and killed nine soldiers, injuring another eleven. Initially, the National Defence Force suspected it might have been a software glitch, but in January 2008 the verdict was in favour of a mechanical problem combined with human negligence. I doubt this might be a question of 'interpretive flexibility'. The sceptic might like to argue that similar issues are not a matter of truth, that there are no right answers to the question whether artificial agents are (or ever could be considered) moral agents: there is no truth to be uncovered, no test that involves identifying whether a system meets or does not meet a set of criteria. Yet this is postmodernism at its worst, a relativistic game that is fine to play in the ivory tower of academia, but should not be exported to real life. Because it is obviously untenable, it ends up fostering bad faith since, when convenient, references to the real and true nature of the artefacts in question (as mere machines) appear to play a decisive role. It is also contradictory, for why worry so much about engineers building artificial agents that could be moral agents if it is really just a matter of interpretation? But ultimately, the real danger is that it seeks to block what Peirce defined as the road of enquiry, by means of a stumbling block of rhetoric and socio-political agendas that promote the sort of head-in-the-sand strategy (if I refuse to see it for long enough it will disappear) that has never worked in the past, and has always increased the trouble left to be resolved. Again, some realism might help, so let me use another old example. In August 2007, the US army deployed to Iraq armed robots known as SWORDS (Special Weapons Observation Remote reconnaissance Direct action System). These agents were armed with M249 machine guns. I would argue that we should consider very carefully whether they are, or can ever become, sources of actions that are morally loaded (i.e. good or evil) and *accountable* (mind, not *responsible*) for foreseeable disasters. We should look at their design, set clear criteria that they should satisfy, investigate the *truth* about their actual specifications and safety measures, and so forth, and perhaps conclude that they should have never been built in the first place, or deployed in that context, or controlled differently, or that they ought to be dismantled as soon as possible. This and similar policies would help us build a better and safer environment. What I strongly doubt is that engaging in some 'interpretive flexibility' exercise might be useful at all. Trying to show that technology is an important component of morality but also that technology is under human control may be good for military propaganda but rather hard to believe in real life. It fosters a false sense of security, especially when you realize that, for example, US military robots ran 30 000 missions in 2006 in Iraq and that the so-called 'surge' involved the deployment of 3000 new robots.[4]

If the ontocentric approach is not mistaken, it still makes IE too theoretical to be of any real use. This is the next objection.

16.6 IE is inapplicable

Objection

IE is too theoretical, metaphysical or philosophical (in the worst sense of these words) to be applicable when we are confronted by very concrete challenges. It does not offer practical guidance.

Reply

As I remarked in the conclusion of Chapter 4, a polarization of theory and practice is sometimes the inevitable price to be paid for any foundationalist project, as it strengthens both. IE is not immediately useful to solve specific ethical problems (including computer ethics problems), but nor does a textbook on Newtonian physics solve your car problems. IE is supposed to provide the conceptual grounds that then

---

[4] Noah Schachtman, 'Military Recruits Thousands More Warbots for New Unmanned Surge', *Wired*, 29 May 2008 <http://www.wired.com/dangerroom/2008/05/in-december-aft/>.

guide problem-solving procedures. Imagine our sceptic complaining that the declaration of human rights is too theoretical, metaphysical, or philosophical to solve the concrete ethical problems she is facing in a specific situation, say in dealing with a particular case of cyber-stalking in the company that employs her. This would be rather short-sighted. The suspicion is that some impatience with conceptual research may betray a lack of understanding of how profound the revolution we are undergoing is, and hence how radical the rethinking of our ethical approaches and principles may need to be in order to cope with it. IE is certainly not the declaration of human rights, but it seeks to obtain a level of generality purporting to provide a foundation for more applied and case-oriented analyses.

As for IE offering no practical guidance once the moral value of all aspects of Being (axiological ecumenism) is accepted, this would be equivalent to saying that, since environmental ethics is based on the value of life and of the absence of suffering, then it offers little help with real-world issues. The truth is exactly the opposite. Having some universal, basic, and robust principles in place helps enormously when it comes to dealing with particular, complex, practical matters. We should not be afraid of respecting any form of reality too much, even if this might be a rather difficult task.

The real question is not whether IE is too theoretical, metaphysical, or philosophical—good foundations for the structure one may wish to see being built inevitably lie well below the surface—but whether it will succeed in providing the robust framework within which practical issues of moral concern may then be more easily identified, clarified, and solved. It is in its actual applications that IE, as an ethics for our global information society and the ever-expanding infosphere, will or will not qualify as a useful approach. Yet building on the foundation provided by IE is a serious challenge, it cannot be an objection. It is encouraging that IE has already been fruitfully applied to deal with a variety of real issues, including the 'tragedy of the digital commons' (Greco and Floridi, 2004), the digital divide (Floridi, 2002), the ethics of computer games (Sicart, 2009) and of computer cheating (Sicart, 2005), environmental issues (York, 2005), software protocols design (Turilli, 2007), information transparency (Turilli and Floridi, 2009), Internet neutrality (Turilli et al., 2012), and trust online (Taddeo, 2009, 2010).

However, even if IE turns out to be applicable, it may still seem to be too demanding. This is the next objection.

16.7 IE is supererogatory

Objection

IE is supererogatory in its demand that we should respect the moral value of any expression of Being/infosphere.

Reply

I have already argued in Sections 6.4.2 and 10.5 that this objection is only superficially convincing. It is actually a consequentialist approach that runs the risk of being supererogatory (see Section 4.7.4). However, it is worth dealing with the objection one last time because it offers the opportunity for two further clarifications.

First, it is important to stress that IE supports a *minimal, overridable,* and *ceteris paribus* sense of ontic moral value. The reader should not be put off by such qualifications, for the idea is really quite intuitive. Environmental ethics, for example, accepts culling as a moral practice and does not indicate as one's duty the provision of a vegetarian diet to wild carnivores. IE is equally reasonable: fighting the decaying of *Being* (metaphysical entropy) is the general approach to be followed, not an impossible and ridiculous struggle against thermodynamics, or the ultimate benchmark for any moral evaluation, as if human beings had to be treated as mere numbers.

Second, age and experience teach us that there is probably nothing more difficult than living a morally good life, even assuming a strong and determined will to be morally good. Now, imagine a sceptic asking what it takes to win a gold medal at the Olympic Games, and then objecting that the answer cannot possibly be correct because it would require too much effort, or too many advanced skills, or unusual capacities and gifts. Moral games are no less difficult. Of course, the sceptic might reply that this is not a good analogy, because a moral life cannot be as difficult to achieve as a gold medal at the Olympics. And it is exactly this reply that unmasks a deeper problem. For it shows that what is at stake is not really a supererogatory issue any more, but rather the mistaken assumption that lies behind the supererogatory objection against IE: that the moral game is a game sufficiently easy to win, and that it must be so because any human being must in principle be able to win it, and that this is the case because some ultimate salvation is at stake, and a game too difficult or even impossible to win would be unfair in itself, or presume an unfair game Designer. This is a very non-Greek and rather Christian silent axiom. It is not Greek, because Greek culture knows only too well the meaning of the tragic: the failure of a good will to do the right thing even when meaning well, as we saw in Chapter 4. Greek eyes do not fear seeing life as intrinsically and sometimes irremediably unfair and unjust. The silent axiom is much more in tune with Christian ethics because the latter presupposes a fair game Designer and Judge and an ultimate *Redde Rationem* ('*redde rationem villicationis tuae*', 'give me an account of your stewardship', *Luke* 16:2). IE finds a compromise between these two positions by seeking to interpret the morally good life as a matter of differential score (see the conclusion to Chapter 15). We cannot avoid doing some evil and this is our tragic predicament, but we can still be good agents if we do more good than evil, and this is our heroic chance. We shall inevitably fail many times, but we could succeed even more times. The question is then: how can we know how to weight in favour of good? This is the way I understand objection number nine. Before it, we need to address another difficulty: even if IE's approach is not supererogatory, it is too hypermoralistic. This is the next objection.

16.8 IE is hypermoralistic

Objection

Getting to IE requires a 'conversion experience'. One has to be willing to see the whole project of one's life through the lens of 'respect for Being' or some such. This will never be more than an esoteric morality; such a view has no 'hooks' into most people (not even to most philosophically minded people). They will not be converted, unless one starts where they are. To be converted is to adopt the point of view of IE as one's own. It is not entirely clear whether IE suffers or must suffer from a hypermoralism—the effort to have one corner of our ethical discourse ('morality' in a narrow sense) dominate all others, 'alienating' us from morality itself—but there is much in the text that should cause concern.

Reply

First, an explanation. In the wider context of the objection, it is clear that the concern about hypermoralism is an ethical concern that urges one to acknowledge that not everything of ethical significance is a narrowly moral concern. It is not merely the complaint that IE can be demanding, and the objection should not be confused with one about supererogation. Nor should the objection of hypermorality be confused with a problem of moral motivation or incentives. Next, a reply. If avoiding alienation from morality means seeking and articulating some harmony between the self and the whole—much like Plato unifies (that is, does not fail to distinguish, but intentionally treats as aspects of a more complex reality) the individual and the social life—then this objection is actually an opportunity for further work. I have stressed in several places that, according to people who know better than I do, IE resonates with spiritual, indeed Buddhist or Shintoist, positions (see also objection 16.20). These are easily interpretable as non-hypermoralistic, poietic ethics, which seek to overcome the polarization between the self and the other, between the agent and the environment, the informational entities and their infosphere. So there is clearly a way forward for IE, which finds in virtue ethics an ally. This is why I have more faith than the sceptic in the fact that 'the whole project of [seeing] one's life through the lens of "respect for Being" . . . will have plenty of hooks into most people'. We have come a long way since it was inconceivable to exercise respect for people different from us, or for animals, or for biological environments. We just need to make the last step. This is not a conversion, it is an evolution.

Even if IE should turn out to be not hypermoralistic, this still leaves unclear how IE measures the intrinsic moral value of things. This is the next objection.

16.9 IE's measure of intrinsic moral value is insufficiently clear and specific

Objection

Overall complexity, or quantity of information, is a poor measure of intrinsic moral value and, besides, IE offers no specific theory of how to measure moral value.

Reply

Neither objection actually applies and we should not be distracted by them. As I clarified in Section 16.2, no quantitative approach to information is in question here. We should not get lost behind false promises of numbers and formulae. When IE refers to more or less evil or moral goodness, to moral thresholds, and to higher or lower degrees of flourishing and well-being, these are *qualitative* assessments that require practical wisdom or *phronesis*, not a pocket calculator. They may be made more precise by some formalization, but cannot be seriously quantified. The objection concerns what sort of metrics one could use in order to determine whether entities (understood informationally) have more or less significant (or negligible) moral value and hence how one could live a morally good life. The reply is quite simple: the more an entity contributes to the well-being of the infosphere the higher its status is in the chain of morally respectable beings. Once again, this is a classic position for which I claim little originality. If there is a God, God is the ultimate respectable entity as the source of all entities. A biological virus must, unfortunately, be destroyed for the sake of the rest of the environment and its flourishing. Between these two boundaries we can aim to imitate God or run the risk of being worse than a virus. People have managed both.

Even if IE's position about the intrinsic moral value of things is clear and specific, it still commits the logical mistake of inferring that if things deserve respect then they have moral value. This is the next objection.

16.10 IE's inference from moral value to moral respect is incorrect

Objection

IE incorrectly infers from the fact that some objects, like rocks and objects of cultural heritage, deserve respect the fact that they therefore have intrinsic value.

Reply

The objection is correct. The reasoning it sketches is blatantly fallacious, so anyone adopting it cannot but fail to be convincing. The trouble is that IE does *not* endorse that reasoning in the first place. The objection confuses a *causal* with an *inferential* reasoning. A quick analogy will help. Suppose you tell me that your car does not start *because* its battery is flat. Next, I take you as saying that *if* your car does not start *then* its battery is flat. I then proceed to show you that you are wrong, for it takes a second to realize that your car may not start for a thousand other reasons (no petrol, for example). Of course, you are not impressed, but complain that I misconstrued what you meant: you said 'because' but I attributed you an 'if...then...' explanation. Now, let's go back to IE. The actual argument seeks to establish that entities deserve respect *because* they have intrinsic value, not that *if* entities deserve respect *then* they have intrinsic value. The latter inference is simply untenable, as the sceptic can easily show, but it is also irrelevant, as anyone may appreciate in the analogy of the car with the flat battery. It is the causal explanation that is at stake and that (not the fallacious inference formulated by the sceptic) leads to the really interesting problem: do non-sentient entities have some minimal, perhaps easily overridable but still intrinsic, value? Without rehearsing the whole discussion, I agree that the answer here can be difficult to grasp. It requires a mental frame rather different from the one that anthropocentric and agent-oriented macroethics have trained us to adopt for millennia. It consists in shifting the burden of proof (a sort of Gestaltic shift) by asking, from a patient-oriented perspective, not 'why should I care, in principle?' but 'what should not be taken care of, in principle?' That is to say, whether there is anything in the universe that is intrinsically and positively worthless ethically and hence rightly disrespectable in *this* particular sense, i.e. insofar as its intrinsic value is concerned (again, something might deserve to be disrespected for other reasons, e.g. instrumentally, symbolically, or for the sake of other entities, as I have clarified in Chapter 9). In short, one line of reasoning in favour of IE's position—not the only one, but the one challenged by the sceptic here—is that, because we have no reasons against the intrinsic value of Being in all its manifestations, we should expand an environmental approach to all entities, including non-sentient beings. The injunction is to treat something as intrinsically valuable and hence worthy of moral respect by default, until 'proven guilty'. The intuitive idea is that a universe without moral evaluators (e.g. humans) would still be morally valuable, and that an ontologically richer universe (consider Moore's 'exceedingly beautiful world' in Moore (1993), pp. 135–6) would be a morally preferable universe compared to an ontologically poorer one.

Even if IE does not make the logical mistake of inferring that if things deserve respect then they have moral value, its negative argument for the intrinsic moral goodness of Being is still incorrect. This is the next objection.

16.11 IE's negative argument for the intrinsic moral goodness of Being is incorrect

Objection

There would be no good reason not to adopt such a higher and more inclusive moral perspective on the goodness of Being if there were, in fact, good objective and independently grounded reasons for adopting such a perspective.

Reply

IE seeks to break the artificial constraints of what may count as morally valuable. One way in which it tries to escape from such ethical chauvinism is by showing that there is no good reason to raise any barrier. To put it simply, this is like arguing:

(a) P is the case because it is not the case that $\neg$ P.

This is classic, elementary logic. Of course, it is also a way of reasoning that one may not wish to endorse for equally good logical reasons; one only needs to recall intuitionistic logic or forms of anti-realism *à la* Dummett. But accepting the logic and not its issuing constraints on the validity of the reasoning is mere inconsistency. Let us now turn to the objection. When the sceptic argues that

(b) there would be no good reason not to adopt such a higher and more inclusive moral perspective if there were, in fact, good objective and independently grounded reasons for adopting such a perspective

a logical mistake is made and a crucial philosophical insight goes missing. The mistake is the following: (b) is not a version of (a) but is rather equivalent to:

(c) if there were good Rs (= good, objective and independently grounded reasons) to adopt P (the thesis concerning the intrinsic moral goodness of Being, or axiological ecumenism) then there would be no good Rs not to adopt P.

Note, however, that while (a) is formally valid, (c) is not (and not because it is expressed subjunctively): the premise could well be true, e.g. there could well be good Rs to adopt P, with the conclusion still being false, that is, while there could also be perfectly good Rs not to adopt P. Indeed, this is a very common scenario in our moral lives, which are full of dilemmas: there are often good reasons both to adopt a position, a decision, a course of action, make a choice, etc. and, on balance, equally good reasons not to do so. This is why IE does not support either (b) or (c).

The missing philosophical insight is connected to (a). One can immediately appreciate that the argument in question is negative or indirect. It consists in reminding historically and showing logically that we have nothing to fear from a holistic attitude towards the value of Being in all its aspects; that it is fine to start from the presupposition that no entity deserves moral disrespect in itself; that anything less than a holistic attitude towards the value of Being would be *prima facie* unjustified.

However, even if IE's negative argument for the intrinsic moral goodness of Being is not incorrect, its universal claims are unclear and possibly contradictory. This is the next objection.

16.12 IE's claim to be universal is unclear and possibly contradictory

Objection

It is not always clear and possibly even contradictory whether IE is universal.

Reply

There are many senses in which a macroethics can qualify as universal, and IE satisfies all of them. Let me clarify how.

First, the objection might be based on the misconception that treating all entities as informational objects, including human beings, means somehow diminishing our 'human dignity'. This is not an unusual complaint, but it is misaddressed, somewhat outdated and definitely unproductive, as we saw in Section 16.3. IE adopts this informational ontology (or better: the corresponding LoA) as a minimal common denominator that unifies all entities. We are not just *inforgs*, but we should not fear considering ourselves as inforgs. This is the first sense in which IE is universal. It is the inclusive sense that the logician will immediately recognize as part of the extensional meaning of a universal quantification: *all entities* are informational in nature, and IE seeks to address the ethical issues that pertain to all of them.

Second, the objection might be based on a mistaken view about how one may choose between different LoAs. It is not just a matter of whimsical preference, personal taste, or subjective inclination of the moment. The reader working in computer science knows only too well that one should never underestimate a crucial component in any use of a LoA, namely its goal or the 'what for?' question. There is a perfectly reasonable LoA, say in terms of shape and topology, at which dad's shoes can be observed and even used as ships; but when Columbus grows up, he will find that ludic LoA useless for the purpose of reaching America. LoAs are always teleological, or goal-oriented. Thus, when observing a building, which LoA one should adopt—architectural, emotional, financial, historical, legal, and so forth—depends on the goal of the analysis. There is no 'right' LoA independently of the purpose for which it is adopted, in the same, relational but non-relativistic sense in which there is no right tool independently of the job that needs to be done. So, the position held by IE is that, when it comes to talking about ethical issues in an ontocentric and more inclusive, non-anthropocentric way, an informational LoA does a good job. This is the real thesis that one may wish to criticize. Unfortunately, it is overlooked by the objection. In IE, it is not a matter of showing that the choice of the right LoA is an ethical imperative in itself, for goals can often be reached equally well in different ways: a hammer or a shoe can both be used to nail a painting to the wall. Nor is the infosphere the highest LoA, for LoAs rarely come ordered in hierarchies, as one is easily reminded by the previous example about the house (asking which LoA is the highest would be missing the point). Indeed, Plato, Berkeley, Spinoza, or perhaps, I dare say, even Heidegger or Buddhist philosophy, all adopt equally abstract LoAs, just to mention a wide selection of different positions. So this is the second, non-relativistic sense in which IE claims to be universal: its analysis is based on the reasonable choice of a plausible and fruitful approach to the sorts of new ethical problem emerging in the information society. Of course, one may disagree on the value of the approach. But the charges of relativism (any LoA is a good LoA) and absolutism (there is only one right LoA, the highest) could not be more misplaced.

In order to grasp the other senses in which IE is universal, it is now worth asking what exactly is being universalized when we request a macroethics to be universal. Here is what one may mean.

(1) Universality as *universal applicability* of ethics to all entities concerned by the moral discourse. This is one of the senses in which one speaks of universality as a matter of 'validity' or 'scope, range, or applicability'. We have seen not only that IE satisfies it but also that it does so better than many other macroethics, since it endorses a wider scope of ethical concerns.

(2) Universality as *universal impartiality* of ethics. This second sense is strictly related to (1), insofar as it qualifies the sort of universal applicability in question (the applicability is *impartial*). In this case too, IE can only be said to be more impartial than many other macroethics that, for example, discriminate between living and non-living beings, or between human/rational and non-human/non-rational agents.

(3) Universality as *universal acceptability* of ethics by everyone involved, who shares it without coercion. This might be what one means when referring to universal values, shared by all, and to a macroethics' strategy of convincing agents to accept or to follow it. This third sense requires some disambiguation. Insofar as it is a matter of *empirical description*, the most universal macroethics in the first (applicability) and second (impartiality) sense above may still fail to be universally acceptable in the third (acceptability) sense if, for example, there exist a group of agents who are determined to reject it in principle. In this sense, the existence of the Ku Klux Klan would undermine the universality of human rights. True, so this is not what can be at stake here. Insofar as it is a matter of *normative prescription*—an ethics ought to be universally acceptable by anyone without coercion—then (3) reduces to a combination of (1) and (2). In other words, it is because a macroethics is universally applicable and impartial that it might rightly aspire to gain uncoerced acceptance. But then it follows that any theory that satisfies (1) and (2) is strategically well positioned (or at least as well positioned as any other) to satisfy (3) as well. And we have seen that IE does satisfy (1) and (2), so IE is not challenged by (3), or at least no more than other macroethics.

(4) Universality as *universal inclusivity* of ethics. This holistic sense is different from (1) or (2) insofar as it refers to the capacity of a macroethics not only of applying to the specific agents and patients involved in a particular moral action, but also of widening its consideration to ever-larger circles of interested parties as stakeholders, who may be taken into account when evaluating a moral action. Now, if any macroethics seeks to be inclusive, this is certainly IE, which I have described as an extension of environmental ethics that looks at a wider and more inclusive environment. Indeed, we shall see, in the next section, that critics have moved the objection that IE runs the risk of being too *inclusive*.

(5) Universality as *anti-relativism* of macroethics. This last sense has already been discussed above, where we have seen how IE succeeds in being non-relativistic without falling into the trap of being authoritarian or absolute.

To summarize, in any of the aforementioned senses in which a macroethics might be requested to be universal, IE turns out to be in a rather satisfactory position. Indeed, one should acknowledge that it performs better than many other theories.

However, even if universality claims concerning IE are neither unclear nor contradictory, IE's egalitarianism is still untenable. This is the next objection.

16.13 IE's egalitarianism is untenable

Objection

IE is committed to an untenable egalitarianism in the valuation of informational entities. From the point of view of IE, a work of Shakespeare is as valuable as a piece of pulp fiction, and a human being as valuable as a vat of toxic waste. Any source of additional worth lies beyond the scope of IE, because IE only assigns worth to things *qua* informational entities. IE tells us that we should be equally protective of human beings and vats of toxic waste, or of any other information object, and that we have an (albeit overridable) duty to contribute to the improvement and flourishing of pieces of lint and human excrement. At best, this suggests that IE gives us very little guidance in making moral choices. At worst, it suggests that IE gives us the wrong kind of guidance.

Reply

The trouble with this objection is twofold. First, when defending the intrinsic value of all aspects of Being, understood as informational entities, the point at stake is not some daft idea about the intrinsic value of Shakespeare vs. airport novel, or chocolate vs. excrement. The actual issue is whether Goodness and Being (capitals intended) might be two sides of the same concept, as Evil and non-Being might be. Without disturbing Eastern traditions within Buddhism, Hinduism, or Shintoism—which I understand attribute intrinsic value both to sentient and to non-sentient realities—the reader sufficiently acquainted with the history of Western philosophy may easily list classic thinkers, including Plato, Aristotle, Plotinus, Augustine, Aquinas, and Spinoza, who have elaborated and defended in various ways this fundamental equation. For Plato, for example, Goodness and Being are intimately connected. Plato's universe is value-ridden at its very roots: value is there from the start, not imposed upon it by a rather late-coming, new mammalian species of animals, as if before evolution had the chance of hitting upon *Homo sapiens* the universe were a value-neutral reality, devoid of any moral worth. By and large, IE proposes the same line of reasoning, by updating it in terms of an informational ontology, whereby Being is understood informationally and non-Being in terms of metaphysical entropy. Note that this is not a defence of IE but an explanation of why the objection fails to apply. Although keeping company with Plato or Spinoza, for example, might be reassuring, it is not an insurance against being mistaken. But it is a rectification of the incorrect remark that IE stands rather alone in its defence of its axiological ecumenism.

Second, the objection misses the crucial importance of the minimalist approach defended by IE, which invites one to consider every entity as ethically valuable in itself and deserving some moral respect *to begin with*, exactly in the same way as environmental ethics invites us to approach any form of life as worth preserving, *if possible*. Nobody in his right mind would ever argue that this is equivalent to saying that a spider's and a human life are equally worthy of respect. Culling, for example, might be considered an ethical duty in environmental ethics. And even in Buddhism, killing animals is a minor offence (*pāyantika*) compared to the much more serious offence (*pārājika*) represented by killing a human being. Likewise, in IE, the destruction of entities might easily be not only inevitable but also mandatory. Again, IE is not about respecting a single grain of sand as much as one respects the whole earth. It is about fixing the threshold below which something should be morally disrespectable in itself and rightly so. With a Cartesian analogy, the sceptic's mistake lies in thinking that, if one argues that all physical things are extended, then one is arguing that they are all of the same size. Of course they are not, and nobody could reasonably argue that they are. To revert to IE, the view that all entities are *at least minimally and overridably* valuable in themselves should not be confused with the view that they all share the same value. Contrary to what the sceptic seems to think, the adverbs play a crucial role here. It is often the case that one philosopher's use of Ockham's razor is another philosopher's chainsaw massacre.

However, even if IE's egalitarianism is defensible, it is based on the naturalistic fallacy. This is the next objection.

16.14 IE commits the naturalistic fallacy

Objection

By developing an ontocentric ethics, IE commits the naturalistic fallacy.

Reply

An ontocentric approach is indeed threatened with the naturalistic fallacy. The latter, in order to be applicable, must presuppose a value-empty or value-neutral reality, from which then not a single drop of morality could be squeezed, on pain of contradiction. The 'no ought from is' principle, with its Humean roots, is perfectly fine. If Being (or reality or nature or indeed the infosphere) is interpreted as being entirely and absolutely devoid of any moral value—if it is simply meaningless to say that 'to be is to be good'—then any moral value, any goodness, and the corresponding ethical orientations that we long for, must come from elsewhere. A drained and dry container cannot fill itself. But if the ontic source, from which we seek to draw some moral guidance, is not empty, if, following Plato and Spinoza, for example, we acknowledge that Being and Goodness are intrinsically intertwined well before any metaphysical or ethical discourse attempts to rescind them, then trying to extract values and the corresponding moral lessons from Being becomes a very natural but not fallacious process. One may try to find guidance and inspiration in the life of the universe without committing any logical fallacy. Now, it seems to me that IE has the merit of reviving, if not establishing, this ontocentric perspective. I am happy to concede that perhaps it takes a spiritualistic form of naturalism to find the approach attractive. Any strictly and uncompromising materialistic view of the world, like Hume's, will struggle with the possibility that Being might be morally pervaded by, and overflowing with, Goodness. And any existentialist view, like Heidegger's, will be too anthropocentric, self-referential, nihilistic, and too reluctant to de-centralize the human condition to be truly enlightening. This is why it is very fruitful to read Spinoza in terms of a naturalistic philosopher closer to the Greeks.

However, even if IE does not commit the naturalistic fallacy, its account of intrinsic value in general is still incorrect. This is the next objection.

16.15 IE's account of intrinsic value is incorrect

Objection

Why should the correct account of intrinsic value be a general, minimalist, homogeneous account, as argued by IE?

Reply

The 'because' is in the pudding. Less metaphorically and more explicitly, we encounter here a twofold confusion. First, there is no 'the correct account'. This approach belongs to a non-pluralist and hence inevitably intolerant way of doing ethics that IE seeks to overcome. There are, however, 'correct accounts' that may complement and reinforce each other, like stones in an arch. The second confusion concerns precisely what makes them 'correct'. Suppose someone says that he is a good driver. Although one might require him to produce a driving licence, nobody would demand a syllogism. If pushed, one would eventually test the person's skills by having him actually drive a car. The reader acquainted with Wittgenstein's distinction between *saying* and *showing* will find this familiar. Now, IE tries to *show* that there is a way of conceptualizing Being informationally in such a way as to build a minimalist and homogeneous account of all entities. IE also tries to *show* that this is *a* correct account.

However, even if IE's account of intrinsic value in general is correct, IE remains highly counterintuitive. This is the next objection.

16.16 IE is counterintuitive

Objection

If we are to start valuing things as intrinsically valuable that we do not already value as such, we need good reasons to do so. Since people do not normally seem to assign intrinsic value to informational entities, IE needs to provide strong arguments for us to start valuing them as such.

Reply

The objection is mistaken for several reasons. First, what people normally assign intrinsic value to is a matter of sociology (*description*) not of ethics (*prescription*), and moving from one to the other means committing an obvious fallacy. Second, history is full of 'people' who failed to assign intrinsic value to at least some human beings (e.g. barbarians, black people, children, foreigners, handicapped, homosexuals, indigenous, immigrants, Jews, slaves, women ... the list is long), but it would be odd to argue that they are (or even were) right until proved wrong. Third, *Vox populi vox Dei* (literally: the voice of the people is God's voice) has never been a decent argument but, if one really likes to stick to the alleged 'wisdom of crowds', why not choose the 'right crowd'? For example, I remarked on more than one occasion that several philosophical schools, as well as many Buddhist, Christian, Hindu, Taoist, or Shinto cultures, attribute intrinsic value both to sentient and to non-sentient realities. Why not consult those cultures instead, if one really wants to rely on what people actually value? Finally, the logical mistake is with the initial argument itself. For its rationale is a conservative and cautious attitude that might be fine when talking about potential moral risks of evil, but is out of place when engaging with the morally good. Consider its form: if we start ϕ-ing things as intrinsically ϕ-able that we do not already ϕ as such, we need good reasons to do so. Now, it makes a big and quite obvious difference whether we replace ϕ with negative (hate, destroy, despise, discriminate, etc.) or positive attitudes (love, admire, cherish, protect, etc.). For example, 'if we start hating things as intrinsically hateable that we do not already hate as such, we need good reasons to do so' might sound reasonable. But 'if we start loving things as intrinsically lovable that we do not already love as such, we need good reasons to do so' is definitely questionable, for love does not bear very much accountancy, as Paul of Tarsus would remark. In other words, we should not fear respecting the whole universe too much. Rather, 'respect and take care of all entities for their own sake, if you can' is the injunction. Or, as Augustine nicely put it, *dilige, et quod vis fac* (love/respect and do what you wish).[5] Note that Augustine uses the Latin 'diligere', not 'amare', a term that precisely refers to love as careful respect. This is in keeping with the emphasis in IE that informational ontocentrism is a naturalistic philosophy that closely resonates with Spinoza's, Plato's, Confucius', and Buddhist thought (among others) in its affirmation of the intrinsic moral worth of the *cosmos* as such.

However, even if IE's position is not counterintuitive, its reliance on the method of abstraction is a mistake. This is the next objection.

16.17 IE's adoption of LoA is mistaken

Objection

If informational entities are to possess intrinsic value, they cannot be observer-dependent, as indicated by the adoption of a level of abstraction, because for an object to possess intrinsic value it must possess one or more properties that bestow intrinsic value upon it, such as the property of being rational, being capable of suffering, or being an informational entity. Such properties have to be objective and inalienable properties of the object in question, not subjective or contingent ones, because otherwise the assigned value is at best extrinsic, that is, resulting from the attribution of contingent roles or subjective meanings to objects.

Reply

What lies behind this objection is a conceptual confusion. When we adopt a Level of Abstraction at which we observe things in terms of their chemical composition, for example, this does not make water contingently and subjectively H_2O. The same applies to the adoption of an information-theoretical LoA. A LoA is an interface that takes advantage of the constraints and affordances offered by the system under observation, in view of a specific goal. Once this is grasped, the second step is to realize that IE tries to move from a materialist ontology to an informational one. As stressed above, this is a matter of metaphysics, not science, so anyone suggesting that information science, as a science for describing the universe, is not on a par with chemistry or physics, commits a category mistake, using Ryle's terminology. Once the ground is cleared of all these confusions and errors, it is obvious that object-oriented programming, as introduced in Section 6.1, is provided only as a means of helping make sense of an informational ontology. Compare this to Plato's or Descartes' use of geometry to make sense of their metaphysical views. Nobody ever complained that the development of, say, non-Euclidean geometry or topology undermined their ontologies, or Plato's analogy of the line, for this is merely irrelevant.

---

[5] Augustine, 'Homily on the First Epistle of St John', Eng. trans. available in Augustine (1984).

However, even if IE's reliance on the method of abstraction is correct, its use to argue that artificial agents may qualify as autonomous moral agents is still mistaken. This is the next objection.

16.18 IE's interpretation of artificial agents as autonomous moral agents is mistaken

Objection

IE confuses, on the one hand, something being autonomous and moral at the level of abstraction and, on the other hand, something being autonomous and moral writ large.

Reply

The method of abstraction may seem difficult to grasp, but its main lesson is simple enough. When applied to agents, it can be fruitfully used in order to understand autonomy and morality in terms of self-regulating agency. The interested reader might check how, in Chapter 7, artificial agents are shown to be autonomous moral agents exactly and precisely at the required level of abstraction. The real issue is whether a better analysis, i.e. a better level of abstraction, may be provided. Hand-waving or feet-stomping is not an alternative, while trying to explain a concept (autonomous agency) by referring to more obscure concepts (freedom, intentionality, and so forth) is simply falling into the classic fallacy of *obscurum per obscurius*.

However, even if IE's inclusive attitude towards artificial agents is correct, like other forms of environmental ethics, it remains a very conservationist approach. This is the next objection.

16.19 IE is too conservationist

Objection

IE supports a morally conservationist attitude, according to which we would be required not to modify, improve, or in any way interfere with the natural course of things.

Reply

Wrong. As indicated in Section 4.7.2, IE is not just a green but a *blue* ethics like virtue ethics (in that section I borrowed the expression from 'blue-print'): IE is fundamentally proactive, in a way similar to *restorationist* or *interventionist ecology*. The *homo poieticus* has the duty to look after the world and make it a better place. The unavoidable challenge lies precisely in understanding how reality can be improved, better shaped, or helped to develop in the best way. A gardener transforms the environment for the better, but that is why he needs to be very knowledgeable. Prune your fruit trees or your roses in the

328   THE ETHICS OF INFORMATION

wrong season or not at all, and the results will be very disappointing. IE has no bias in principle against abortion, eugenics, GM food, human cloning, enhancement or plastic surgery, animal experiments and other highly controversial, yet technically and scientifically possible, ways of transforming or 'enhancing' reality. But it is definitely opposed to any associated ignorance of the consequences of such radical transformations and any delegation of the responsibility of our choices to the gods.

Even if IE's approach is not conservationist, it is too spiritualistic. This is the next objection.

16.20 IE is pantheistic or panpsychistic

Objection

IE offers a Stoic pantheistic ethics that endows everything in the universe with a moral significance and status through a pre-determined divine rational order in which everything is ontologically interconnected and of which everything forms an ontic part, no matter how big or small. This pantheism or panpsychism is untenable.

Reply

IE is compatible with, and may be associated with, religious and spiritual beliefs, including a Buddhist (Herold, 2005), Confucian, Shintoist, or a Judeo-Christian view of the world. In the latter case, the reference to *Genesis* 2:15 readily comes to one's mind. *Homo poieticus* is supposed 'to tend (*'abad*) and exercise care and protection over (*shamar*)' God's creation. 'Stewardship' is a much better way of rendering this stance towards reality than 'dominion'. Or consider the very complex concept of *kami* in the Shinto faith. The Japanese word refers, loosely speaking, to the divine spirits, forces, or essence inhabiting any aspect of Being. In a simple article by the BBC aimed at beginners like myself, one reads:

> Shinto is based on belief in, and worship of, *kami*. The best English translation of kami is 'spirits', but this is an oversimplification of a complex concept—kami can be elements of the landscape or forces of nature.... Kami can refer to beings or to a quality which beings possess.... So the word is used to refer to both the essence of existence or beingness which is found in everything, and to particular things which display the essence of existence in an awe-inspiring way.... Kami as a property is the sacred or mystical element in almost anything. It is in everything and is found everywhere, and is what makes an object itself rather than something else. The word means *that which is hidden*. Kami have a specific life-giving, harmonising power, called *musubi*, and a truthful will, called *makoto* (also translated as *sincerity*). Not all kami are good—some are thoroughly evil. The idea that kami are the same as God stems in part from the use of the word kami to translate the word 'God' in some 19th century translations of the Bible into Japanese. This caused a great deal of confusion even among Japanese: the Shinto theologian Ueda Kenji estimated in 1990 that nearly 65% of entering students now associate the Japanese term kami with some version of the Western concept of a supreme being.... The concept of kami is hard to explain. Shintoists would say that this is because human beings are simply incapable of forming a true understanding

of the nature of kami.... The term kami is sometimes applied to spirits that live in things, but it is also applied directly to the things themselves—so the kami of a mountain or a waterfall may be the actual mountain or waterfall, rather than the spirit of the mountain or waterfall.[6]

The temptation of appropriating such a wonderful concept as *kami* to describe IE as an ethics of respect and care for *kami* is very strong. I shall barely resist it in order to stress that IE is based on an immanent, if perhaps rather spiritual, philosophy. *Homo poieticus* has a vocation for responsible stewardship in the world. Unless some other form of intelligence is discovered in the universe, we cannot presume to share this burden with any other being. *Homo poieticus* should certainly not entrust his responsibility for the flourishing of *Being* to some transcendent power. As the Enlightenment has taught us, the religion of reason can be immanent. If the full, rational responsibilization of humanity is then consistent with a spiritual or religious view, this can only be a welcome conclusion, not a premise.

CONCLUSION

One of the highest honours that a research and scholarly community can bestow on its members is to pay sustained and careful attention to their work. In philosophy, this attention ultimately translates into constructive criticism.

Criticism is in the very nature of any philosophical investigation. With its intrinsically open questions (Floridi, 2011a), philosophy invites dialogue in the form of objections and replies, suggestions for revisions, and proposals for further improvements or very different alternatives. The scientist may find this process of 'creative destruction' unfamiliar: she might dislike it as at best fruitless, at worst counterproductive, in any case suspiciously symptomatic of a lack of clear criteria—through which progress might be assessed—and hard data, by which the same progress might be anchored and constrained. That scientist may not be in error about the facts, but she would certainly be mistaken about their interpretation, for she would be confusing the different directions in which philosophy and science move. Scientists build, whereas philosophers dig.

In the process of building, one cannot help but construct every higher step upon a lower step. It is trivial to remark that there is no second floor without a first and that the solidity of the whole construction depends heavily on the reliability of every layer. Trust and team-work are everything, getting things right vital. That is why scientific revolutions happen rarely but, when they do, they are as dangerous as major earthquakes.

In the process of digging, on the other hand, every single shovelful helps. So philosophers are more akin to individual explorers of the depths, and are more likely to proceed by removing rather than augmenting, reminding one of Michelangelo's

---

[6] 'What are kami?', BBC Religions, 4 September 2009 <http://www.bbc.co.uk/religion/religions/shinto/beliefs/kami_1.shtml>.

definition of sculpture, the art of 'taking away', as opposed to the art of 'adding on' characteristic of painting. As we all know, the higher one wishes to build, the more deeply (or better: profoundly) one needs to explore. Yet philosophers not only search for the deepest and firmest ground on which our understanding may rest more safely; they also—or perhaps I should say mainly, in these anti-foundationalist days—seek to extract precious conceptual resources that, once unearthed, purified, and carefully processed, may help humanity to make sense of an ever-changing reality. We do not pass slabs, like collaborative Wittgensteinian constructors, we go back into the darkness of Plato's cave to help ourselves and others. Our hands are roughened and dirty. Forget about Athena, our god is Hephaestus.

The result is that, in the philosophical underground, where everything is so dimly lit and hidden, one often hears other explorers shouting curses and advice, eurekas and warnings. These are not sterile expressions of emotional states; these are signs of passion and interest in the intellectual work being done. It is with this general picture of a collaborative enterprise in mind that I have tried to address in this chapter the criticisms moved against IE. I believe IE both to provide a reliable foundation for a wider ethical discourse and to offer valuable resources to make sense of our time and the information revolution that characterizes it. But I am also aware that it is a field that we have only just started to explore. I am afraid that, in many cases, I might have failed to do justice to the value of the theoretical analyses offered by so many colleagues from whom I have learnt so much. To my justification, I may point out that IE is a very rich and still largely unexploited mine. So, to the reader who may find some of the theses in this book less than satisfactory, the invitation is to join us. The advice is that there is plenty of rewarding work still to be done.

Epilogue

> The goodness, or badness of actions, does not arise from hence, that the epithet, interested, or disinterested, may be applied to them, ... but from their being what they are.
>
> Joseph Butler, *Fifteen Sermons Preached at the Rolls Chapel* (1914), p. xvii.

There are places, like the small village where I live, that are difficult to find. They lie in remote locations, not clearly indicated on the map, few people have ever heard of them, and hardly anyone can tell you how to get there. There are places, like the university where I work, which are difficult to reach. It is so big that, if you are driving following a GPS, its postcode will take you miles away from the campus, to a mail deposit. Sometimes, I fear that the philosophy of information that I have been working on combines the geographical problems of my home and working places: difficult to find and hard to reach. I hope this second volume on the ethics of information helps to map some less tortuous paths that, if followed, should enable the reader to get to the philosophy of information that I have in mind, and alert the same reader to some wrong turns, potential pitfalls, and misleading road signs that have side-tracked more than one fellow traveller. Of course, indicating more clearly how to reach a place does not mean that the place itself is worth visiting. I believe that the philosophy of information is the philosophy *of* our time properly conceptualized *for* our time, but then you might expect this level of commitment on my side. I also hope that the journey to reach it will be rewarding, but on this I can only rely on the traveller's experience. What I may say is that the view from here is very interesting and shows an immense conceptual space still virgin. If you join me, you will see.

Let me now conclude this exploration of information ethics by quoting a famous passage in one of Einstein's letters that summarizes well the perspective advocated by IE understood as a macroethics.

Some five years prior to his death, Einstein wrote a letter to Robert S. Marcus, the Political Director of the World Jewish Congress, who was grieving over the loss of his young son:

February 12, 1950
Dear Mr. Marcus:
A human being is a part of the whole, called by us 'Universe', a part limited in time and space. He experiences himself, his thoughts and feelings as something separated from the rest—a kind of optical delusion of his consciousness. The striving to free oneself from this delusion is the one issue of true religion. Not to nourish the delusion but to try to overcome it is the way to reach the attainable measure of peace of mind.[1]

There is a unique ethical duty that follows from humanity's unique semanticizing role. The epistemic responsibility involved in the design of a meaningful reality is not just an epistemological task, placed on our shoulders as individual epistemic agents. It is also, and probably more importantly, a social and ethical obligation that we have towards each other. If we and no one else make reality meaningful to ourselves and to others; if there is no other source of meaning in the universe but us; if our 'semantic currency' is not backed up by some God standard; then there is only an immanent semantics, which is up to us to design, develop, protect, and share. This is our call. From it, it follows that each human life becomes valuable, and something to be cherished, as a precious source of sense-making. It would be a logical mistake to read such call solipsistically (and here is where I distance myself from Descartes and Kant), for the following reason. Semanticization is an information process. But then, semanticization is a social process, to which we may contribute only a bit, but from which we all benefit enormously. Most, indeed almost (yet not) all the sense we can give to our lives is due to the sense-making activities of millions of other people. Hell is not the other, but the death of the other, for that is the drying up of the main source of meaning. As any old person knows, solitude is a social choice, made possible by the presence of others, but loneliness is a desperate condition due to the absence of any other and their helpful semanticization of our lives. In the same way as data and rules are the relations representing the constraining affordances for our behaviour (not only epistemic), our semanticization of them is both an epistemic and ethical task that we can fulfil only as social agents. Civilization is both an epistemic and an ethical concept for a multi-agent system. Wittgenstein was right: there is no private game of semanticization.

From this perspective, in the previous chapters I have argued that the agent-related *behaviour* and the patient-related *status* of all entities *qua* informational entities can be morally significant, over and above the instrumental function that may be attributed to them by other ethical approaches, and hence that they can contribute to determining, normatively, our ethical duties and rights. IE's position, like that of any other macroethics, is not devoid of problems. But it can interact with other macroethical theories and contribute an important new perspective: a process or action may be morally good or bad irrespective of its sources, consequences, motives, universality, or

---

[1] In the past I mistakenly quoted a slightly different text circulated by *The New York Times* (29 March 1972) and *The New York Post* (28 November 1972). The correct citation is published in Alice Calaprice (ed.), *The New Quotable Einstein* (Princeton, NJ: Princeton University Press, 2005), p. 206.

virtuous nature, but depending on how it affects the infosphere, Einstein's 'whole'. An ontocentric ethics provides an insightful perspective. Without IE's contribution, our understanding of moral facts in general, not just of ICT-related problems in particular, would be less complete. Our struggle to escape from our anthropocentric and solipsistic condition, be that Plato's cave or Einstein's delusion, will be more successful if we can take a patient-oriented, informational perspective to the universe and its value.

References

Adam, A. (2005). 'Delegating and Distributing Morality: Can We Inscribe Privacy Protection in a Machine?', *Ethics and Information Technology*, 7(4), 233–42.
Adams, M. M. and Adams, R. M. (1990). *The Problem of Evil* (Oxford: Oxford University Press).
Alexander, C. (1964). *Notes on the Synthesis of Form* (Cambridge, MA: Harvard University Press).
Alfino, M. and Pierce, L. (1997). *Information Ethics for Librarians* (Jefferson, NC: McFarland & Co.).
Allen, A. L. (2011). *Unpopular Privacy: What Must We Hide?* (Oxford: Oxford University Press).
Allen, C., Varner, G., and Zinser, J. (2000). 'Prolegomena to Any Future Artificial Moral Agent', *Journal of Experimental & Theoretical Artificial Intelligence*, 12, 251–61.
Allgrove, B. (2004). 'Legal Personality for Artificial Intellects: Pragmatic Solution or Science Fiction?' (MPhil Thesis, University of Oxford).
Allhoff, F. (2007). *Nanoethics: The Ethical and Social Implications of Nanotechnology* (Hoboken, NJ: Wiley-Interscience).
Allo, P. (ed.) (2010). *Metaphilosophy*, 41(3), Special issue on *Luciano Floridi and the Philosophy of Information*.
Allo, P. (ed.) (2011). *Putting Information First* (Oxford: Wiley-Blackwell).
Alpaydin, E. (2010). *Introduction to Machine Learning* (2nd edn., Cambridge, MA: MIT Press).
Alterman, A. (2003). '"A Piece of Yourself": Ethical Issues in Biometric Identification', *Ethics and Information Technology*, 5(3), 139–50.
Anderson, M. and Anderson, S. L. (2011). *Machine Ethics* (New York, NY: Cambridge University Press).
Anderson, S. L. (1990). 'Evil', *Journal of Value Inquiry*, 24(1), 43–53.
Andrade, F. et al. (2004). 'Software Agents as Legal Persons', in Camarinha-Matos, L. M. (ed.), *Virtual Enterprises and Collaborative Networks* (Dordrecht: Kluwer Academic Publishers).
Andrade, F. et al. (2007). 'Contracting Agents: Legal Personality and Representation', *Artificial Intelligence and Law*, 15(4), 357–73.
Ariáes, P. and Duby, G. (1987). *A History of Private Life*, 5 vols. (Cambridge, MA: Belknap Press of Harvard University Press).
Arnold, A. and Plaice, J. (1994). *Finite Transition Systems: Semantics of Communicating Systems* (Paris: Masson; Englewood Cliffs, NJ: Prentice Hall).
Asimov, I. (1956). 'The Dead Past', *Astounding Science Fiction*, 81(2), 6–46.
Augustine (1984). *Selected Writings* (New York: Paulist Press).
Baird Callicott, J. (1980). 'Animal Liberation: A Triangular Affair', *Environmental Ethics*, 2, 311–38. Reprinted with a new Preface in Elliot (1995).
Barandiaran, X. E., Paolo, E. D., and Rohde, M. (2009). 'Defining Agency: Individuality, Normativity, Asymmetry, and Spatio-Temporality in Action', *Adaptive Behavior—Animals, Animats, Software Agents, Robots, Adaptive Systems*, 17(5), 367–86.
Barfield, W. (2005). 'Issues of Law for Software Agents within Virtual Environments', *Presence: Teleoperators and Virtual Environments*, 14(6), 741–8.

Becker, F. and Sims, W. (2000). *Offices That Work: Balancing Cost, Flexibility, and Communication* (New York: Cornell University International Workplace Studies Program).
Bedau, M. A. (1996). 'The Nature of Life', in Boden, M. A. (ed.), *The Philosophy of Life* (Oxford: Oxford University Press), 332–57.
Benn, I. (1985). 'Wickedness', *Ethics and Information Technology*, 95(4), 795–810.
Benn, P. (1998). *Ethics* (London: UCL Press).
Benn, S. I. (1975). 'Privacy, Freedom, and Respect for Persons', in Wasserstrom, R. (ed.), *Today's Moral Problems* (New York: Macmillan).
Benson, J. (2000). *Environmental Ethics: An Introduction with Readings* (London: Routledge).
Binkley, T. (1998). 'Computer Art', in Kelly, M. (ed.), *Encyclopedia of Aesthetics* (Oxford: Oxford University Press), vol. 1, 412–14.
Biocca, F. (2001). 'Inserting the Presence of Mind into a Philosophy of Presence: A Response to Sheridan and Mantovani and Riva', *Presence: Teleoperators and Virtual Environments*, 10(5), 546–57.
Bloustein, E. (1964). 'Privacy as an Aspect of Human Dignity: An Answer to Dean Prosser', *New York University Law Review*, 39, 962–1007.
Bohn, J. et al. (2004). 'Social, Economic, and Ethical Implications of Ambient Intelligence and Ubiquitous Computing', *Journal of Human and Ecological Risk Assessment*, 10(5), 763–86.
Boltuc, P. (ed.) (2008). *American Philosophical Association Newsletter on Computers and Philosophy*, 7(2)–8(1), Special Issue on *Luciano Floridi's Ethics and Philosophy of Information*.
Boman, M. (1997). *Conceptual Modelling* (Hemel Hempstead: Prentice Hall).
Bond, A. H. and Gasser, L. (eds.) (1988). *Readings in Distributed Artificial Intelligence* (San Mateo, CA: Morgan Kaufmann).
Bracken, C. C. and Skalski, P. D. (eds.) (2010). *Immersed in Media: Telepresence in Everyday Life* (New York: Routledge).
Brandt, D. and Henning, K. (2002). 'Information and Communication Technologies: Perspectives and Their Impact on Society', *AI & Society*, 16(3), 210–23.
Brey, P. (2005). 'Freedom and Privacy in Ambient Intelligence', in Brey, P., Grodzinsky, F., and Introna, L. (eds.), *Ethics of New Information Technology—Proceedings of the Sixth International Conference of Computer Ethics: Philosophical Enquiry (Cepe2005)* (Enschede: CEPTES University of Twente), 91–100.
Buchanan, E. A. (1999). 'An Overview of Information Ethics Issues in a World-Wide Context', *Ethics and Information Technology*, 1(3), 193–201.
Bunge, M. (1977). 'Towards a Technoethics', *The Monist*, 60, 96–107.
Butler, J. (1914), *Fifteen Sermons Preached at the Rolls Chapel*. With an introduction, analyses, and notes by W.R. Matthews (London: G. Bell).
Bynum, T. W. (1992). 'Human Values and the Computer Science Curriculum', in Bynum, T. W., Maner, W., and Fodor, J. (eds.), *Computing and Human Values* (New Haven, CT: Research Center on Computing & Society, Southern Connecticut State University).
Bynum, T. W. (1998). 'Global Information Ethics and the Information Revolution', in Bynum, T. W. and Moor, J. H. (eds.), *The Digital Phoenix: How Computers Are Changing Philosophy* (Oxford: Blackwell), 274–89.
Bynum, T. W. (2000). 'A Very Short History of Computer Ethics', *APA Newsletter on Philosophy and Computers*, 99(2) <http://web.cs.wpi.edu/~hofri/Readings/bynum_sh0rt_hist.html>.

Bynum, T. W. (2001a). 'Computer Ethics: Basic Concepts and Historical Overview', in Zalta, E. N. (ed.), *The Stanford Encyclopedia of Philosophy*.

Bynum, T. W. (2001b). 'Computer Ethics: Its Birth and Its Future', *Ethics and Information Technology*, 3(2), 109–12.

Bynum, T. W. (2010). 'Philosophy in the Information Age', *Metaphilosophy*, 41(3), 420–42.

Bynum, T. W. and Rogerson, S. (1996). 'Global Information Ethics: Introduction and Overview', *Science and Engineering Ethics*, 2(2), 131–6.

Bynum, T. W. and Rogerson, S. (2004). *Computer Ethics and Professional Responsibility* (Oxford: Blackwell).

Capurro, R. (2006). 'Towards an Ontological Foundation of Information Ethics', *Ethics and Information Technology*, 8(4), 175–86.

Carus, A. W. (2007). *Carnap and Twentieth-Century Thought: Explication as Enlightenment* (Cambridge: Cambridge University Press).

Cassirer, E. (1953). *Substance and Function, and Einstein's Theory of Relativity* (New York: Dover; London: Constable & Co.).

Cavalier, R. J. (2005). *The Impact of the Internet on Our Moral Lives* (Albany, NY: State University of New York Press).

Cavoukian, A. (2009). *Privacy by Design* (Ottawa: IPC Publications).

Chen, S. and Ravallio, M. (2008–2009). 'The Developing World Is Poorer Than We Thought, but No Less Successful in the Fight against Poverty', Policy Research Working Paper No. 4703 <http://go.worldbank.org/HAG6SG9G30>.

Chopra, S. and Dexter, S. (2008). *Decoding Liberation: The Promise of Free and Open Source Software* (New York: Routledge).

Christensen, B. (2009). 'Can Robots Make Ethical Decisions?', *Live Science*, 16 September 2009 <http://www.livescience.com/5729-robots-ethical-decisions.html>.

Churchland, P. M. (1999). 'Densmore and Dennett on Virtual Machines and Consciousness', *Philosophy and Phenomenological Research*, 59(3), 763–7.

Clarke, R. (1994). 'Asimov's Laws of Robotics: Implications for Information Technology', *Computer*, 27, 57–66.

Coates, J. F. (1982). 'Computers and Business—A Case of Ethical Overload', *Journal of Business Ethics*, 1(3), 239–48.

Cohen, J. (2000). 'Examined Lives: Informational Privacy and the Subject as Object', *Stanford Law Review*, 52, 1373–437.

Coleman, K. G. (1999). 'Responsible Computers', in D'Atri, A. et al. (eds.), *Proceedings of the 4th ETHICOMP International Conference on the Social and Ethical Impacts of Information and Communication Technologies* (Rome: LUISS Guido Carli Centro di Ricerca sui Sistemi Informativi).

Coleman, K. G. (2001). 'Android Arete: Toward a Virtue Ethic for Computational Agents', *Ethics and Information Technology*, 3(4), 247–65.

Coroama, V., Bohn, J., and Mattern, F. (2004). 'Living in a Smart Environment—Implications for the Coming Ubiquitous Information Society', IEEE International Conference on Systems, Man & Cybernetics 2004, The Hague, The Netherlands, 10–13 October, 5633–8 <http://www.informatik.uni-trier.de/~ley/db/conf/smc/smc2004.html>.

Crane, A. and Matten, D. (2004). *Business Ethics: A European Perspective: Managing Corporate Citizenship and Sustainability in the Age of Globalization* (Oxford: Oxford University Press).

Crane, A. and Matten, D. (2007). *Business Ethics: Managing Corporate Citizenship and Sustainability in the Age of Globalization* (2nd edn., Oxford: Oxford University Press).

Dahrendorf, R. (1967). *Society and Democracy in Germany* (1st edn., Garden City, NY: Doubleday).

Daniels, N. (1996). *Justice and Justification: Reflective Equilibrium in Theory and Practice* (Cambridge: Cambridge University Press).

Danielson, P. (1992). *Artificial Morality: Virtuous Robots for Virtual Games* (London / New York: Routledge).

Davidsson, P. and Johansson, S. J. (2005). 'On the Metaphysics of Agents', *Proceedings of the Fourth International Joint Conference on Autonomous Agents and Multiagent Systems* (The Netherlands: ACM), 1299–300.

De George, R. T. (2003). *The Ethics of Information Technology and Business* (Malden, MA / Oxford: Blackwell).

De George, R. T. (2006). 'Information Technology, Globalization and Ethics', *Ethics and Information Technology*, 8(1), 29–40.

Deleuze, G. and Guattari, F. (1994). *What Is Philosophy?* (New York: Columbia University Press).

Demir, H. (ed.) (2010). *Knowledge, Technology and Policy*, 23(1–2), Special Issue on *Luciano Floridi's Philosophy of Technology: Critical Reflections*.

Demir, H. (ed.) (2012). *Luciano Floridi's Philosophy of Technology* (New York: Springer).

Dennet, D. (1997). 'When Hal Kills, Who's to Blame?', in Stork, D. (ed.), *Hal's Legacy: 2001's Computer as Dream and Reality* (Cambridge, MA: MIT Press), 351–65.

Densmore, S. and Dennett, D. (1999). 'The Virtues of Virtual Machines', *Philosophy and Phenomenological Research*, 59(3), 747–61.

Dijkstra, E. W. (1968). 'Structure of the 'the'-Multiprogramming System', *Communications of the ACM,* 11(5), 341–6.

Dixon, B. A. (1995). 'Response: Evil and the Moral Agency of Animals', *Between the Species*, 11 (1–2), 38–40.

Donaldson, T. (1982). *Corporations and Morality* (Englewood Cliffs, NJ: Prentice Hall).

Dreyfus, H. L. (2001). *On the Internet* (London: Routledge).

Drury, M. O. C. (1981). 'Some Notes on Conversations with Wittgenstein', in Rhees, R. (ed.), *Ludwig Wittgenstein: Personal Recollections* (Totowa, NJ: Rowman and Littlefield), 71–171.

Einstein, A. (1954). *Ideas and Opinions* (New York: Crown Publishers).

Eisenstadt, M. and Vincent, T. (2000). *The Knowledge Web: Learning and Collaborating on the Net* (London: Kogan Page).

Elliot, R. (1995). *Environmental Ethics* (Oxford: Oxford University Press).

Ellis, R. (1996). *Data Abstraction and Program Design: From Object-Based to Object-Oriented Programming* (2nd edn., London: UCL Press).

Ennals, J. R. (1994). *Information Technology and Business Ethics* (Kingston upon Thames: Kingston Business School).

Epstein, R. G. (1997). *The Case of the Killer Robot: Stories About the Professional, Ethical, and Societal Dimensions of Computing* (New York / Chichester: Wiley).

Ermann, M. D. and Shauf, M. S. (2003). *Computers, Ethics and Society* (3rd edn., New York / Oxford: Oxford University Press).

Ess, C. (2002). 'Computer-Mediated Colonization, the Renaissance, and Educational Imperatives for an Intercultural Global Village', *Ethics and Information Technology*, 4(1), 11–22.

Ess, C. (2006). 'Ethical Pluralism and Global Information Ethics', *Ethics and Information Technology*, 8(4), 215–26.

Ess, C. (ed.) (2005). *Ethics and Information Technology*, 7(1), Special Issue on *Privacy and Data Privacy Protection in Asia*.

Ess, C. (ed.) (2008). *Ethics and Information Technology*, 10(2–3), Special Issue on *Luciano Floridi's Philosophy of Information and Information Ethics: Critical Reflections and the State of the Art*.

European Group on Ethics in Science and New Technologies (2012). *Opinion no. 26: Ethics of Information and Communication Technologies* (Luxembourg: Publications Office of the European Union) <http://ec.europa.eu/bepa/european-group-ethics/docs/publications/ict_final_22_february-adopted.pdf>.

Ewin, R. E. (1991). 'The Moral Status of the Corporation', *Journal of Business Ethics*, 10(10), 749–56.

Fagin, R. et al. (1995). *Reasoning About Knowledge* (Cambridge, MA / London: MIT Press).

Fischer, P. M. et al. (1991). 'Brand Logo Recognition by Children Aged 3 to 6 Years. Mickey Mouse and Old Joe the Camel', *The Journal of the American Medical Association*, 266(22), 3145–8.

Floridi, L. (1994). 'Scepticism and the Search for Knowledge: A Peirceish Answer to a Kantian Doubt', *Transactions of the Charles S. Peirce Society*, 30(3), 543–73.

Floridi, L. (1995a). 'Cupiditas Veri Videndi: Pierre Villemandy's Dogmatic Versus Cicero's Sceptical Interpretation of Man's Desire to Know', *British Journal for the History of Philosophy*, 3(1), 29–56.

Floridi, L. (1995b). 'Internet: Which Future for Organized Knowledge, Frankenstein or Pygmalion?', *International Journal of Human-Computer Studies*, 43, 261–74.

Floridi, L. (1999a). 'Information Ethics: On the Philosophical Foundations of Computer Ethics', *Ethics and Information Technology*, 1(1), 33–52.

Floridi, L. (1999b). *Philosophy and Computing: An Introduction* (London / New York, NY: Routledge).

Floridi, L. (ed.) (1999c). *Etica & Politica*, 1(2), Special Issue on *Computer Ethics* <http://www2.units.it/etica/1999_2/index.html>.

Floridi, L. (2002). 'Information Ethics: An Environmental Approach to the Digital Divide', *Philosophy in the Contemporary World*, 9(1), 39–45.

Floridi, L. (2003). 'On the Intrinsic Value of Information Objects and the Infosphere', *Ethics and Information Technology*, 4(4), 287–304.

Floridi, L. (2004). 'Information', in Floridi, L. (ed.), *The Blackwell Guide to the Philosophy of Computing and Information* (Oxford / New York: Blackwell), 40–61.

Floridi, L. (2005a). 'The Ontological Interpretation of Informational Privacy', *Ethics and Information Technology*, 7(4), 185–200.

Floridi, L. (2005b). 'The Philosophy of Presence: From Epistemic Failure to Successful Observability', *Presence: Teleoperators and Virtual Environments*, 14(6), 656–67.

Floridi, L. (2005c). 'Information Ethics, Its Nature and Scope', *Computers and Society*, 35(2), 1–3.

Floridi, L. (2006a). 'Four Challenges for a Theory of Informational Privacy', *Ethics and Information Technology*, 8(3), 109–19.

Floridi, L. (2006b). 'Information Technologies and the Tragedy of the Good Will', *Ethics and Information Technology*, 8(4), 253–62.

Floridi, L. (2007a). 'A Look into the Future Impact of Ict on Our Lives', *The Information Society*, 23(1), 59–64.

Floridi, L. (2007b). 'Computers', in Wolf, C., Ryberg, J., and Petersen, T. S. (eds.), *New Waves in Applied Ethics* (Basingstoke / New York: Palgrave MacMillan).

Floridi, L. (2008a). 'Foundations of Information Ethics', in Tavani, H. T. and Himma, K. (eds.), *Information and Computer Ethics* (Hoboken, NJ: John Wiley & Sons), 3–23.

Floridi, L. (2008b). 'Information Ethics: Its Nature and Scope', in Van den Hoven, J. and Weckert, J. (eds.), *Moral Philosophy and Information Technology* (Cambridge: Cambridge University Press), 40–65.

Floridi, L. (2008c). 'Artificial Intelligence's New Frontier: Artificial Companions and the Fourth Revolution', *Metaphilosophy*, 39(4–5), 651–5.

Floridi, L. (2008d). 'The Method of Levels of Abstraction', *Minds and Machines*, 18(3), 303–29.

Floridi, L. (2008e). 'Understanding Information Ethics—Replies to "Commentaries on Floridi"', *APA Newsletter on Philosophy and Computers*, 8(2), 4–11 <http://www.apaonline.org/APAOnline/Publications/Newsletters/Past_Newsletters/Vol08/Vol._08_Fall_2008_Spring_2009_.aspx>.

Floridi, L. (2008f). 'Information Ethics: A Reappraisal', *Ethics and Information Technology*, 10(2–3), 189–204.

Floridi, L. (2008g). 'A Defence of Informational Structural Realism', *Synthese*, 161(2), 219–53.

Floridi, L. (2009a). 'Against Digital Ontology', *Synthese*, 168(1), 151–78.

Floridi, L. (2009b). *Infosfera—Etica e filosofia nell'età dell'informazione* (Torino: Giappichelli).

Floridi, L. (2010a). 'Levels of Abstraction and the Turing Test', *Kybernetes*, 39(3), 423–40.

Floridi, L. (2010b). 'Network Ethics: Information and Business Ethics in a Networked Society', *Journal of Business Ethics*, 90(4), 649–59.

Floridi, L. (2010c). *Information—A Very Short Introduction* (Oxford: Oxford University Press).

Floridi, L. (ed.) (2010d). *The Cambridge Handbook of Information and Computer Ethics* (Cambridge: Cambridge University Press).

Floridi, L. (2011a). *The Philosophy of Information* (Oxford: Oxford University Press).

Floridi, L. (2011b). 'A Defence of Constructionism: Philosophy as Conceptual Engineering', *Metaphilosophy*, 42(3), 282–304.

Floridi, L. (ed.) (2011c). *The Construction of Personal Identities Online*, 21(4), Special Issue on *Minds and Machines*.

Floridi, L. (2012a). 'Acta—the Ethical Analysis of a Failure, and Its Lessons', *ECIPE Working Papers*, 04/2012.

Floridi, L. (2012b). 'Degenerate Epistemology', *Philosophy & Technology*, 25(1), 1–3.

Floridi, L. (2012c). 'Big Data and Their Epistemological Challenge', *Philosophy & Technology*, 25(4), 435–7.

Floridi, L. (2012d). 'Hyperhistory and the Philosophy of Information Policies', *Philosophy & Technology*, 25(2), 129–31.

Floridi, L. (2012e). 'Steps Forward in the Philosophy of Information', *Etica Politica/Ethics Politics*, 14(1), 304–10.

Floridi, L. (2012f). 'The Road to the Philosophy of Information', in Demir, H. (ed.), *Luciano Floridi's Philosophy of Technology—Critical Reflections* (New York: Springer), 245–71.

Floridi, L. (2012g). 'Turing's Three Philosophical Lessons and the Philosophy of Information', *Philosophical Transactions*, A(370), 3536–42.

Floridi, L. (forthcoming-a). 'The Ethics of Distributed Morality', *Science and Engineering Ethics*.

Floridi, L. (forthcoming-b). 'Turing Test and the Method of Levels of Abstraction', in Cooper, S. B. and Leeuwen, J. v. (eds.), *Alan Turing—His Work and Impact* (Oxford: Elsevier Science).

Floridi, L. and Sanders, J. W. (1999). 'Entropy as Evil in Information Ethics', *Etica & Politica*, 1(2), Special Issue on *Computer Ethics* <http://www2.units.it/etica/1999_2/index.html>.

Floridi, L. and Sanders, J. W. (2001). 'Artificial Evil and the Foundation of Computer Ethics', *Ethics and Information Technology*, 3(1), 55–66.

Floridi, L. and Sanders, J. W. (2002). 'Computer Ethics: Mapping the Foundationalist Debate', *Ethics and Information Technology*, 4(1), 1–9.

Floridi, L. and Sanders, J. W. (2004a). 'Mapping the Foundationalist Debate', in Spinello, R. and Tavani, H. (eds.), *Readings in Cyberethics* (2nd edn., Boston, MA: Jones and Bartlett), 81–95.

Floridi, L. and Sanders, J. W. (2004b). 'The Method of Abstraction', in Negrotti, M. (ed.), *Yearbook of the Artificial. Nature, Culture and Technology. Models in Contemporary Sciences* (2nd edn., Berne: Peter Lang), 177–220.

Floridi, L. and Sanders, J. W. (2004c). 'On the Morality of Artificial Agents', *Minds and Machines*, 14(3), 349–79.

Floridi, L. and Sanders, J. W. (2005). 'Internet Ethics: The Constructionist Values of Homo Poieticus', in Cavalier, R. (ed.), *The Impact of the Internet on Our Moral Lives* (New York: State University of New York Press).

Floridi, L. and Taddeo, M. (eds.) (forthcoming). *The Ethics of Information Warfare* (Berlin / New York: Springer).

Flynn, D. J. and Fragoso Diaz, O. (1996). *Information Modelling: An International Perspective* (London: Prentice Hall).

Foot, P. (1967). 'The Problem of Abortion and the Doctrine of the Double Effect', *Oxford Review*, 5, 5–15.

Forester, T. and Morrison, P. (1994). *Computer Ethics: Cautionary Tales and Ethical Dilemmas in Computing* (2nd edn., Cambridge, MA: MIT Press).

Fox, W. (1995). *Toward a Transpersonal Ecology: Developing New Foundations for Environmentalism* (New York: State University of New York Press).

Franklin, S. and Graesser, A. (1997). 'Is It an Agent, or Just a Program?: A Taxonomy for Autonomous Agents', *Proceedings of the Workshop on Intelligent Agents III, Agent Theories, Architectures, and Languages* (Berlin / New York: Springer), 21–35.

Freeman, R. E. (1984). *Strategic Management: A Stakeholder Approach* (Boston, MA: Pitman).

Freud, S. (1955). 'A Difficulty in the Path of Psycho-Analysis', *The Standard Edition of the Complete Psychological Works of Sigmund Freud. Volume XVII (1917–1919): An Infantile Neurosis and Other Works*, ed. and trans. J. Strachey (London: Hogarth Press), 135–44.

Fried, C. (1970). *An Anatomy of Values: Problems of Personal and Social Choice* (Cambridge, MA: Harvard University Press).

Friedman, M. (1970). 'The Social Responsibility of Business Is to Increase Its Profits', *The New York Times Magazine*, 13 September, pp. 32–3, 122, 126.

Froehlich, T. J. (1997). *Survey and Analysis of Legal and Ethical Issues for Library and Information Services* (Munich: G. K. Saur).

Froomkin, A. M. (2000). 'The Death of Privacy?', *Stanford Law Review*, 52, 1461–543.

Gaita, R. (2004). *Good and Evil: An Absolute Conception* (2nd edn., London: Routledge).
Gantz, J. and Reinsel, D. (2011). *Extracting Value from Chaos*, White paper, Sponsored by EMC—IDC <http://www.emc.com/leadership/programs/digital-universe.htm>.
Gellately, R. (2001). *Backing Hitler: Consent and Coercion in Nazi Germany* (Oxford: Oxford University Press).
Gelven, M. (1983). 'The Meanings of Evil', *Philosophy Today*, 27(3–4), 200–21.
Gibson, J. J. (1979). *The Ecological Approach to Visual Perception* (Boston / London: Houghton Mifflin).
Gide, A. (1960). *The Immoralist* (Harmondsworth: Penguin).
Gips, J. (1995). 'Towards the Ethical Robot', in Ford, K., Glymour, C., and Hayes, P. (eds.), *Android Epistemology* (Cambridge, MA: MIT Press), 243–52.
Goldberg, K. (ed.) (2000). *The Robot in the Garden: Telerobotics and Telepistemology in the Age of the Internet* (Cambridge, MA / London: MIT Press).
Gorniak-Kocikowska, K. (1996). 'The Computer Revolution and the Problem of Global Ethics', *Science and Engineering Ethics*, 2(2), 177–90.
Gotterbarn, D. (1991). 'Computer Ethics: Responsibility Regained', *National Forum: The Phi Beta Kappa Journal*, 71, 26–31.
Gotterbarn, D. (1992). 'The Use and Abuse of Computer Ethics', *Journal of Systems and Software*, 17(1), 75–80.
Gotterbarn, D. W. (2001). 'Software Engineering Ethics', in Marciniak, J. (ed.), *Encyclopedia of Software Engineering* (2nd edn., New York: Wiley-Interscience).
Gow, G. (2005). 'Privacy and Ubiquitous Network Societies—Background Paper', *International Communication Union—Workshop on Ubiquituous Network Societies* <http://www.itu.int/osg/spu/ni/ubiquitous/Papers/Privacy%20background%20paper.pdf>.
Greco, G. M. and Floridi, L. (2004). 'The Tragedy of the Digital Commons', *Ethics and Information Technology*, 6(2), 73–82.
Gresillon, A. (1994). *Éléments de critique Génétique* (Paris: Presses Universitaires de France).
Grodzinsky, F. (2001). 'Revisiting the Virtues: The Practitioner from Within', in Spinello, R. A. and Tavani, H. (eds.), *Readings in Cyberethics* (Sudbury, MA: Jones and Bartlett), ch. 6.
Halpern, J. Y. and Moses, Y. (1990). 'Knowledge and Common Knowledge in a Distributed Environment', *Journal of the Association for Computing Machinery*, 37(3), 549–87.
Hampton, J. (1989). 'The Nature of Immorality', *Social Philosophy and Policy*, 7(1), 22–44.
Hardin, G. (1968). 'The Tragedy of the Commons', *Science*, 162(3859), 1243–8.
Hardin, G. (1998). 'Extensions of "The Tragedy of the Commons"', *Science*, 280(5364), 682–3.
Hauptman, R. (1988). *Ethical Challenges in Librarianship* (Phoenix, AZ: Oryx Press).
Hayes, P. et al. (1992). 'Virtual Symposium on Virtual Mind', *Minds and Machines*, 2(3), 217–38.
Held, D. and McGrew, A. (2001). 'Globalization', in Krieger, J. (ed.), *Oxford Companion to Politics of the World* (2nd edn., Oxford / New York: Oxford University Press).
Held, D. et al. (1999). *Global Transformations: Politics, Economics and Culture* (Cambridge: Polity Press).
Hepburn, R. W. (1984). *'Wonder' and Other Essays: Eight Studies in Aesthetics and Neighbouring Fields* (Edinburgh: Edinburgh University Press).
Herold, K. (2005). 'A Buddhist Model for the Informational Person', *Proceedings of the Second Asia Pacific Computing and Philosophy Conference*, 7–9 January, Bangkok, Thailand.

Hick, J. (1967). 'The Problem of Evil', in Edwards, P. (ed.), *The Encyclopedia of Philosophy* (New York, NY: Macmillan).

Hildebrandt, M. (2008). 'Ambient Intelligence, Criminal Liability and Democracy', *Criminal Law and Philosophy*, 2(2), 163–80.

Hildebrandt, M. (2011). 'Criminal Liability in a Smart Environment', in Duff, R. A. and Green, S. P. (eds.), *Philosophical Foundations of Criminal Law* (Oxford: Oxford University Press), 507–32.

Himma, K. E. (2004a). 'There's Something About Mary: The Moral Value of Things *Qua* Information Objects', *Ethics and Information Technology*, 6(3), 145–59.

Himma, K. E. (2004b). 'The Relationship between the Uniqueness of Computer Ethics and Its Independence as a Discipline in Applied Ethics', *Ethics and Information Technology*, 5(4), 225–37.

Himma, K. E. and Tavani, H. T. (2008). *The Handbook of Information and Computer Ethics* (Hoboken, NJ: Wiley).

Hoare, C. A. R. (1972). 'Notes on Data Structuring', in Dahl, O. J., Dijkstra, E. W., and Hoare, C. A. R. (eds.), *Structured Programming* (London: Academic Press), 83–174.

Hodel, T. B., Holderegger, A., and Lüthi, A. (1998). 'Ethical Guidelines for a Networked World under Construction', *Journal of Business Ethics*, 17(9–10), 1057–71.

Hogan, J. P. (1997). *The Two Faces of Tomorrow* (Riverdale, NY: Baen).

Hongladarom, S. (2006). 'Analysis and Justification of Privacy from a Buddhist Perspective', in Hongladarom, S. and Ess, C. (eds.), *Information Technology Ethics: Cultural Perspectives* (Hershey, Pennsylvania: Idea Publishing).

Hongladarom, S. (2008). 'Floridi and Spinoza on Global Information Ethics', *Ethics and Information Technology*, 10(2), 175–87.

Huizinga, J. (1970). *Homo Ludens: A Study of the Play Element in Culture* (London: Paladin).

Hume, D. (2007). *A Treatise of Human Nature: A Critical Edition* (Oxford: Clarendon).

Ijsselsteijn, W. (2002). 'Elements of a Multi-Level Theory of Presence: Phenomenology, Mental Processing and Neural Correlates', *Presence 2002: 5th Annual International Workshop on Presence, Conference Proceedings* Universidade Fernando Pessoa, Porto, Portugal, 9–11 October, 245–59 <http://ispr.info/presence-conferences/previous-conferences/presence-2002/>.

Ijsselsteijn, W. and Harper, B. (2001). 'Virtually There? A Vision on Presence Research', *PRESENCE-IST 2000-31014*, EC Public Deliverable, December.

Ijsselsteijn, W. A. et al. (2000). 'Presence: Concept, Determinants and Measurement', *Proceedings of the SPIE, Human Vision and Electronic Imaging* (5), 3956–75.

Inness, J. C. (1996). *Privacy, Intimacy, and Isolation* (New York: Oxford University Press).

Introna, L. D. (1997). 'Privacy and the Computer: Why We Need Privacy in the Information Society', *Metaphilosophy*, 28(3), 259–75.

Jamieson, D. (2008). *Ethics and the Environment: An Introduction* (Cambridge: Cambridge University Press).

Johnson, D. G. (1999). 'Sorting out the Uniqueness of Computer-Ethical Issues', *Etica & Politica* 1(2), Special Issue on *Computer Ethics* <http://www2.units.it/etica/1999_2/index.html>.

Johnson, D. G. (2001). *Computer Ethics* (3rd edn., Upper Saddle River, NJ: Prentice Hall).

Johnson, D. G. (2006). 'Computer Systems: Moral Entities but Not Moral Agents', *Ethics and Information Technology*, 8(4), 195–204.

Johnson, D. G. and Miller, K. (2009). *Computer Ethics: Analyzing Information Technology* (4th edn., Upper Saddle River, NJ: Prentice Hall).

Johnson, J. L. (1992). 'A Theory of the Nature of Value of Privacy', *Public Affairs Quarterly*, 6(3), 271–88.

Jonsen, A. R. and Butler, L. H. (1975). 'Public Ethics and Policy Making', *Hastings Center Report*, 5(4), 19–31.

Kabay, M. E. (1998). 'ICSA White Paper on Computer Crime Statistics' <http://groshens.org/Whitepaper/Virus/crime.pdf>.

Kant, I. (2005). 'Groundwork of the Metaphysics of Morals', in Kant, I., *Practical Philosophy*, ed. M. J. Gregor (rev'd edn., Cambridge: Cambridge University Press).

Kekes, J. (1988). 'Understanding Evil', *American Philosophical Quarterly*, 25, 13–24.

Kekes, J. (1990). *Facing Evil* (Princeton, NJ: Princeton University Press).

Kekes, J. (1998a). 'Evil', *Routledge Encyclopedia of Philosophy* (London: Routledge).

Kekes, J. (1998b). 'The Reflexivity of Evil', *Social Philosophy and Policy*, 15(1), 216–32.

Kekes, J. (2005). *The Roots of Evil* (Ithaca, NY / London: Cornell University Press).

Kerr, P. (1996). *The Grid* (New York: Warner Books).

King, D. (1997). *Kasparov v Deeper Blue* (London: Batsford).

King, M. L. (1992). *The Papers of Martin Luther King, Jr* (Berkeley: University of California Press).

Kitiyadisai, K. (2005). 'Privacy Rights and Protection: Foreign Values in Modern Thai Context', *Ethics and Information Technology*, 7(1), 17–26.

Klein, N. (2000). *No Logo: No Space, No Choice, No Jobs* (London: Flamingo).

Koenig, M. E. D., Kostrewski, B. J., and Oppenheim, C. (1981). 'Ethics in Information Science', *Journal of Information Science*, 3, 45–7.

Koops, B. J., Hildebrandt, M., and Jaquet-Chiffelle, D. O. (2010). 'Bridging the Accountability Gap: Rights for New Entities in the Information Society?', *Minnesota Journal of Law, Science & Technology*, 11(2), 497–561.

Kroes, P. and Verbeek, P.-P. (forthcoming). *Moral Agency and Technical Artefacts* (Berlin / New York: Springer).

Langford, D. (1995). *Practical Computer Ethics* (London: McGraw-Hill).

Langford, D. (1999). *Business Computer Ethics* (Harlow: Addison-Wesley).

Langford, D. (2000). *Internet Ethics* (New York: St. Martin's Press).

Laurel, B. (1991). *Computers as Theatre* (Reading, Mass.: Addison-Wesley).

Lauria, R. (2001). 'In Answer to a Quasi-Ontological Argument: On Sheridan's "Toward an Eclectic Ontology of Presence" and Mantovani and Riva's "Building a Bridge between Different Scientific Communities"', *Presence: Teleoperators and Virtual Environments*, 10(5), 557–63.

Leibniz, G. W. (1990). *Theodicy: Essays on the Goodness of God, the Freedom of Man, and the Origin of Evil* (La Salle, IL: Open Court).

Leopold, A. (1949). *A Sand County Almanac; and, Sketches Here and There* (New York: Oxford University Press).

Levi, P. (1959). *If This Is a Man* (London: Orion Press).

Locke, J. (1979). *An Essay Concerning Human Understanding* (Oxford / New York: Clarendon Press).

Lombard, M. and Ditton, T. (1997). 'At the Heart of It All: The Concept of Presence', *Journal of Computer-Mediated Communication*, 3(2) <http://onlinelibrary.wiley.com/doi/10.1111/j.1083-6101.1997.tb00072.x/full>.

Lomborg, B. R. (2009). *Global Crises, Global Solutions* (2nd edn., Cambridge: Cambridge University Press).

Lopes, D. M. (2003). 'Digital Art', in Floridi, L. (ed.), *Blackwell Guide to the Philosophy of Computing and Information* (Oxford: Blackwell), ch. 8.

Lyman, P. and Varian, H. R. (2003). *How Much Information 2003* <http://www.sims.berkeley.edu/how-much-info-2003>.

Maner, W. (1996). 'Unique Ethical Problems in Information Technology', *Science and Engineering Ethics*, 2(2), 137–54.

Maner, W. (1999). 'Is Computer Ethics Unique?', *Etica & Politica*, 1(2) Special Issue on *Computer Ethics* <http://www2.units.it/etica/1999_2/index.html>.

Mantovani, G. and Riva, G. (1999). '"Real" Presence: How Different Ontologies Generate Different Criteria for Presence, Telepresence, and Virtual Presence', *Presence: Teleoperators and Virtual Environments*, 8(5), 540–50.

Mantovani, G. and Riva, G. (2001). 'Building a Bridge between Different Scientific Communities: On Sheridan's Eclectic Ontology of Presence', *Presence: Teleoperators and Virtual Environments*, 10(5), 537–43.

Margulis, S. T. (2003). 'Privacy as a Social Issue and Behavioral Concept', *Journal of Social Issues*, 59(2), 243–61.

Martin, K. and Freeman, R. E. (2004). 'The Separation of Technology and Ethics in Business Ethics', *Journal of Business Ethics*, 53(4), 353–64.

Martin, R. and Barresi, J. (2006). *The Rise and Fall of Soul and Self: An Intellectual History of Personal Identity* (New York, NY: Columbia University Press).

Marturano, A. (2002). 'The Role of Metaethics and the Future of Computer Ethics', *Ethics and Information Technology*, 4(1), 71–8.

Marx, G. T. (2005). 'Some Conceptual Issues in the Study of Borders and Surveillance', in Zureik, E. and Salter, M. B. (eds.), *Global Surveillance and Policing—Borders, Security, Identity* (Cullompton, Devon: Willan Publishing), ch. 2.

Mason, R. (1986). 'Four Ethical Issues of the Information Age', *MIS Quarterly*, 10(1), 5–12.

Mason, R. O., Mason, F. M., and Culnan, M. J. (1995). *Ethics of Information Management* (Thousand Oaks, CA / London: Sage Publishing).

Mather, K. (2005). 'Object Oriented Goodness: A Response to Mathiesen's "What Is Information Ethics?"', *Computers and Society*, 34(4) <http://dl.acm.org/citation.cfm?id=1111637&dl=ACM&coll=DL&CFID=308019942&CFTOKEN=32520655>.

Mathiesen, K. (2004). 'What Is Information Ethics?', *Computers and Society*, 34(1), 6 <http://dl.acm.org/citation.cfm?doid=1050305.1050312>.

Mattelart, A. (2001). *Histoire de la société de l'information* (Paris: Éditions de la Découverte).

May, L. (1983). 'Vicarious Agency and Corporate Responsibility', *Philosophical Studies*, 43(1), 69–82.

Mealing, S. (1997). *Computers and Art* (Exeter: Intellect).

Medvidovic, N., Taylor, R. N., and E. James Whitehead, J. (1996). 'Formal Modeling of Software Architectures at Multiple Levels of Abstraction', *Proceedings of the California Software Symposium*, 17 April, Los Angeles, CA, 28–40.

Michie, D. (1961). 'Trial and Error', in Garratt, A. (ed.), *Penguin Science Surveys* (Harmondsworth: Penguin), 129–45.

Microsoft-Research (2005). 'Towards 2020 Science' <http://research.microsoft.com/en-us/um/cambridge/projects/towards2020science/>.

Mills, E. (2005). 'Google Balances Privacy, Reach', *CNet News.com*, 14 July <http://news.cnet.com/Google-balances-privacy,-reach/2100-1032_3-5787483.html>.

Milo, R. D. (1984). *Immorality* (Princeton: Princeton University Press).

Minsky, M. (1980). 'Telepresence', *Omni Magzine*, 2(9), 45–51.

Minsky, M. L. (1986). *The Society of Mind* (New York: Simon & Schuster).

Mintz, A. P. (ed.) (1990). *Information Ethics: Concerns for Librarianship and the Information Industry: Proceedings of the Twenty-Seventh Annual Symposium of the Graduate Alumni and Faculty of the Rutgers School of Communication, Information and Library Studies, 14 April 1989* (Jefferson, NC / London: McFarland).

Mitchell, M. (1998). *An Introduction to Genetic Algorithms* (Cambridge, MA / London: MIT Press).

Mitchell, T. M. (1997). *Machine Learning* (Int'l edn., New York, NY / London: McGraw-Hill).

Moody, G. (2001). *Rebel Code: The Inside Story of Linux and the Open Source Revolution* (Cambridge, MA: Perseus Publishing).

Moor, J. H. (1985). 'What Is Computer Ethics?', *Metaphilosophy*, 16(4), 266–75.

Moor, J. H. (1997). 'Towards a Theory of Privacy in the Information Age', *ACM SIGCAS Computers and Society*, 27, 27–32.

Moor, J. H. (2001a). 'The Status and Future of the Turing Test', *Minds and Machines*, 11(1), 77–93.

Moor, J. H. (2001b). 'The Future of Computer Ethics: You Ain't Seen Nothin' Yet!', *Ethics and Information Technology*, 3(2), 89–91.

Moore, A. D. (ed.) (2005). *Information Ethics: Privacy, Property and Power* (Seattle, WA / London: University of Washington Press).

Moore, G. E. (1993). *Principia Ethica*, ed. and introd. Thomas Baldwin (revised edn., Cambridge: Cambridge University Press).

Motwani, R. and Raghavan, P. (1995). *Randomized Algorithms* (Cambridge: Cambridge University Press).

Moya, L. J. and Tolk, A. (2007). 'Towards a Taxonomy of Agents and Multi-Agent Systems', *Proceedings of the 2007 Spring Simulation Multiconference—Volume 2* (Norfolk, VA: Society for Computer Simulation International), 11–18.

Murray, J. H. (1997). *Hamlet on the Holodeck: The Future of Narrative in Cyberspace* (New York: Free Press).

Naess, A. (1973). 'The Shallow and the Deep, Long-Range Ecology Movement', *Inquiry*, 16, 95–100.

Nagle, D. B. (2006). *The Household as the Foundation of Aristotle's Polis* (Cambridge: Cambridge University Press).

Nakada, M. and Tamura, T. (2005). 'Japanese Conceptions of Privacy: An Intercultural Perspective', *Ethics and Information Technology*, 7(1), 27–36.

Naresh, S. (1999). 'Ethical Norms for the Information Society', *Proceedings of the First Session of Unesco's Comest* (Paris: UNESCO), 169–77.

Narveson, J. (2002). 'Collective Responsibility', *The Journal of Ethics*, 6(2), 179–98.

Nash, R. (1989). *The Rights of Nature: A History of Environmental Ethics* (Madison, WI: University of Wisconsin Press).

Nature (2006). '2020—Future of Computing', 440 <http://www.nature.com/nature/focus/futurecomputing/>.

Negroponte, N. (1995). *Being Digital* (New York, NY: Knopf).

Nelson, B. L. (2006). *Law and Ethics in Global Business: How to Integrate Law and Ethics into Corporate Governance around the World* (London: Routledge).

Neurath, O. (1959). 'Protocol Sentences', in Ayer, A. J. (ed.), *Logical Positivism* (Glencoe, IL: The Free Press), 199–208.

Nissenbaum, H. (1998). 'Protecting Privacy in an Information Age: The Problem of Privacy in Public', *Law and Philosophy*, 17(5–6), 559–96.

Nissenbaum, H. (2010). *Privacy in Context: Technology, Policy, and the Integrity of Social Life* (Stanford, CA: Stanford Law Books).

Norton, B. G. (1989). 'The Cultural Approach to Conservation Biology', in Western, D. and Pears, M. C. (eds.), *Conservation in the Twenty-First Century* (Oxford: Oxford University Press).

O'Reilly, T. (2005). 'What Is Web 2.0: Design Patterns and Business Models for the Next Generation of Software', 30 September <http://www.oreillynet.com/pub/a/oreilly/tim/news/2005/09/30/what-is-web-20.html>.

Orwell, G. (1970). *The Collected Essays, Journalism, and Letters of George Orwell: Volume 3: As I Please, 1943–1945* (London: Penguin Books).

Pacuit, E., Parikh, R., and Cogan, E. (2006). 'The Logic of Knowledge-Based Obligation', *Synthese*, 149(2), 57–87.

Pagallo, U. (2012). 'On the Principle of Privacy by Design and Its Limits: Technology, Ethics, and the Rule of Law', in Gutwirth, S. et al. (eds.), *European Data Protection: In Good Health?* (Berlin / London: Springer), 331–46.

Pagallo, U. and Bassi, E. (2010). 'The Future of EU Working Parties' "the Future of Privacy" and the Principle of Privacy by Design', *3rd International Seminar on Information Law 2010*, Corfu, Greece, 25–26 June.

Papert, S. (1993). *Mindstorms: Children, Computers, and Powerful Ideas* (2nd edn., Cambridge, MA: Perseus).

Parker, D. B. (1979). *Ethical Conflicts in Computer Science and Technology* (Arlington, VA: AFIPS Press).

Parker, D. B. (1982). 'Ethical Dilemmas in Computer Technology', in Hoffman, W. M. and Moore, J. M. (eds.), *Ethics and the Management of Computer Technology* (Cambridge, MA: Oelgeschlager, Gunn & Hain).

Parker, D. B., Swope, S., and Baker, B. N. (1990). *Ethical Conflicts in Information and Computer Science, Technology, and Business* (Wellesley, MA: QED Information Sciences).

Parnas, D. C. (1972). 'On the Criteria to Be Used in Decomposing Systems into Modules', *Communications of the ACM*, 15(12), 1053–8.

Patton, J. W. (2000). 'Protecting Privacy in Public? Surveillance Technologies and the Value of Public Places', *Ethics and Information Technology*, 2(3), 181–7.

Patton, M. F. J. (1988). 'Tissues in the Profession: Can Bad Men Make Good Brains Do Bad Things?', *Proceedings and Addresses of the American Philosophical Association*, 61(3) <http://www.mindspring.com/~mfpatton/Tissues.htm>.

Peirce, C. S. (1931–1958). *Collected Papers*, 8 vols. (Cambridge, MA: Harvard University Press).

Perry, J. (2008). *Personal Identity* (2nd edn., Berkeley, CA / London: University of California Press).

Phillips, R. (2003). *Stakeholder Theory and Organizational Ethics* (1st edn., San Francisco: Berrett-Koehler).

Piaget, J. (1977). *The Essential Piaget* (London: Routledge and Kegan Paul).

Plantinga, A. (1977). *God, Freedom, and Evil* (Grand Rapids, MI: Eerdmans).

Pogue, D. (1999). *Palmpilot: The Ultimate Guide* (2nd edn., Beijing / Cambridge, MA: O'Reilly).

Pollock, J. (2008). 'What Am I? Virtual Machines and the Mind/Body Problem', *Philosophy and Phenomenological Research*, 76(2), 237–309.

Proust, M. (1982). *Remembrance of Things Past. Volume 1: Swann's Way*, trans. C. K. S. Montcrieff and T. Kilmartin (London: Vintage).

Putnam, H. (1960). 'Minds and Machines', in Hook, S. (ed.), *Dimensions of Mind* (New York, NY: New York University Press).

Queneau, R. (2008). *Exercises in Style*, trans. B. Wright (Richmond: Oneworld Classics).

Quine, W. V. (1992). 'Structure and Nature', *Journal of Philosophy*, 89(1), 5–9.

Rachels, J. (1975). 'Why Privacy Is Important', *Philosophy and Public Affairs*, 4, 323–33.

Rawls, J. (1999). *A Theory of Justice* (rev'd edn., Cambridge, MA: Belknap Press of Harvard University Press).

Raymond, E. S. (2001). *The Cathedral and the Bazaar: Musings on Linux and Open Source by an Accidental Revolutionary* (rev'd edn., Beijing / Cambridge, MA: O'Reilly).

Reiman, J. H. (1976). 'Privacy, Intimacy, and Personhood', *Philosophy and Public Affairs*, 6(1), 26–44.

Rolston III, H. (1985). 'Duties to Endangered Species', *BioScience*, 35, 718–26. Reprinted in Elliot (1995).

Rosenfeld, R. (1995a). 'Can Animals Be Evil?: Kekes' Character-Morality, the Hard Reaction to Evil, and Animals', *Between the Species*, 11(1–2), 33–8.

Rosenfeld, R. (1995b). 'Reply', *Between the Species*, 11(1–2), 40–1.

Rowlands, M. (2000). *The Environmental Crisis: Understanding the Value of Nature* (Basingstoke: Macmillan).

Rumbaugh, J. (1991). *Object-Oriented Modeling and Design* (Englewood Cliffs, NJ / London: Prentice Hall International).

Russell, S. J. and Norvig, P. (2010). *Artificial Intelligence: A Modern Approach* (3rd, int'l edn., Boston / London: Pearson).

Sacau, A. et al. (2003). 'Presence in Computer-Mediated Environments: A Short Review of the Main Concepts, Theories, and Trends', *Proceedings of IADIS International Conference e-Society 2003*, Lisbon, Portugal, 3–6 June.

Sachs, J. D. (2004). 'Seeking a Global Solution', *Nature*, 430(7001), 725–6.

Saponas, T. S. et al. (2007). 'Devices That Tell on You: Privacy Trends in Consumer Ubiquitous Computing', *Proceedings of 16th USENIX Security Symposium* (Boston, MA: USENIX Association).

Scanlon, T. (1975). 'Thomson on Privacy', *Philosophy and Public Affairs*, 4, 315–22.

Schachter, M. (2003). *Informational and Decisional Privacy* (Durham, NC: Carolina Academic Press).

Schechtman, M. (1996). *The Constitution of Selves* (Ithaca, NY / London: Cornell University Press).

Schmid, M. (1998). *Processes of Literary Creation: Flaubert and Proust* (Oxford: Legenda).

Schmitz, K. L. (1978). 'Entitative and Systemic Aspects of Evil', *Dialectics and Humanism*, 5, 149–61.

Schuemie, M. J. et al. (2001). 'Research on Presence in Virtual Reality: A Survey', *CyberPsychology & Behavior*, 4(2), 183–200.

Seltzer, W. and Anderson, M. (2001). 'The Dark Side of Numbers: The Role of Population Data Systems in Human Rights Abuses', *Social Research*, 68(339–71), 2.

Severson, R. J. (1997). *The Principles of Information Ethics* (Armonk, NY: M.E. Sharpe).

Shaheen, S. A., Guzman, S., and Zhang, H. (2010). 'Bikesharing in Europe, the Americas, and Asia—Past, Present, and Future', *Transportation Research Record—Journal of the Transportation Research Board*, 2143, 159–67.

Sheridan, T. (1999). 'Descartes, Heidegger, Gibson, and God: Toward an Eclectic Ontology of Presence', *Presence: Teleoperators and Virtual Environments*, 8(5), 551–9.

Sicart, M. (2005). 'On the Foundations of Evil in Computer Game Cheating', *Proceedings of the Digital Games Research Association's 2nd International Conference—Changing Views: Worlds in Play*, 16–20 June, Vancouver, BC, Canada.

Sicart, M. (2009). *The Ethics of Computer Games* (Cambridge, MA / London: MIT Press).

Simon, J. (2012). 'Epistemic Responsibility in Entangled Socio-Technical Systems', *Proceedings of AISB/IACAP World Congress—Alan Turing*, 2–6 July, Birmingham, UK.

Siponen, M. (2004). 'A Pragmatic Evaluation of the Theory of Information Ethics', *Ethics and Information Technology*, 6(4), 279–90.

Sloman, A. and Chrisley, R. L. (2003). 'Virtual Machines and Consciousness', *Journal of Consciousness Studies*, 10(4–5), 133–72.

Slote, M. (2000). 'Virtue Ethics', in Follette, H. L. (ed.), *The Blackwell Guide to Ethical Theory* (Oxford: Blackwell), 325–47.

Smart, J. J. C. and Williams, B. A. O. (1987). *Utilitarianism: For and Against* (Cambridge: Cambridge University Press).

Smith, M. M. (1996). *Information Ethics: An Hermeneutical Analysis of an Emerging Area in Applied Ethics*, Ph.D. Thesis, The University of North Carolina at Chapel Hill.

Smith, M. M. (1997). 'Information Ethics', *Annual Review of Information Science and Technology*, 32, 339–66.

Smith, M. M. (2002). 'Global Information Ethics: A Mandate for Professional Education', *Proceedings of the 68th IFLA Council and General Conference*, 18–24 August, Glasgow, UK.

Solove, D. J. (2002). 'Conceptualizing Privacy', *California Law Review*, 90, 1087–155.

Sorabji, R. (2006). *Self: Ancient and Modern Insights About Individuality, Life, and Death* (Chicago: University of Chicago Press).

Spinello, R. A. (1997). *Case Studies in Information and Computer Ethics* (Upper Saddle River, NJ: Prentice Hall).

Spinello, R. A. (2003). *Case Studies in Information Technology Ethics and Policy* (2nd edn., Upper Saddle River, NJ: Prentice Hall).

Spinello, R. A. (2005). 'Trespass and Kyosei in Cyberspace', in Spinello, R. A. and Tavani, H. T. (eds.), *Intellectual Property Rights in a Networked World: Theory and Practice* (Hershey, PA: Idea Group Inc.).

Spinello, R. A. (2011). *Cyberethics: Morality and Law in Cyberspace* (4th edn., Sudbury, MA: Jones & Bartlett Learning).

Spinello, R. A. and Tavani, H. T. (2001). *Readings in Cyberethics* (Boston: Jones and Bartlett Publishers).

Spinoza, B. (2000). *Ethics* (Oxford: Oxford University Press).

Stichler, R. N. and Hauptman, R. (eds.) (1998). *Ethics, Information, and Technology: Readings* (Jefferson, NC / London: McFarland).

Stone, C. D. (2010). *Should Trees Have Standing?: Law, Morality, and the Environment* (3rd edn., New York, NY / Oxford: Oxford University Press).

Sumner, L. W. (1996). *Welfare, Happiness, and Ethics* (Oxford: Oxford University Press).

Sunstein, C. R. (2001). *Republic.Com* (Princeton, NJ: Princeton University Press).

Sycara, K. P. (1998). 'Multiagent Systems', *AI Magazine*, 19(2), 79–92.

Taddeo, M. (2009). 'Defining Trust and E-Trust: Old Theories and New Problems', *Journal of Technology and Human Interaction*, 5(2), 23–35.

Taddeo, M. (2010). 'Modelling Trust in Artificial Agents, a First Step toward the Analysis of E-Trust', *Minds and Machines*, 20(2), 243–57.

Taddeo, M. and Floridi, L. (2007). 'A Praxical Solution of the Symbol Grounding Problem', *Minds and Machines*, 7(4), 369–89.

Tavani, H. T. (2002). 'The Uniqueness Debate in Computer Ethics: What Exactly Is at Issue, and Why Does It Matter?', *Ethics and Information Technology*, 4(1), 37–54.

Tavani, H. T. (2003). *Ethics and Technology: Ethical Issues in an Age of Information and Communication Technology* (New York: John Wiley & Sons).

Tavani, H. T. (2010). 'The Foundationalist Debate in Computer Ethics', in Floridi, L. (ed.), *The Cambridge Handbook of Information and Computer Ethics* (Cambridge: Cambridge University Press), 251–70.

Tavani, H. T. (2011). *Ethics and Technology: Controversies, Questions, and Strategies for Ethical Computing* (3rd edn., Hoboken, NJ: John Wiley & Sons).

Taylor, P. W. (1981). 'The Ethics of Respect for Nature', *Environmental Ethics*, 3(3) 197–218. Reprinted in M. E. Zimmerman (ed.), *Environmental Philosophy* (Englewood Cliffs, NJ: Prentice Hall, 1993), 67–88.

Taylor, P. W. (2011). *Respect for Nature: A Theory of Environment Ethics* (25th anniversary edn., Princeton, NJ: Princeton University Press).

Thompson, J. (1996). *Models of Value: Eighteenth-Century Political Economy and the Novel* (Durham, NC: Duke University Press).

Thomson, J. (1975). 'The Right to Privacy', *Philosophy and Public Affairs*, 4, 295–314.

Thomson, J. J. (1976). 'Killing, Letting Die, and the Trolley Problem', *The Monist*, 59, 204–17.

Turilli, M. (2007). 'Ethical Protocols Design', *Ethics and Information Technology*, 9(1), 49–62.

Turilli, M. and Floridi, L. (2009). 'The Ethics of Information Transparency', *Ethics and Information Technology*, 11(2), 105–12.

Turilli, M., Vaccaro, A., and Taddeo, M. (2012). 'Internet Neutrality: Ethical Issues in the Internet Environment', *Philosophy & Technology*, 25(2), 133–51.

Turing, A. M. (1936). 'On Computable Numbers, with an Application to the Entscheidungsproblem', *Proceedings of the London Mathematical Society*, 2(42), 230–65.

Turing, A. M. (1950). 'Computing Machinery and Intelligence', *Mind*, 59(236), 433–60.

Turkle, S. (1995). *Life on the Screen: Identity in the Age of the Internet* (New York: Simon & Schuster).

Vaccaro, A. (2006). 'Privacy, Security, and Transparency: ICT-Related Ethical Perspectives and Contrasts in Contemporary Firms', *Social Inclusion: Societal and Organizational Implications for Information Systems*, 208, 245–58.

Vaccaro, A. (2008). 'Information Ethics as Macroethics: Perspectives for Further Research', *American Philosophical Association Computers and Philosophy Newsletter*, 7(2), 16–17.

Vaccaro, A. and Madsen, P. (2009). 'Corporate Dynamic Transparency: The New ICT-Driven Ethics?', *Ethics and Information Technology*, 11(2), 113–22.

Van den Hoven, J. (1995). 'Equal Access and Social Justice: Information as a Primary Good', *ETHICOMP95: An International Conference on the Ethical Issues of Using Information Technology*, 28–30 March, De Montfort University, Leicester, UK.

Van den Hoven, J. and Weckert, J. (2008). *Information Technology and Moral Philosophy* (Cambridge: Cambridge University Press).

Verbeek, P.-P. (2011). *Moralizing Technology: Understanding and Designing the Morality of Things* (Chicago, IL: University of Chicago Press).

Veryard, R. (1992). *Information Modelling: Practical Guidance* (New York, NY / London: Prentice Hall).

Volkman, R. (2003). 'Privacy as Life, Liberty, Property', *Ethics and Information Technology*, 5(4), 199–210.

Walker, K. (2000). 'Where Everybody Knows Your Name: A Pragmatic Look at the Costs of Privacy and the Benefits of Information Exchange', *Stanford Technology Law Review*, 2, 1–50.

Wallace, K. A. (1999). 'Anonymity', *Ethics and Information Technology*, 1(1), 23–35.

Wallach, W. and Allen, C. (2009a). *Moral Machines: Teaching Robots Right from Wrong* (Oxford / New York: Oxford University Press).

Wallach, W. and Allen, C. (2009b). *Moral Machines: Teaching Robots Right from Wrong* (Oxford / New York: Oxford University Press).

Warnock, G. J. (1971). *The Object of Morality* (London: Methuen).

Warren, S. and Brandeis, L. D. (1890). 'The Right to Privacy', *Harvard Law Review*, 4(5), 193–220.

Weckert, J. (2001). 'Computer Ethics: Future Directions', *Ethics and Information Technology*, 3(2), 93–6.

Weckert, J. (2007). *Computer Ethics* (Aldershot / Burlington, VT: Ashgate).

Weckert, J. and Adeney, D. (1997). *Computer and Information Ethics* (Westport, CT: Greenwood Press).

Weinert, F. (2009). *Copernicus, Darwin, and Freud: Revolutions in the History and Philosophy of Science* (Oxford: Blackwell).

Wellman, M. P., Greenwald, A., and Stone, P. (2007). *Autonomous Bidding Agents: Strategies and Lessons from the Trading Agent Competition* (Cambridge, MA: MIT Press).

Westin, A. F. (1968). *Privacy and Freedom* (1st edn., New York: Atheneum).

White, L. J. (1973). 'Continuing the Conversation', in Barbour, I. G. (ed.), *Western Man and Environmental Ethics* (Reading, MA: Addison-Wesley), pp. 55–64.

Wiener, N. (1950). *The Human Use of Human Beings: Cybernetics and Society* (Boston, MA: Houghton Mifflin).

Wiener, N. (1954). *The Human Use of Human Beings: Cybernetics and Society* (rev'd edn., Boston, MA: Houghton Mifflin).

Wiener, N. (1961). *Cybernetics: Or Control and Communication in the Animal and the Machine* (2nd edn., New York: MIT Press).

Wiener, N. (1964). *God and Golem, Inc.: A Comment on Certain Points Where Cybernetics Impinges on Religion* (Cambridge, MA: MIT Press).

Williams, S. (2010). *Free as in Freedom (2.0): Richard Stallman and the Free Software Revolution* (2nd edn., Boston, MA: Free Software Foundation).

Wittgenstein, L. (2001). *Philosophical Investigations: The German Text with a Revised English Translation* (3rd edn., Oxford: Blackwell).

Wong, K. (2000). 'The Development of Computer Ethics: Contributions from Business Ethics and Medical Ethics', *Science and Engineering Ethics*, 6(2), 245–53.

Wood, D. J. and Logsdon, J. M. (2008). 'Educating Managers for Global Business Citizenship', in Swanson, D. L. and Fisher, D. G. (eds.), *Advancing Business Ethics Education* (Charlotte, NC: Information Age Publishing), 265–84.

Wood, D. J. et al. (2008). 'Dialogue: Toward Superior Stakeholder Theory', *Business Ethics Quarterly*, 18(2), 153–90.

Woodbury, M. C. (2003). *Computer and Information Ethics* (Champaign, IL: Stipes).

Wooldridge, M. J. (2009a). *An Introduction to Multiagent Systems* (2nd edn., Chichester: Wiley).

Wooldridge, M. J. (2009b). *An Introduction to Multiagent Systems* (2nd edn., Hoboken, NJ / Chichester: Wiley).

Woolf, V. (1965). *The Years* (Orlando, FL: Harcourt).

Woolf, V. (1992). *A Woman's Essays: Selected Essays*, 2 vols. (Annotated edn., London: Penguin).

York, P. F. (2005). *Respect for the World: Universal Ethics and the Morality of Terraforming*, Ph.D. Thesis, The University of Queensland.

Zahorik, P. and Jenison, R. L. (1998). 'Presence as Being-in-the-World', *Presence: Teleoperators and Virtual Environments*, 7(1), 78–89.

Zimmerman, M. E. (2005). *Environmental Philosophy: From Animal Rights to Radical Ecology* (4th edn., Upper Saddle River, NJ: Pearson/Prentice Hall).

Index

a4a 9
a2a 9
accessibility 22, 231, 273
accountability 23, 154, 157
  see also responsibility
accuracy 22
adaptability 140–1
  see also agent
advertising 23
agent
  advantage of extending class of 157–9
  business 280
  definition of 140–1
  empowering the information agent 236
  entities that are not agents 141–2
  evaluation of 151
  global 205
  identification of 151–2
  level of abstraction 139–48
  logical relations between classes of 136
  Menace 142–5
  moral 134, 146–8
  morality of artificial 134–60, 166, 180–93, 327
  nature of 138–46
  standard vs. non-standard theories 135–8
  taxonomy of 187
akrasia 196
anagnorisis 225
animals 151, 157
  see also agent anonymity 233
appliances
  augmenting 15–17, 233
  enhancing 15–17, 233
artificial intelligence 39
Asimov's laws of robotics 163, 234
autonomy 140, 190
  see also agent
availability 22, 273
axiology
  critique of Kantian axiology 114–15
  five objections to 124–33
  patient-oriented approach 115–22
  see also information

Being 64, 69, 75, 85, 98, 131–3, 308
  and flourishing 111
  intrinsic moral goodness of 319–20
  non-Being 65, 75
  ontic trust 302–3
  uniformity of 65

bioengineering 82
business 277–91
  concentric ring model of 282–3
  relational analysis of 282
  whirlpool model of 283–4
  see also business ethics
business ethics 277–91
  ethical questions, the what, how, and impact 284–5
  from information ethics to business ethics 279–80, 287–90
  informational analysis of business 280–4
  normative pressure points 285–7
Bynum, T. W. 87, 89, 92, 177–8

Cassandra 199–200, 205
  predicament of 201
  see also Good Will
censorship 25, 156–7
change 67–8
children 151, 157
  see also agent
cloud computing 11
communication problem 130–1
community
  return of the digital 236–40, 248
  virtual 174
computer ethics
  as applied ethics 94–7
  information ethics as foundation of 97–9
  theoretical and uniqueness debate 92–4
  see also ethics
conceptual design 2
confidentiality 245–6
  see also informational privacy
consequentialism 61, 77, 80
  and comparative problem 78
  and mathematical problem 78
  and monotonic problem 78
  and supererogatory problem 78
contractualism 61, 301
conservationism 327–8
constructionism 2, 76
  ethics 164–6
  on the web 175
  values 161–79
contents control 25
  see also monitoring
Copenhagen Consensus 206–8
Copernicus 14

354   INDEX

Daily Me 202
Darwin 14
death 82–4
deception 23
deontologism 61, 77, 81
Descartes, R. 14, 38–9, 106
digital arts 172–3
digital divide 22
digital gaze 223–5
disinformation 23
distributed morality 137, 261–76
  basic idea of 262
  and enabling infraethics 272–4
  examples of 267–9
  harnessing the power of 269–71
  new ethical scenario with 265–7
  old ethical scenario without 263–5
dynamic source of action/interaction 42
  see also presence

ecopoiesis 162, 168–9
egalitarianism 322–3
egopoiesis 162, 164–6, 218
egology
  as synchronic individualization 215–17
  from egology to ecology of self 226–7
  its two branches 213–15
  see also self
embodiment 221–2
endurantism 213
entropy 65–6, 71, 98
  informational 66
  metaphysical 65, 67, 130, 202
  thermodynamic concept of 66
ethics
  action-oriented 61–2
  after the information revolution 1–18
  agent-oriented 61–2
  as informational resources 21–3, 24
  as a macroethics 25–8, 56, 74–9, 94, 109
  as a microethics 26, 56, 95, 109
  augmented 206
  biocentric 64, 69, 76, 98
  blue 76
  business, *see* business ethics
  code of 155–6
  computer 63–4, 85
  defence of information ethics 306–30
  e-nvironmental 18, 53–85, 300
  foundationalist debate 86–101
  four ethical principles 70–4
  global information 292–305
  globalizing 296
  nature of 19–28
  naturalized semantic information 25
  non-standard and IE 76
  normative aspect of 70–4

  of the informational environment 24–5
  of informational products 23–4
  ontocentric 64, 65–70, 98, 326
  patient-oriented 61–4, 65–70, 110
  situated action 162
  unified model of 20–1
  virtue 61, 77, 81, 83, 164–8
  see also macroethics; foundationalist debate
eudokia 196–8
evil 72, 75, 128–30, 203, 269, 284
  artificial 181–93
  deflatory interpretation of existence of 183–5
  evolution of 185–7
  nature of 180–3
  see also ethics; distributed morality
exabyte 5
exaflood 5

filtering 25
foundationalist debate 86–101
  conservative approach 94–7
  innovative approach 97–9
  no resolution approach 88–90
  professional approach 90–2
  radical approach 93–4
foundationalist problem 54–7
fourth revolution 13–14
freedom of expression 25
Freud 14
futuristic thermostats 145–6

global-communication ethics
  vs. global-information ethics 296–7
  see also ethics
global-information ethics
  advantages of 299–300
  cost of 300–3
  and problem of the lion 297–9
  see also ethics
globalization 292–305
  processes of 293–6
Golden Rule 59, 69
Good Will
  empowering individual 205
  modelling of 194–7
  tragedy of 194–209
goodness 85
  non-monotonic nature of 72–4
  resilience of 72–4
Gotterbarn, D. 88–91
Gutenberg 4

hacking 24
homo poieticus 161–79
Hume, D. 216, 217
hyperhistorical predicament 3
hypermoralism 316–17

identity
  and biometrics 246–9
  personal 213, 215, 243–4
  see also informational privacy
inapplicability 313–14
infoglut 22
inforgs
  evolution of 14–18
information
  axiological analysis of 112–33
  ethics 63–4, 122–4, 292–305, 306–30
  life cycle of 3–4
  naturalized semantic 25
  role of in ethics 109–12
  societies 4
  see also business ethics; ethics
Information Communication Technologies (ICTs)
  and ethical problems 58–60
  and IT-heodicean problem 188, 200–1
  as re-ontologizing technologies 6–9, 15, 20
  as technologies of the self 221–5
informational entity 125–30
informational privacy 228–60
  and computer ethics 229–31
  as a function of ontological friction 231–2
  four challenges for a theory of 249–56
  individualism and the anthropology of 251–3
  in the re-ontologized infosphere 235–46
  non-Western approaches to 249–51
  ontological interpretation of and its value 242–6
  ownership-based interpretation 241–2, 244
  personal identity and biometrics 246–9
  public, passive, and active 256–7
  reductionist interpretation 240
  scope and limits of 253–6
infosphere 6–7, 9, 14, 26, 29, 70, 98, 132–3
  and informational privacy 235–46
  global 3
  intrinsic value of 102–33, 324–5
  metaphysics of 10–13
  poiesis in the 169–75
infraethics
  see distributed morality
intellectual property 24, 26
interactivity 140
  see also agent
interfaces 169–70
ITentities 8

Johnson, D.G. 95

Kant, I. 38, 81, 82, 83, 110, 136, 216–17
  and Good Will 197

  critique of Kantian axiology 114–20
  and value of infosphere 102–3

level of abstraction 27, 29–52, 104, 111, 214, 307–8, 320
  definition of 31–4
  and Menace 144
  method of 29–52
  mistaken adoption of 326–7
  on very idea of 30–1
  and ontological commitment 34–6
  and telepresence 51
  see also agent
liability 23
libel legislation 23
local space of observation 43–51
  see also presence
logic of realization 225–6
Lucretius 198, 204
  see also Good Will

macroethics 58–64, 74–9, 94
  and ICTs' ethical problems 58–60
  informational model 61–3
  proactive 162–4
  reactive 162–4
  see also ethics
memories 223
Menace 142–5, 153
  see also noughts and crosses
Mephistopheles 67
mindless morality 148–52
  freedom objection 149–50
  intentional objection 149
  responsibility objection 150–2
  teleological objection 148
MINMAX principle 59
Miranda 198
  see also Good Will
misinformation 23, 25
monitoring 26
Moor, J. H. 86–7, 178
moral action 147–8
  intrinsic value of object-oriented model of 103–9
  see also infosphere
morality
  concrete application of 157
  distributed 261–76
  threshold 152–3
multi-agent system 138, 212–13
  see also distributed morality; business ethics

naturalistic fallacy 323–4
non-informational privacy 257–8
  see also informational privacy

356 INDEX

nonsubstantialism 183–5
noughts and crosses 142–5

object-oriented programming 103–9, 116
observables 31–2
Oedipus 199, 205
  see also Good Will
onlife experience 8
ontic divide 132
ontic trust 300–3
ontological equality principle 68–70
ontological friction 7
  and the difference between old and new ICTs 232–5
  see also informational privacy
ontology
  need for 124–5, 299
open source software 25, 170–2
organizations 146

panpsychism, see pantheism
pantheism 328–9
patients, see agents
pedagogical methodology, see foundationalist debate
perception 37–8, 223–5
perdurantism 213
physis 303–5
Piaget, J. 176
piracy 25
plagiarism 23
Plato
  and the problem of the chariot 211–13, 218
poiesis 160–75, 243
  see also ecopoiesis; egopoiesis
pornography
  virtual 44–6
pragmatic rules of communication 23
presence 29
  as epistemic failure 37–41
  as successful observation 41–4
  backward/forward 48
  objection against presence as successful observation 46–9
  strong 44
  weak 44
  see also level of abstraction
principle
  of ontic solidarity 131
  of ontic uniformity 131
  of reflective respect 130
privacy 24
  assessing theories of 240–2
  see also informational privacy
profit 280, 288–90
  see also business ethics

propaganda 23
Proust, M. 215

Quine, V. W. 51–2

Radio Frequency Identification 8
reality
  nature of 10
reductionism 308–9
relativism 32–3, 312–13
reliability 22
re-ontology 6
remote space of observation 43–51
  see also presence
responsibility 25, 154, 157–9, 168, 182, 196
  see also mindless morality
Resource Product Target model 26–7

scandalous, see tragic
security 24
self
  construction of 173–4
  ICTs as technologies of 221–5
  informational nature of 210–27
  three membranes model 217–21
  see also ecopoiesis; egopoiesis
Shannon, C. E. 66, 195, 220
SmartPaint 146
social contract 301–3
  see also ontic trust
space 222
static property-bearer 42
  see also presence
stewardship 168, 328–9
supererogation 78, 314–16

techne 303–5
telepistemics 47
  see also presence
telepresence 29, 36–51, 295
  absence 40–1
  nested 40
  see also presence
testimony 23, 25
theodicean problem 185–7, 200–1
  see also evil
three membranes model 217–21
  see also self
time 222–3
tragic, the 194–209
  escaping from 204–8
  and the scandalous 197–200
  see also Good Will
transition system 34
triple A 22
trustworthiness 22, 25

2P2Q hypothesis 230–1, 234, 235
Turing, A. 4, 7, 134, 139
twin earth 131

universality 320–2
utilitarianism 61

vandalism 24, 80–1
variable 31

typed variable 31, 42
see also observables
veil of ignorance 22
virtualization 11–12, 174
virtue ethics, *see* ethics

wastefulness 74–9, 290
webbots 145

zettabyte era 5–6

Printed and bound by CPI Group (UK) Ltd, Croydon, CR0 4YY